T0305693

Circular Economy Applications for Water Security

In arid and semi-arid regions, where water demand exceeds water availability, water security is becoming a significant concern not only related to water availability but also to rigorous and costly requirements to remove conventional and emerging contaminants from effluents discharging into drinking water sources or as water reuse becomes an alternate water supply for communities in these regions. Water and wastewater treatment demands a great amount of energy and resources, highlighting the need for novel applications of the circular economy concept. *Circular Economy Applications for Water Security* examines knowledge gaps, avenues of future research, and challenges related to the potential of enhanced underutilized/waste materials as a transition to circular economy applications for ensuring the proper quality of water.

- This book includes fundamental information and practical examples that help to better understand the concepts included.
- The circular economy concept is helpful to incept sustainability in the water treatment processes.
- Every chapter includes the identification of knowledge gaps, avenues for further research, and challenges that guide readers toward real state-of-the-art analysis.
- Contributors are experts in their areas and will commit to explaining concepts in a user-friendly way without missing scientific rigor.

Circular Economy Applications for Water Security

Edited by
Erick R. Bandala

CRC Press
Taylor & Francis Group
Boca Raton London New York

CRC Press is an imprint of the
Taylor & Francis Group, an **informa** business

Designed cover image: Shutterstock

First edition published 2025
by CRC Press
6000 Broken Sound Parkway NW, Suite 300, Boca Raton, FL 33487-2742

and by CRC Press
4 Park Square, Milton Park, Abingdon, Oxon, OX14 4RN

CRC Press is an imprint of Taylor & Francis Group, LLC

ISBN: 978-1-032-57786-9 (hbk)
ISBN: 978-1-032-57790-6 (pbk)
ISBN: 978-1-003-44100-7 (ebk)

DOI: 10.1201/9781003441007

Typeset in Times
by codeMantra

Contents

PART I Fundamentals

PART II Case Studies

About the Editor

Dr. Erick R. Bandala is Director Research & Development at DASCO Inc. He holds a PhD in Engineering, an MSc in Organic Chemistry, and a BE in Chemical Engineering. His research interests in Environmental Engineering include: a) the water-energy-food nexus; b) water reliability; c) international water, sanitation, and hygiene (IWASH); d) climate change adaptation technologies for water reliability; and e) synthesis, characterization, and application of novel materials for environmental restoration. He is the author or co-author of 150 peer-reviewed papers in international journals (2022 average impact factor 5.8, >5,390 citations, h-index 41), five books, 30 book chapters, and 65 works published in the proceedings of international conferences. He is the Chief Editor of *Water in Air, Soil and Water Research*, and Associate Editor of *Frontiers in Environmental Science*.

Contributors

Thusalini Asharp
Department of Agricultural Engineering
University of Jaffna
Jaffna, Sri Lanka

Marcia Regina Assalin
Institute of Chemistry
UNICAMP
Campinas, Brazil

Erick R. Bandala
DASCO Inc.
Centennial, Colorado

Piotr F. Borowski
Department of Production Management
    and Engineering
Vistula University
Warsaw, Poland

Yaneth A. Bustos-Terrones
CONAHCYT-TecNM-Technological
    Institute of Culiacan
Division of Graduate Studies and
    Research
Culiacan, Mexico

Alain S. Conejo-Dávila
CIATEC
León, Mexico

Gabriel Contreras-Zarazua
CIATEC
León, Mexico

Thyerre Santana da Costa
Institute of Chemistry
UNICAMP Campinas, Brazil

Jorge del Real-Olvera
Environmental Technology
Center of Research and
    Assistance in Technology
    and Design of the State
    of Jalisco
Guadalajara, Mexico

Nelson Durán
Laboratory of Urogenital
    Carcinogenesis and Immunotherapy
UNICAMP
Campinas, Brazil

Catalina Ferat-Toscano
Division of Civil and Geomatics
    Engineering
National Autonomous University of
    Mexico
Mexico City, Mexico

Yolanda G. Garcia-Huante
Departamento de Ciencias Basicas
Instituto Politecnico Nacional
Mexico City, Mexico

Tania-Ariadna García-Mejía
Instituto de Ingenieria
Universidad Nacional Autonoma de
    Mexico
Mexico City, Mexico

M.A. Gomez-Gallegos
Departamento de Ingenieria Civil y
    Ambiental
Universidad de las Americas Puebla
Puebla, Mexico

Ashantha Goonetilleke
School of Civil and Environmental
  Engineering
Queensland University of Technology
  (QUT)
Brisbane, Australia

Leonel Hernández-Mena
Department of Environmental
  Technology
Center of Research and Assistance in
  Technology and Design of the State
  of Jalisco
Guadalajara, Mexico

Ajay Kumar
Department of Engineering and
  Innovation
The Open University
Milton Keynes, United Kingdom

Maelson Cardoso Lacerda
Institute of Chemistry
UNICAMP
Campinas, Brazil

Yukun Ma
School of Environment
Beijing Normal University
Beijing, China

José-Alberto Macías-Vargas
Instituto de Ingenieria
Universidad Nacional Autonoma de
  Mexico
Mexico City, Mexico

Wendy N. Medina-Esparza
Departamento de Medio Ambiente y
  Energia
Advanced Materials Research Center
Chihuahua, Mexico

Gabriela E. Moeller-Chavez
Academic Direction of Environmental
  Technology Engineering
Polytechnic University of the State of
  Morelos (UPEMOR)
Jiutepec, Mexico

Claudia Montoya-Bautista
Instituto de Ingenieria
Universidad Nacional Autonoma de
  Mexico
Mexico City, Mexico

Brenda S. Morales-Verdín
CIATEC
León, Mexico

Kannan Nadarajah
Department of Agricultural
  Engineering, Faculty of Agriculture
University of Jaffna
Jaffna, Sri Lanka

Alberto Ordaz
Department of Bioengineering
Tecnologico de Monterrey
Monterrey, Mexico

Natyeli A. Ortiz-Tirado
CIATEC
León, Mexico

Irwing Ramirez
School of Engineering and Innovation
The Open University
Milton Keynes, United Kingdom

Rosa-María Ramírez-Zamora
Instituto de Ingenieria
Universidad Nacional Autónoma de
  México, Instituto de Ingeniería
Coordinación de Ingeniería Ambiental
Ciudad de México, México

Kattia A. Robles-Estrada
Departamento de Materiales
CINVESTAV
Mexico City, Mexico

Ernestina Moreno Rodriguez
Departamento de Ingenieria Quimica
Universidad de las Americas Puebla
Puebla, Mexico

Oscar M. Rodríguez-Narvaez
Dirección de investigación y soluciones
 tecnológicas
Centro de Innovación Aplicada en
 Tecnologías Competitivas (CIATEC)
León, México

Lewis S. Rowles
Department of Civil Engineering and
 Construction
Georgia Southern University
Statesboro, Georgia

J.L. Sanchez-Salas
Department of Chemistry and
 Biological Sciences, Sciences School
Universidad de las Américas Puebla
Puebla, Mexico

Palistha Shrestha
Division of Hydrologic Sciences
Desert Research Institute
Reno, Nevada

Loveciya Sunthar
Department of Agricultural Engineering
University of Jaffna
Jaffna, Sri Lanka

Ljubica Tasic
Institute of Chemistry, Biological
 Chemistry Laboratory
Universidade Estadual de Campinas,
 UNICAMP
Campinas, Brazil

Karen Valencia-García
Instituto de Investigacion en Materiales
Universidad Nacional Autonoma de
 Mexico
Mexico City, Mexico

Déborah L. Villaseñor-Basulto
Departamento de Quimica
Universidad de Guanajuato
Guanajuato, Mexico

Buddhi Wijesiri
Faculty of Engineering
School of Civil and Environmental
 Engineering, Faculty of Engineering
Queensland University of Technology
 (QUT)
Queensland, Australia

Preface

In 1993, I got my first academic job when I joined a research group in Water Quality and Water Treatment. Several things have changed in the past three decades since I got that first position. Only the year before, in 1992, the United Nations (UN) published Agenda 2021 after the UN Conference on Sustainable Development in Rio de Janeiro, setting the basis for sustainable development. Less than a decade later, in 2000, the global representative agreed to create the Millennium Development Goals (MDGs), a compendium of eight goals to be achieved from 2000 to 2015 that centered on extreme poverty and hunger eradication, achieving universal primary education, promoting gender equality, reducing child mortality, improving maternal health, combating HIV/AIDS, reducing the prevalence of malaria, developing a global partnership for development, and ensuring environmental sustainability.

At the end of the 2015 deadline for the MDGs, the UN announced the adoption of the 2030 Agenda with the Sustainable Development Goals (SDGs) for a "shared blueprint for peace and prosperity for people and the planet." For many of us, the SDGs were the natural extension of the MDGs, splitting the eight original MGDs into detailed, specific, and more ambitious actions that were missed in the first iteration. The SDGs included no poverty; zero hunger; health and well-being; quality education; gender equality; clean water and sanitation; affordable and clean energy; decent work and economic growth; industry, innovation, and infrastructure; reduced inequalities and sustainable cities and communities; responsible consumption and production; climate action; life below water; life on land; and peace, justice, and strong institutions.

The most interesting feature of the SDGs was that they provided opportunities for crosscutting issues and generating synergies between them. In the case of water treatment/water quality, the change of mindset along the timeline of all these historical milestones has been mind-blowing, to say the least. Along with many other fields of knowledge, we have not only leaned toward maximizing the use of resources and minimizing waste, generating great synergy between SDG 6 (clean water and sanitation) and SDG 12 (responsible consumption and production), but also other more intermingled relationships, such as the nexus between food (SDG 2), energy (SDG 7), and water (SDG 6).

All the changes happening during the three decades I have dedicated to academic work in water treatment/quality are both interesting and challenging, and therefore, they deserve a detailed analysis, so that we can regroup, gather our thoughts, and design a path forward. This book is the work of enthusiasts and experts in the field aimed at exploring one interesting, feasible, affordable pathway to contribute to the achievement of the collective goals, SDG 6. I have made the commitment to gather a group of bright minds and ask them to prepare contributions that reflect a way to ensure access to clean water and proper sanitation for everyone while ensuring the unitary operations involved will not generate a worse environmental mess and will support poverty reduction, prevent hunger, encourage good health and well-being, provide for quality education, and encourage gender equality.

You, esteemed reader, will be the best judge of our efforts in delivering useful, up-to-date information that we hope will generate new ideas and ignite the enthusiasm to continue exploring the concepts included in the next chapters of this book. I hope you will have as much fun reading our work as I had editing it for you.

Sincerely,

Erick R. Bandala, PhD, PH
Director, Research & Development
DASCO Inc. (www.dascoinc.com)
9000 E. Nichols Ave. Suit 205, Centennial, Colorado 89119-7363
Tel: (720) 459 3146
e-mail: ebandala@dascoinc.com

Introduction

In 2015, the United Nations General Assembly approved the 2030 Agenda for Sustainable Development, describing an action plan that benefits people, the planet, and community prosperity. The objectives of the 2030 agenda include strengthening universal peace, eradicating poverty, and moving the human collective toward sustainable development [1]. As its main outcome, a collection of 17 interlinked objectives, the Sustainable Development Goals (SDGs), were proposed to enhance social, environmental, and economic sustainability [2]. The SDG focused on ensuring clean water and sanitation for all, SDG 6, is probably one of the most fundamental and is devoted to securing access to safe drinking water, sanitation, and hygiene as a basic human need for health and well-being [3].

Climate change has aggravated the imbalance between water demand and supply, as well as other environmental stressors such as water contamination from potentially toxic pollutants or degradation from point- and diffuse-source pollution from anthropogenic (e.g., industrial and agricultural) activities [4]. To fulfill the SDG 6 objectives, a broad assortment of technologies have been developed intended to generate safe drinking water and/or remove pollutants from wastewater influents to avoid detrimental environmental impacts and enhance wastewater reuse [5]. As continued population growth has increased the global demand for safe water, the production of clean drinking water or reclamation of wastewater requires treatment processes that also generate unavoidable by-products [6]. The materials used for water treatment usually have a very short life and may only be used once or a few times to minimize the loss of their treatment properties, which severely impacts investment capital and operational expenditures [7]. This practice is not only outdated and incongruent with the goal of responsible consumption and production (e.g., SDG 12), but it also highlights the need to quickly adapt to new information and changing trends.

Circular economy (CE) is a relatively new concept coined to promote resource efficiency and sustainability by minimizing the use of natural resources while supporting economic growth [8]. CE increases waste minimization, and it has been consistently applied to enhance the chances of achieving different SDGs [9]. However, relatively little is known about applying the concept of CE in water and wastewater treatment besides the evident applications of wastewater reuse for either indirect potable use or agricultural, nutrient, and energy recovery applications [10]. Limiting the CE concept to water reuse and water/energy recovery seriously restricts innovative approaches, such as the recovery of biomolecules and organic/inorganic compounds, valorization of sewage sludge for agricultural reuse, production of carbon-based and metallic materials with enhanced adsorptive or catalytic capabilities [11], and application of underused/waste materials to generate adsorbents or catalysts to remove organic pollutants by adsorption or advanced oxidation processes [12].

This book was conceived as a state-of-the-art compendium in which the fundamental concepts of CE, SDGs, and life cycle analysis are explained and case studies

DOI: 10.1201/9781003441007-1

with applications are discussed by experts from different regions of the world (e.g., Africa, America, Asia, Europe, and Oceania), including their views on the most pressing knowledge gaps and interesting avenues for future research.

REFERENCES

1. Rodriguez-Anton, J. M., Rubio-Andrada, L., Celemín-Pedroche, M. S. & Alonso-Almeida, M. D. M. Analysis of the relations between circular economy and sustainable development goals. *Int. J. Sustain. Dev. World Ecol.* **26**, 708–720 (2019).
2. Rodríguez-Antón, J. M., Rubio-Andrada, L., Celemín-Pedroche, M. S. & Ruíz-Peñalver, S. M. From the circular economy to the sustainable development goals in the European Union: an empirical comparison. *Int. Environ. Agreements Polit. Law Econ.* **22**, 67–95 (2022).
3. Chen, T. L. et al. Implementation of green chemistry principles in circular economy system towards sustainable development goals: Challenges and perspectives. *Sci. Total Environ.* **716**, 136998 (2020).
4. Baskar, A. V. et al. Recovery, regeneration and sustainable management of spent adsorbents from wastewater treatment streams: A review. *Sci. Total Environ.* **822** (2022).
5. Padhye, L. P., Bandala, E. R., Wijesiri, B., Goonetilleke, A. & Bolan, N. Hydrochar: A promising step towards achieving a circular economy and sustainable development goals. *Front. Chem. Eng.* **4**, 867228 (2022).
6. Nguyen, M. D., Thomas, M., Surapaneni, A., Moon, E. M. & Milne, N. A. Beneficial reuse of water treatment sludge in the context of circular economy. *Environ. Technol. Innov.* **28**, 102651 (2022).
7. Saidulu, D., Srivastava, A. & Gupta, A. K. Enhancement of wastewater treatment performance using 3D printed structures: A major focus on material composition, performance, challenges, and sustainable assessment. *Journal Environ. Manag.* **306**, 11461 (2022).
8. Dolgen, D. & Alpaslan, N. M. Reuse of wastewater from the circular economy (CE) perspective. In *A Sustainable Green Future, Perspectives on Energy, Economy, Industry, Cities and Environment* (ed. Oncel, S. S.) 643 (Springer, 2023). doi:10.1007/978-3-31-24942-6.
9. Gonzalez-Gonzalez, R. B., Iqbal, M. N., Bilal, M. & Parra-Saldivar, R. (Re)-thinking the bio-prospect of lignin biomass recycling to meet Sustainable Development Goals and circular economy aspects. *Curr. Opin. Green Sustain. Chem.* **38**, 100699 (2022).
10. Kumar, P. G., Kanmani, S., Kumar, P. S. & Vellingiri, K. Efficacy of simultaneous advanced oxidation and adsorption for treating municipal wastewater for indirect potable reuse. *Chemopshere* **321**, 138115 (2023).
11. Guerra-Rodríguez, S., Oulego, P., Rodríguez, E., Singh, D. N. & Rodríguez-Chueca, J. Towards the implementation of circular economy in the wastewater sector: Challenges and opportunities. *Water (Switzerland)* **12**, 1431 (2020).
12. Rodriguez-Narvaez, O. M., Peralta-Hernandez, J. M., Goonetilleke, A. & Bandala, E. R. Biochar-supported nanomaterials for environmental applications. *J. Ind. Eng. Chem.* **78**, 21–33 (2019).

Part I

Fundamentals

1 Sustainable Developing Goals

SDG 6 and Its Significance for Sustainable Growth

Erick R. Bandala

1.1 INTRODUCTION

The 17 Sustainable Development Goals (SDGs) agreed on by the United Nations are devoted to ensuring peace and prosperity and safeguarding the environment from human-induced pollution and climate change [1]. The SDGs are considered by some to be the last remaining opportunities to achieve sustainable growth, despite significant controversy about their progress, which some authors suggest is lagging behind the stipulated 2030 target because of continued environmental degradation, increasing social problems (e.g., poverty, inequality, and inequity), and financial and economic instability [2]. Although weighing the priority of the different SDGs is complicated, SDG 6 has been systematically reported to have a catalytic effect on other goals linked to water [3]. Because water permeates all aspects of our daily lives, securing water availability with proper quality and in the necessary quantity is essential not only for survival but also for maintaining healthy ecosystems, which in turn translates to community well-being [2]. Access to water with the proper quality promotes healthy communities and generates economic development (not to mention the different trans-sectorial interdependencies that involve water as a key natural resource, such as the water–food–energy nexus), which highlights its role in the sustainable development framework [4]. Therefore, clean water availability is critical to achieving sustainable development and is fundamental to the successful completion of all the proposed SDGs.

Achieving SDG 6 requires innovative, cost-effective technologies that prevent pollutants from entering water sources, remove pollutants from contaminated water bodies, and protect drinking water from pathogenic microorganisms, which require significant effort and economic resources that need to be used wisely and with great efficiency [5]. As water and wastewater regulations become more stringent, the need to incorporate advanced treatments into conventional water/wastewater treatment plants increases, which is frequently associated with increased economic demands. The environmental impacts of upstream and/or downstream water/wastewater treatment processes are an underexplored part of the water supply chain [6]. Single-use materials needed to ensure water quality contribute to the

DOI: 10.1201/9781003441007-3

5

significant environmental footprint of water treatment processes, and therefore, finding cost-effective alternatives to reduce that footprint must be explored to allow sustainable and continuous implementation in economically challenged regions [7]. An interesting alternative for tackling this problem is to look at underused or waste materials that, with the proper treatment, could become as effective as other materials already on the market.

The goals of this chapter are to summarize the main requirements related to achieving SDG 6, identify the knowledge gaps that need attention to meet these global commitments, and suggest avenues for future research that may help fill the identified gaps.

1.2   BRIEF ANALYSIS OF SDG 6 ECONOMIC NEEDS

Although a detailed economic analysis is included in Chapter 2, this section provides a broad overview of the economic needs already identified for the successful achievement of SDG 6 to prove a point about the need for cost-effective materials for water treatment and sanitation.

There is significant controversy about the cost estimation for successfully accomplishing SDG 6, with estimates varying widely across different studies from ca. $30 billion to $1.1 trillion per year [8]. Using a simple unit costs approach to include community wells and latrines, access to handwashing stations with soap and piped water, and fecal waste treatment for safely managed water, sanitation, and hygiene (WASH), a study from 2016 estimated universal access to basic WASH services required $15–$50 billion annually [9]. In a more recent study, the Organization for Economic Cooperation and Development (OECD) estimated a much higher annual investment need of $0.9 trillion [10], which included maintaining and upgrading urban water and sanitation infrastructure [11]. Beyond drinking water and sanitation, however, investments are required to enhance water resources management infrastructure to ensure the impacts of climate change are addressed. Projected water demands and energy use, as well as behavioral change and population growth scenarios, need to be considered to estimate investment needs for SDG 6 and account for climate change. Besides the water withdrawal and distribution infrastructure required to meet increased demand, infrastructure needs to ensure water supply (e.g., pollution control, wastewater recycling, and desalination), reduce water demand (e.g., improved household, agricultural, and industrial water-use efficiency), and allow for integrated water resources management. Estimates of the total annual investment to meet SDG 6 targets derived by this method range from $445 to $885 billion [12]. Any study conducting a comprehensive needs estimate for SDG 6 targets must include full capital and operational expenditures to secure access to drinking water and sanitation, agricultural and industrial water pollution control, scarcity scenarios, regulations, and integrated water management. After all these costs have been included, an estimate of $1.12 trillion has been suggested, which is the highest estimated investment gap for SDG 6 and one of the highest for any SDG [13]. If only the basic WASH access targets are considered for SDG 6, the estimated investment gap is $46–50 billion, which is comparable to recent results that looked at similar targets [8].

The significant disparity in economic estimates related to the investment costs of achieving SDG 6 is a knowledge gap that requires urgent attention because we need to have a clear idea of the economic burden associated with water security and provide decision-makers and stakeholders with the most accurate numbers so that proper decisions can be made based on accurate, reliable information.

1.3 WATER-TREATMENT-RELATED COSTS

As stated in Section 1.2, making safe water available for all involves significant effort and demands attention to identifying where technology contributions may help reduce the economic burden. The investment and operational costs related to water treatment depend on a variety of parameters but, for the scope of this book, they are simplified to (i) water source, (ii) water use, (iii) degree of water pollution, and (iv) treatment method [14]. The following subsections are devoted to further analyzing these basic parameters.

1.3.1 WATER SOURCE

Depending on the origin of the water (e.g., surface water, groundwater, reclaimed water, seawater, or brackish water), its characteristics will vary based on a wide variety of components that may need to be removed before it can be considered safe for human consumption. For example, groundwater has the most constant quality over time because the filtering capacity of soil helps eliminate suspended solids and, in some cases, microorganisms [15]. Some dissolved solids (e.g., fluoride, arsenic, and carbonates) that require further treatment may also be added to water as it passes through soil layers [16]. Surface water encompasses most of the water usage worldwide, and it is a significantly variable source (in quantity and quality) over time because it is easily impacted by weather conditions and/or anthropogenic activities [17]. Reclaimed water refers to wastewater that has been submitted to the required treatment process, which allows it to be used for beneficial purposes. Usually because of social acceptance and a lack of proper treatment, reclaimed water only constitutes a small amount of water usage compared with surface water or groundwater [18]. Seawater and brackish water are alternative water sources that have attracted significant interest in recent years, mainly in countries where accessing conventional water sources (e.g., surface water and groundwater) is difficult. The high content of dissolved solids (>35 g/L in the case of seawater) makes its treatment necessary before human use can be proposed [19].

Different organizations and scholars as well as regional, national, and local regulators have attempted to provide guidelines on the required treatment depending on the water source [20,21]. Although commendable, these attempts have proven to be insufficient to address all the different sources (conventional and alternative) and water quality situations faced by the water users in need of determining the best approach to deal with their local water-related challenges. Therefore, generating general guidelines to help water professionals address these different challenges in a dynamic, adaptable way is another knowledge gap worth exploring. Although the task may appear complex, a variety of artificial intelligence (AI) applications, such as machine learning algorithms, have been suggested to be useful for collecting, interpreting, and compiling water-related data that may serve this purpose [22].

1.3.2 WATER USE

The different ways water is used dictate the water treatment required to ensure it will be fit for its intended use [23]. Water use (e.g., public supply, irrigation, or industry) correlates with the water source characteristics described above and the size of the treatment facilities, which also contribute to investment and operational costs [14]. For example, reclaimed water is a well-known and highly interesting alternative source for irrigation and some industrial uses, but other nonconventional water sources (e.g., stormwater and rainwater) have also been identified with the potential to help alleviate water shortages, mainly in arid and semiarid locations [23].

Water use is affected by the quality of the water source, which is closely related to climate change stressors because higher temperatures affect several of the chemical and biological processes that occur in water bodies [24,25]. However, little is known about the impacts of stressors such as climate change on water use from other conventional water sources (such as groundwater or reclaimed water). Although there is a significant amount of information available on the impacts of climate change on water availability, less is known about the impacts of these stressors on water quality [26], and even less is known about their impacts on nonconventional sources (e.g., stormwater and rainwater) [27]. Information about how climate change or other stressors may affect water use from different sources or for different applications is another significant knowledge gap that requires attention. Generating scenarios that provide information on the effects of stressors (e.g., climate change, population growth, and land-use changes) on water use for both conventional and nonconventional water sources is key to generating the fundamental regulations and adaptation technologies to achieve SDG 6.

1.3.3 WATER POLLUTION DEGREE

Characterizing the pollution degree of different water sources to identify proper treatment is a paramount but very complex task. Depending on the source type, the degree of water pollution may vary spatially and temporally, which poses additional challenges for water professionals [14]. For example, the type and load of conventional pollutants included in wastewater vary significantly depending on whether it is municipal (e.g., gray water or black water) or industrial effluent, with the latter being significantly affected by the industrial process generating the wastewater effluent (e.g., extraction, transformation, or service sector) [28]. The presence of nonconventional pollutants in wastewater effluents—and eventually surface water, groundwater, brackish water, and seawater—has generated significant public concern because of the known and unknown effects these pollutants have on human health [29]. Emerging pollutants such as microplastics, antibiotics, and polyfluorinated alkyls in different ecological spheres (even pristine sites expected to be unaltered by anthropogenic activities) have created the need for novel treatment technologies capable of dealing with these undesirable substances [30,31].

1.4   ENVIRONMENTAL FOOTPRINT OF WATER TREATMENT PROCESSES

Although water and wastewater treatments help protect human health and the environment, the processes involved in these unitary operations also have negative environmental impacts, such as generating greenhouse gases or by-products that require proper disposal [32]. For example, desalination of seawater or brackish water as an alternative water source for human consumption or agricultural applications in arid regions has become crucial to combatting global water shortages. However, the discharge of the rejected hypersaline brine steam, the high-power demand of the membrane process, and the generation of waste materials also contribute to the significant environmental footprint of the desalination process [33].

Water desalination is not the only water treatment process capable of generating waste materials as one of its main outcomes. Several other processes (both conventional and emerging) are known to produce a secondary effluent of collateral derivatives, such as sludge, exhausted adsorbents, used nanoparticles, or consumed electrodes that may generate significant environmental impacts if improperly disposed of [34]. The need for alternative materials produced from waste or underserved sources that can be used in water and/or wastewater treatment processes and/or with characteristics that allow their secondary use in future applications with additional value added is a highly interesting avenue for future research. These types of materials would help reduce the environmental footprint of current water treatment processes and increase their sustainable application. Many of these concepts are further developed in the chapters included in the second part of this book and schematically shown in Figure 1.1. As shown in Figure 1.1, the whole idea is to identify waste or underserved materials related to or resulting from human activities to generate value-added products that may be used to remove water and/or wastewater pollutants. After pollutants have been removed, the treated water/wastewater effluents can be safely used or reused for other applications that may generate raw materials for human use, which closes the circular economy cycle and provides other potential benefits.

1.5   SDG 6-RELATED CHALLENGES

Several challenges are threatening SDG 6 achievement in time to complete the tasks by the 2030 proposed deadline. This section discusses some of the most significant challenges, but chapters in the second part of this book will also expand on these and other specific challenges identified by the contributors.

1.5.1   CLIMATE CHANGE

Growing populations and the increasing vulnerability of urban areas to extreme weather events, climate variations, and warming trends are stressors that likely will be outside of the range of historical extremes [35–37] and are expected to cause changes not only in mean temperature but also in extreme weather episodes. The exposure of cities to extreme heat appears to be more critical under climate change

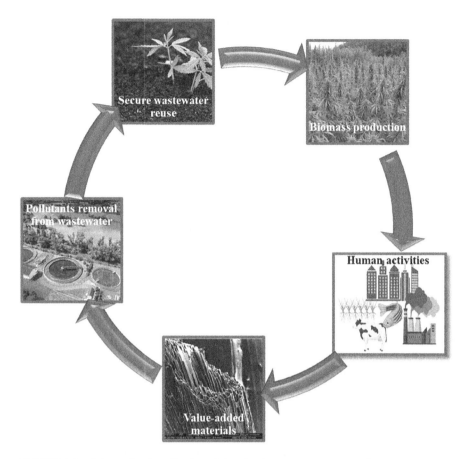

FIGURE 1.1   Schematic of application of the circular economy concept for water security.

scenarios and is exacerbated by the high solar absorption of urban environments (i.e., urban heat island [UHI]) [38]. Decision-makers require more information on local options and trade-offs to take adaptation actions to improve resilience in natural resources management and infrastructure design, as well as the well-being of residents [39]. Exposure to extreme heat is a very interesting example. Urban adaptation can significantly reduce the impact of heat events through vegetation cover, permeable surfaces, green space, reflective rooftops, cooling corridors, and shade from building heights and trees, as well as by taking advantage of wind and breezeways in zoning. Urban communities in arid or semiarid locations face significant challenges in adaptation options because of resource constraints related to water. For example, Las Vegas, Nevada, is one of the most vulnerable cities to climate change because it relies on Colorado River water, its location in the warming and drying Southwest, and the prevalence of extreme heat events [40]. It is also the most rapidly warming urban area in the United States [41] and is ranked as the location with the most intense UHI effect [42]. The region is also facing extreme water scarcity because of a 12-year drought that has affected water flow in the Colorado River basin and reduced the water level in Lake Mead to a historic low. One suggested approach to deal with such a complex problem is to find secure alternative water sources that can be used instead of freshwater to increase greenery in urban communities. This requires assessing the water needs of urban greenery by creating smart desert urban forestry (defined as a collection of trees and plants in a city, town, or suburb where the required sensing array is deployed to measure all the variables related to its development) and assessing the requirements to build and maintain it, as well as the benefits it provides to community well-being.

The water scarcity crisis faced by Las Vegas and other urban communities in the southwestern United States has become so critical that the Southern Nevada Water Authority recently declared a ban on the use of natural turf grass within the city and a program is currently in place to offer economic incentives to community members to switch out turf grass in front and backyards for rocks or pebbles, which may exacerbate UHI and increase heat wave events. Heat waves are projected to become more frequent and intense because of climate change [43]. People in under-resourced neighborhoods are expected to struggle to cope with the health effects associated with extreme weather events [44], of which heat waves are the deadliest [45]. Different studies have identified the health impacts of heat waves, which include respiratory illness, cognitive effects, heat stroke and exhaustion, and increased hospitalizations for all causes [44]. Extreme heat events pose particular risks for vulnerable populations, especially those in urban areas [46] and those lacking adequate shelter [47–49]. Morbidity and mortality associated with extreme heat exposure are increasing overall, but they most profoundly impact vulnerable groups with limited resources [50].

Urban desert regions with high concentrations of vulnerable populations face intensifying risks for heat-related morbidity and mortality. A shortage of affordable housing usually forces trade-offs that threaten public health in extreme weather events [51]. In 2021, one in five Nevada households reported keeping their homes at unsafe or unhealthy temperatures to reduce energy costs [52], which increased the risk of heat-related illness [53]. The state also has one of the highest rates of unsheltered homelessness [54], which is defined as sleeping on the streets, in parks, or vehicles. Over half of Nevada's homeless population is unsheltered [54] and nearly two-thirds of individuals experiencing homelessness are unsheltered on a given night [55], and therefore face urgent heat-related risks given the extreme desert climate and limited water resources [46]. The UHI contributes to elevated temperatures in metropolitan areas with large numbers of inadequately housed and unsheltered individuals who experience enduring exposure to extreme heat. The lack of affordable energy and housing costs threatens to worsen health impacts as climate change intensifies.

Dwindling water supply because of increased drought risk and pollution further hinders the well-being of inadequately housed and homeless individuals. Sanitation and hygiene pose major challenges for these groups because of limited access to clean water [56]. Furthermore, unsheltered homelessness can contribute to the contamination of existing water sources [57], compounding shortages in areas prone to drought conditions. Water shortages directly impede the well-being of inadequately housed and homeless individuals, as well as opportunities to leverage heat mitigation strategies such as increased vegetation cover. Several challenges impede efforts to mitigate climate-related risks for inadequately housed and unsheltered individuals.

Marginalized populations frequently mistrust and are hesitant to engage with service providers and social systems [58,59]. Those unable to meet housing and energy needs may be hard to reach due to residential mobility or other barriers, such as a lack of transportation. Extreme financial hardship forces difficult trade-offs that may appear illogical to providers and policymakers [60,61], and traditional modes of service and resource delivery can lead to disparities in access and benefit [59]. Failure to accommodate the unique needs of this population increases the public health consequences of extreme heat as climate change worsens and heat waves intensify. Innovative approaches are needed to overcome health risks faced by inadequately housed and homeless populations in the context of rising temperatures and dwindling water supplies. Restructuring outdoor and communal spaces is a crucial leverage point for protecting the large unsheltered homeless population. Increased canopy cover to offset UHI can directly reduce the risk of heat-related morbidity and mortality in Las Vegas, as well as promote more distal indicators of well-being associated with access to green space [62,63]. Securing enough water to accomplish all these activities is necessary for implementing all these measurements and ensuring the well-being of the community. However, conventional water sources cannot be included in the equation and the need for alternative sources is fundamental, which may also imply the need for fit-for-use treatment to generate the amount and quality of water needed to expand urban greenery and generate some relief for the inhabitants of these urban communities during hot weather events. This information, however, is missing and represents another interesting avenue for future research that requires urgent attention.

1.5.2 ENDOCRINE-DISRUPTING CHEMICALS

Another SDG 6 critical challenge is securing access to safe water. Climate change, industrialization, high rates of urbanization, and population growth have resulted in many countries suffering from water crises, especially in arid and semiarid areas [64]. Countries in different regions of the world have also been struggling over regional water availability and it is anticipated that these struggles may result in conflicts over shared water resources [20]. Considering the adverse consequences of the water crisis, countries have increasingly been trying to cope with the problem of water availability by implementing sustainable water management plans and looking for alternative water sources [64]. In recent years, more countries have been considering water reuse to supplement freshwater sources [20]. Water reuse decreases the pressure on freshwater resources, reduces the pollution being discharged into water bodies, and can be a reliable source compared with other water resources that are directly dependent on rainfall [23]. Global wastewater discharge is estimated to be 400 billion m^3/year, but with the recent developments in wastewater treatment technologies, the worldwide volume of recycled water in the 2010–2015 period increased from 33.7 million m^3/d to 54.5 million m^3/d. Another report suggested that 7.1 billion m^3/year of treated wastewater is reused for irrigation and at least 20 million hectares in 50 countries are irrigated with wastewater [65]. Wastewater production is expected to reach 470 billion m^3/year by 2030 and 574 billion m^3/year by 2050, which could provide a dependable and sustainable way to address water scarcity in dry areas [66].

Currently, global annual agricultural water consumption exceeds 2,700 billion m³, accounting for 70% of global freshwater withdrawal, and it is projected to reach 3,200 billion m³ by the middle of the century [67]. Agricultural production is closely connected to water resources, particularly in arid and semiarid areas where agricultural production largely depends on irrigation [68]. As population growth increases, the high agricultural demand for water may lead to two-thirds of the global population (approximately 6 billion) facing a water crisis related to food production, which will only be exacerbated by climate change [67]. There are 275 million hectares (ha) of land consisting of irrigated agriculture worldwide, of which approximately 15% (40 million ha) could be irrigated using wastewater [69]. Of the total global water withdrawal, 11% was for municipal water demand, and 32% was discharged as agricultural wastewater and drainage [69]. Most of the wastewater recycled for agriculture is municipal wastewater, which suggests the need to change water reuse policies and plans from municipal wastewater management to the sustainable management of municipal and agricultural wastewater [64].

One of the barriers to water reuse is the potential adverse effects on human and environmental health [70]. Even today, treated and untreated wastewater reuse is poorly controlled. Over 80% (and over 95% in developing countries) of the world's wastewater is released into the environment without treatment [31]. The current challenge is to make reuse circuits shorter, safer, and economically sustainable. Although recycled water reduces pressure on freshwater sources, implementing properly treated and safe water reuse practices is the ultimate goal [64]. One of the basic needs for safely using recycled water is ensuring its quality [71]. Without proper treatment, wastewater can be loaded with organic matter, suspended solids, nitrate, phosphate, heavy metals, pathogens, and endocrine-disrupting chemicals (EDCs; e.g., estrogens and hormones) [65]. Table 1.1 shows the concentration levels of select EDCs in different water types. These contaminants cause severe soil, crop, and groundwater problems, as well as affect the health of farmers and consumers [31].

TABLE 1.1
Concentration of Selected EDCs in Different Water Types

EDC	Concentration, ng/L			
	Surface Water	Groundwater	Drinking Water	Wastewater
17b-Estradiol	<0.1–6.0 [72–74]	13–80 [75]	0.2–2.1 [73]	1.5–650 [73,76]
Estrone	<0.1–17 [73,74]	–	0.2–0.6 [73]	<0.1–19 [73,77]
Ethynylestradiol	0.1–5.1 [73]	–	0.15–0.5 [73]	0.1–8.9 [73,78]
PFOA	143 [79]	Up to 21,200 [80]	0.0051–2.0 [81]	8.4* [82]
Estriol	1.0–2.5 [74]	–	–	5.0–7.3 [74]
Sulfamethoxazole	10 µg/L [83]	0.2–1,100 [84]	0.1–32 mg/L [85]	10–1,500 [85]
Bisphenol A	0.5–250 [73,74,78]	3–1,410 [86]	0.5–44 [73,86]	4.8–258 [73,78]
Nonylphenol	100–15,000 [87]	200–760 [88]	2.5–2,700 [73]	18–770 [73,78]

* U.S. national mean concentration.

EDCs have gained notoriety because they are known to interfere with the reproduction and development of aquatic wildlife [89,90]. In addition, climate change scenarios predict that changes in environmental conditions may influence the fate and transport of, and human exposure to, EDCs by reducing the capacity of water bodies to dilute contaminants and increasing the need for wastewater reuse [91]. Concentrations of EDCs in water are expected to increase because of drought and decreased removal efficiency under anoxic conditions, resulting in concerning scenarios for sites where their presence has been reported [92] because the release of EDCs into aquatic ecosystems affects fish health and reproduction [93]. A recent national-scale study examined the extent of effluent-dominated streams and rivers in the United States and identified exceedances of levels of concern for EDCs in 60% of the cases, particularly under low-flow conditions when a lack of precipitation limits pollutant dilution [94].

Exposure to EDCs is linked with decreased fertility, changed sexual behavior, and amplification of abnormalities and cancers in humans and laboratory animals [95]. Because conventional treatments in wastewater treatment plants cannot eliminate EDCs, these contaminants are released into the receiving water bodies. For example, 40% of all oral contraceptives consumed, which act as EDCs, end up in sewage streams and are released into the environment [96]. As a result, chemicals with endocrine activity have been detected in groundwater, recycled water, and wastewater effluents and have also been reported in drinking water sources [97–100]. There is a lack of environmental toxicity information related to the presence of EDCs in water, but the available information on individual estrogens shows that EDCs have severe effects on gonad development and reproduction [101–104]. Previous studies all around the country [105–107] have reported a link between the presence of EDCs and the observed endocrine disruptive activity in fish species. Exposure to EDCs has been identified at Lake Mead that is associated with the discharge of treated wastewater [108,109]. Some of these studies have shown endocrine disruptive activity in common fish (*Cyprinus carpio* and *Micropterus salmoides*) in the Lower Colorado River [110,111]. As mentioned before, conventional water and/or wastewater treatments were not designed to eliminate EDCs from water effluents, which makes their presence in water effluents and the lack of cost-efficient treatments for their removal a highly interesting knowledge gap that is worth attention. The presence of EDCs and other highly persistent, refractory, toxic pollutants dramatically reduces opportunities for applying the circular economy concept to water security and suggests the significance of research devoted to generating new materials with specific characteristics that will enhance the efficacy of eliminating such undesirable constituents in water. Exhaustive examples of these types of materials will be discussed in depth in the following chapters of this book by experts in their field.

1.6   CONCLUSIONS

SDGs are an interesting challenge and probably, as some authors have proposed in the past, the last real call for a sustainable future. Despite the current controversy over the achievability of the goals it is, I guess, a challenge we better address with the most of our energy and enthusiasm to try to ensure the future we will leave for further generations.

Among SDGs, SDG 6 remains to be successfully concluded and contributions, either big or small, should be welcome to help arrive at the finish line as expected and within the timeline. This chapter was intended as a collection of general information and get the readers in the context of the problem while setting the mood for the information and discussions in the next chapters of this book.

It is my desire that you, esteemed reader, may find the information above and in the next chapters useful and encouraging to bring your expertise and contribute with us to show that the needs of the many certainly should outweigh the needs of the few which is, in my humble opinion, the essence of the SDGs global effort and the endgame of this book.

REFERENCES

1. United Nations. The Sustainable Development Goals Report 2022. Web page, N/A https://sdgs.un.org/goals/goal6 (2022).
2. Padhye, L. P., Bandala, E. R., Wijesiri, B., Goonetilleke, A. & Bolan, N. Hydrochar: A promising step towards achieving a circular economy and sustainable development goals. *Front. Chem. Eng.* **4**, 867228 (2022).
3. Germann, V. et al. Development and evaluation of options for action to progress on the SDG 6 targets in Austria. *J. Environ. Manage.* **25**, 116487 (2023).
4. Zhang, J., Wang, S., Pradhan, P., Zhao, W. & Fu, B. Mapping the complexity of the food–energy–water nexus from the lens of sustainable development goals in China. *Resour. Conserv. Recycl.* **183**, 106357 (2022).
5. Bustos-Terrones, Y. A., Norman, L., Perez-Estrada, L. A., El Nemr, A. & Bandala, E. R. Advanced physico-chemical technologies for water detoxification and disinfection. *Front. Environ. Sci.* **11**, 1132758 (2023).
6. Mehmeti, A. & Canaj, K. Environmental assessment of wastewater treatment and reuse for irrigation: A mini-review of LCA studies. *Resources* **11**, 94 (2022).
7. Razman, K. K., Hanifiah, M. M. & Mohammad, A. W. An overview of LCA applied to various membrane technologies: Progress, challenges, and harmonization. *Environ. Technol. Innov.* **27**, 102803 (2022).
8. Kulkarni, S. et al. Investment needs to achieve SDGs: An overview. *PLoS Sustain. Transform.* **1**, e0000020 (2022).
9. Hutton, G. & Varaghese, M. *The Costs of Meeting the 2030 Sustainable Development Goal Targets on Drinking Water, Sanitation, and Hygiene* (The World Bank, 2016).
10. Filho, W. L. et al. The economics of the UN Sustainable Development Goals: Does sustainability make financial sense? *Discov. Sustain.* **3**, 20 (2022).
11. OECD. *Water and Cities: Ensuring Sustainable Futures* (OECD, 2015). doi:10.2166/9781780407609.
12. McCollum, D. L. et al. Energy investment needs for fulfilling the Paris Agreement and achieving the Sustainable Development Goals. *Nat. Energy* **3**, 589–599 (2018).
13. Strong, C., Kuzma, S., Vionnet, S. & Reig, P. Achieving Abundance: Understanding the Cost of a Sustainable Water Future. https://www.wri.org/research/achieving-abundance-understanding-cost-sustainable-water-future (2020).
14. Bhojwani, S., Topolski, K., Mukherjee, R., Sengupta, D. & El-Hawagi, M. M. Technology review and data analysis for cost assessment of water treatment systems. *Sci. Total Environ.* **651**, 2749–2761 (2019).
15. Laney, B., Rodriguez-Narvaez, O. M., Apambire, B. & Bandala, E. R. Water defluoridation using sequentially coupled *Moringa oleifera* seed extract and electrocoagulation. *Groundw. Monit. Remediat.* gwmr.12396 (2020) doi:10.1111/gwmr.12396.

16. McMahon, P. B., Brown, C. J., Johnson, T. D., Belitz, K. & Lindsey, B. D. Fluoride occurrence in United States groundwater. *Sci. Total Environ.* **732**, 139217 (2020).

17. Duran-Encalada, J. A., Paucar-Caceres, A., Bandala, E. R. & Wright, G. H. The impact of global climate change on water quantity and quality: A system dynamics approach to the US-Mexican transborder region. *Eur. J. Oper. Res.* **256**, 567–581 (2017).

18. Ricart, S., Rico, A. M., & Ribas, A. Risk-Yuck factor nexus in reclaimed wastewater for irrigation: Comparing farmers' attitudes and public perception. *Water (Switzerland)* **11**, 187 (2019).

19. Ghazi, Z. M. et al. An overview of water desalination systems integrated with renewable energy sources. *Desalination* **542**, 116063 (2022).

20. Shoushtarian, F. & Negahban-Azar, M. Worldwide regulations and guidelines for agriculturalwater reuse: A critical review. *Water* **12**, 971 (2020).

21. WHO. *Guidelines for Drinking-Water Quality* (World Health Organization, 2022).

22. Nasir, N. et al. Water quality classification using machine learning algorithms. *J. Water Process Eng.* **48**, 102920 (2022).

23. Goonetilleke, A. et al. Stormwater reuse, a viable option: Fact or fiction? *Econ. Anal. Policy* **56**, 14–17 (2017).

24. Quevedo-Castro, A., Bustos-Terrones, Y. A., Bandala, E. R., Loaiza, J. G. & Rangel-Peraza, J. G. Modeling the effect of climate change scenarios on water quality for tropical reservoirs. *J. Environ. Manage.* **322**, 116137 (2022).

25. Quevedo-Castro, A. et al. Temporal and spatial study of water quality and trophic evaluation of a large tropical reservoir. *Environments* **6**, 61 (2019).

26. Stigter, T. Y., Miller, J., Chen, J. & Re, V. Groundwater and climate change: threats and opportunities. *Hydrogeol. J.* **31**, 7–10 (2023).

27. Wijesiri, B., Bandala E., Liu, A. & Goonetilleke, A. A framework for stormwater quality modelling under the effects of climate change to enhance reuse. *Sustainability* **12**, 10463 (2020).

28. Rodriguez-Narvaez, O. M., Peralta-Hernandez, J. M., Goonetilleke, A. & Bandala, E. R. Biochar-supported nanomaterials for environmental applications. *J. Ind. Eng. Chem.* **78**, 21–33 (2019).

29. Zeidman, A., Rodriguez-Narvaez, O. M., Moon, J. & Bandala, E. R. Removal of antibiotics in aqueous phase using silica-based immobilized nanomaterials: A review. *Environ. Technol. Innov.* **20**, 101030 (2020).

30. Rodríguez-Narvaez, O. M., Goonetilleke, A., Perez, L. & Bandala, E. R. Engineered technologies for the separation and degradation of microplastics in water: A review. *Chem. Eng. J.* **414**, 128692 (2021).

31. Rodriguez-Narvaez, O. M., Peralta-Hernandez, J. M., Goonetilleke, A. & Bandala, E. R. Treatment technologies for emerging contaminants in water: A review. *Chem. Eng. J.* **323**, 361–380 (2017).

32. Maktabifard, M. et al. Comprehensive evaluation of the carbon footprint components of wastewater treatment plants located in the Baltic Sea region. *Sci. Total Environ.* **806**, 150436 (2022).

33. Shokri, A. & Fard, M. S. A comprehensive overview of environmental footprints of water desalination and alleviation strategies. *Int. J. Environ. Sci. Technol.* **20**, 2347–2374 (2023).

34. Bhowmik, D. et al. Multitudinous approaches, challenges and opportunities of bioelectrochemical systems in conversion of waste to energy from wastewater treatment plants. *Clean. Circ. Bioecon.* **4**, 100040 (2023).

35. Simolo, C., Brunetti, M., Maugeri, M. & Nanni, T. Evolution of extreme temperatures in a warming climate. *Geophys. Res. Lett.* **38**, L16701 (2011).

36. Trenberth, K. E. Framing the way to relate climate extremes to climate change. *Clim. Change* **115**, 283–290 (2012).

37. City of Las Vegas. City of Las Vegas, 2050 Master Plan. Report (2020).
38. Fischer, E. M. & Schar, C. Consistent geographical patterns of changes in high-impact European heatwaves. *Nat. Geosci.* **3**, 398–403 (2010).
39. Rosenzweig, C. et al. Enhancing climate resilience at NASA centers: A collaboration between science and stewardship. *Bull. Am. Meteorol. Soc.* **95**, 1351–1363 (2014).
40. Dame, U. of N. Notre Dame Global Adaptation Initiative. Web page N/A. https://gain.nd.edu (2021).
41. USGCRP. Fourth National Climate Assessment: Report-in-Brief. vol. II https://nca2018.globalchange.gov/ (2018).
42. Smith, E., Pattni, K., Saladino, C. & Brown, W. E. The Urban Heat Island Effect in Nevada. https://digitalscholarship.unlv.edu/bmw_lincy_env/1 (2020).
43. Intergovernmental Panel of Climate Change. Climate Change 2021: The Physical Science Basis (2021).
44. Gronlund, C. J., Zanobetti, A., Schwartz, J. D., Wellenius, G. A. & O'Neill, M. S. Heat, heat waves, and hospital admissions among the elderly in the United States. *Environ. Health Perspect.* **122**, 1187–1192 (2014).
45. Martínez-Austria, P. F. & Bandala, E. R. Heat waves: Health effects, observed trends and climate change. In Philip John Sallis (ed.), *Extreme Weather* 107–123 (2018). doi:10.5772/intechopen.75559.
46. Bandala, E. R. et al. Extreme heat and mortality rates in Las Vegas, Nevada: Inter-annual variations and thresholds. *Int. J. Environ. Sci. Technol.* **16**, 7175–7186 (2019).
47. Kinay, P., Morse, A. P., Villanueva, E. V., Morrisey, K. & Staddon, P. L. Direct and indirect health impacts of climate change on the vulnerable elderly population in East China. *Environ. Rev.* **27**, 295–303 (2019).
48. Svoboda, R. B. Health of the homeless and climate change. *J. Urban Heal.* **86**, 654–664 (2009).
49. Chambers, J. Global and cross-country analysis of exposure of vulnerable populations to heatwaves from 1980 to 2018. *Clim. Change* **163**, 539–558 (2020).
50. Bandala, E. R., Brune, N. & Kebede, K. Assessing the effect of extreme heat on workforce health in the southwestern USA. *Int. J. Environ. Sci. Technol.* (2022) doi:10.1007/s13762-022-04180-1.
51. National Low Income Housing Coalition. The Gap: A Shortage of Affordable Homes. https://reports.nlihc.org/sites/default/files/gap/Gap-Report_2021.pdf (2021).
52. U.S. Census Bureau. Household Pulse Survey Data Tables. Web page N/A https://www.census.gov/programs-surveys/household-pulse-survey/data.html (2021).
53. Cardoza, J. E. et al. Heat-related illness is associated with lack of air conditioning and pre-existing health problems in Detroit, Michigan, USA: A community-based participatory co-analysis of survey data. *Int. J. Environ. Res. Public Health* **17**, 5704 (2020).
54. Henry, M., Watt, R., Mahathey, A., Ouellette, J. & Sitler, A. The 2019 Annual Homeless Assessment Report (AHAR) to Congress Part I: Point-in-Time Estimates of Homelessness. https://www.huduser.gov/portal/sites/default/files/pdf/2019-AHAR-Part-1.pdf (U.S. Department of Housing and Urban Development, 2019).
55. Southern Nevada Homelessness Continuum. Homelessness Census Comprehensive Report. 2019. https://helphopehome.org/wp-content/uploads/2019/09/2019-Homeless-Census-Narratives-and-Methodology-Final-2.0.pdf (2021).
56. Yoon, J., Klassert, C. & Selby, P. A coupled human–natural system analysis of freshwater security under climate and population change. *Proc. Natl. Acad. Sci.* **118**, 431118 (2021).
57. Verbyla, M. E., Calderon, J. S. & Flanigan, S. An assessment of ambient water quality and challenges with access to water and sanitation services for individuals experiencing homelessness in riverine encampments. *Environ. Eng. Sci.* **38**, 389–401 (2021).

58. Magwood, O., Leki, V. Y. & Kpade, V. Common trust and personal safety issues: A systematic review on the acceptability of health and social interventions for persons with lived experience of homelessness. *PLoS One* **14**, e0226306 (2019).
59. Ramsay, N., Hossain, R., Moore, M., Milo, M. & Brown, A. Health care while homeless: Barriers, facilitators, and the lived experiences of homeless individuals accessing health care in a Canadian Regional Municipality. *Qual. Health Res.* **29**, 1839–1849 (2019).
60. Beatty, T. K. M., Blow, L. & Crossley, T. F. Is there a "heat-or-eat" trade-off in the UK? *J. R. Stat. Soc. Ser. A, Stat. Soc.* **177**, 281–294 (2014).
61. Mullainathan, S. & Shafir, E. Scarcity: *Why Having Too Little Means So Much* (Times Books, 2013).
62. Koprowska, K., Kronenberg, J., Kuzma, I. B. & Laszkiewicz, E. Accessibility and attractiveness of urban green spaces to people experiencing homelessness. *Geoforum* **113**, 1–13 (2020).
63. van den Berg, M. et al. Health benefits of green spaces in the living environment: A systematic review of epidemiological studies. *Urban For. Urban Green.* **14**, 806–816 (2015).
64. Eslamian, S. *Urban Water Reuse Handbook* (CRC Press, 2016).
65. Ungureanu, N., Vladut, V. & Voicu, G. Water scarcity andwastewater reuse in crop irrigation. *Sustainability* **12**, 9055 (2020).
66. Qadir, M. et al. Global and regional potential of wastewater as awater, nutrient and energy source. *Nat. Resour. Forum* **44**, 40–51 (2020).
67. He, G. et al. Impact of food consumption patterns change on agricultural water requirements: An urban–rural comparison in China. *Agric. Water Manag.* **243**, 106504 (2021).
68. Pisinaras, V., Paraskevas, C. & Panagopoulos, A. Investigating the effects of agriculturalwater management in a Mediterranean Coastal Aquifer under current and projected climate conditions. *Water* **13**, 108 (2021).
69. WWP. The United Nations World Water Development Report 2017-Wastewater: The Untapped Resource, The United Nations Environmental Programme (2017).
70. Deviller, G., Lundy, L. & Fatta-Kassinos, D. Recommendations to derive quality standards for chemical pollutants in reclaimed water intended for reuse in agricultural irrigation. *Chemosphere* **240**, 124911 (2020).
71. Khan, N. Natural ecological remediation and reuse of sewage water in agriculture and its effects on plant health. *Sewage* **1**, 1 (2018).
72. Dorabawila, N. & Gupta, G. Endocrine disrupter – estradiol – in Chesapeake Bay tributaries. *J. Hazard. Mater.* **120**, 67–71 (2005).
73. Kuch, H. M. & Ballschmiter, K. Determination of endocrine-disrupting phenolic compounds and estrogens in surface and drinking water by HRGC-(NCI)-MS in the picogram per liter range. *Environ. Sci. Technol.* **35**, 3201–3206 (2001).
74. Cargouet, M., Perdiz, D., Mouatassin-Souali, A., Tamisier-Karolak, S. & Levi, Y. Assessment of river contamination by estrogeniccompounds in Paris area (France). *Sci. Total Environ.* **324**, 55–66 (2004).
75. Wicks, C., Kelley, C. & Peterson, E. Estrogen in a karstic aquifer. *Groundwater* **42**, 384–389 (2004).
76. Kolodziej, E. P., Harter, T. & Sedlak, D. L. Dairy wastewater, aquaculture, and spawning fish as sources of steroid hormones in theaquatic environment. *Environ. Sci. Technol.* **38**, 6377–6384 (2004).
77. Pawlowski, S. et al. Combined in situ and in vitro assessment of the estrogenic activity of sewage and surfacewater samples. *Toxicol. Sci.* **75**, 57–65 (2003).
78. Heisterkamp, I., Ganrass, J. & Ruck, W. Bioassay-directed chemicalanalysis utilizing LC-MS: a tool for identifying estrogenic compounds in water samples. *Anal. Bioanal. Chem.* **378**, 709–715 (2004).

79. Galloway, J. E. et al. Evidence of air dispersion: HFPO-DA and PFOA in Ohio and West Virginia surface water and soil near a fluoropolymer production facility. *Environ. Sci. Technol.* **54**, 7175–7184 (2020).

80. Xu, B. et al. PFAS and their substitutes in groundwater: Occurrence, transformation and remediation. *J. Hazard. Mater.* **412**, 125159 (2021).

81. Garnik, L. et al. An evaluation of health-based federal and state PFOA drinking water guidelines in the United States. *Sci. Total Environ.* **761**, 144107 (2021).

82. Thompson, K. A. et al. Poly- and perfluoroalkyl substances in municipal wastewater treatment plants in the United States: Seasonal patterns and meta-analysis of long-term trends and average concentrations. *Environ. Sci. Technol. Water* **2**, 690–700 (2022).

83. Shimabuku, K. K. et al. Biochar sorbents for sulfamethoxazole removal from surface water, stormwater, and wastewater effluent. *Water Res.* **96**, 236–245 (2016).

84. Barber, L. B. et al. Fate of sulfamethoxazole, 4-nonylphenol, and 17β-estradiol in groundwater contaminated by wastewater treatment plant effluent. *Environ. Sci. Technol.* **43**, 4843–4850 (2009).

85. Bizi, M. Sulfamethoxazole removal from drinking water by activated carbon: Kinetics and diffusion process. *Molecules* **25**, 4656 (2020).

86. Rudel, R. A., Melly, S. J., Geno, P. W., Sun, G. & Brody, J. G. Identification of alkylphenols and other estrogenic phenolic com-pounds in wastewater, septage and groundwater, on Cape Cod, Massachusetts. *Environ. Sci. Technol.* **32**, 861–869 (1998).

87. Petrovic, M., Eljarrat, E., Lopez de Alda, M. J. & Barcelo, D. Endocrine disrupting compounds and other emerging contaminants in the environment: A survey on new monitoring strategies and occurrence data. *Anal. Bioanal. Chem.* **378**, 549–562 (2004).

88. Ahel, M., Schaffner, C. & Giger, W. Behavior of alkyphenolpolyethoxylate surfactants in the aquatic environment III: Occurrence and elimination of their persistent metabolites during infiltration of river water to groundwater. *Water Resour.* **30**, 37–46 (1996).

89. Ramírez-Sánchez, I. M., Tuberty, S., Hambourger, M. & Bandala, E. R. Resource efficiency analysis for photocatalytic degradation and mineralization of estriol using TiO2 nanoparticles. *Chemosphere* **184**, 1270–1285 (2017).

90. Yost, E. E., Hawkins, M. B., & Kullman, S. W. Bridging the gap from screening assays to estrgenic effects in fish: Potential roles of multiple estrogen receptor subtypes. *Environ. Sci. Technol.* **48**, 5211–5219 (2014).

91. Nadal M., Marques M., Mari M., & Domingo, J. L. Climate change and environmental concentrations of POPs: A review. *Environ. Res.* **143**, 177–185 (2015).

92. Sprenger C., Lorenzen G., Hulshoff I., Grutzmacher G., Ronghang M., & Pekdeger A. Vulnerability of back filtration system to climate change. *Sci. Total Environ.* **409**, 655–663 (2011).

93. Spearow J. L., Kota R. S., & Ostrach D. J. Environmental contaminant effects on juvenile striped bass in the San Francisco Estuary, California, USA. *Environ. Toxicol. Chem.* **30**, 393–402 (2011).

94. Svecova, H. et al. De facto reuse at the watershed scale: Seasonal changes, populationcontributions, instreamflows and water quality hazards of human pharmaceuticals. *Environ. Pollut.* **268**, 115888 (2021).

95. Gadupudi, C. K., Rice, L., Xiao, L. & Kantamaneni, K. Endocrine disrupting compounds removal methods from wastewater in the United Kingdom: A review. *Science* **3**, 11 (2021).

96. Wise A., Brien K. O., & Woodruff T. Are oral contraceptives a significant contributor to the estrogenicity of drinking water? *Environ. Sci. Technol.* **45**, 51–60 (2011).

97. Caldwell D. J., Mastrocco F., Nowak E., Jonhson J., Yekel H., Pfeiffer D., Hoyt M., Duplessie B. M., & Anderson P. D. An assessment of potential exposure and risk from estrogens in drinking water. *Environ. Health Perspect.* **118**, 338–344 (2010).

98. Tourad E., Roig B., Sumpter J. P., & Coetsier C. Drug residues and endocrine disruptors in drinking water: Risk for humans? *Int. J. Hyg. Environ. Health* **214**, 437–441 (2011).
99. Bradley, P. M. et al. Expanded target-chemical analysis reveals extensive mixed-organic-contaminant exposure in U.S. Streams. *Environ. Sci. Technol.* **51**, 4792–4802 (2017).
100. Glassmeyer, S. T. et al. Nationwide reconnaissance of contaminants of emerging concern in source and treated drinking waters of the United States. *Sci. Total Environ.* **581–582**, 909–922 (2017).
101. Kidd, K. A., Blanchfield, P. J., Mills, K. H., Palace, V. P., Evans, R. E., Lazorchak, J. M., & Flick, R. W. Collapse of a fish population after exposure to a synthetic estrogen. *PNAS* **104**, 8897–8901 (2007).
102. Fischer, H., Kloep, F., Wilzcek, S., & Pusch, M. A river's liver: Microbial processes within the hyporheic zone of a large lowland river. *Biogeochemistry* **76**, 349–371 (2005).
103. Koplin, D. W., Furlong, E. T., Meyer, M. T., Thurman, E. M., Zaugg, S. D., Barber, L. B., & Buxton, H. T. Pharmaceuticals, hormones and other organic wastewater contaminants in U.S. streams, 1999–2000: A national reconnaissance. *Environ. Sci. Technol.* **36**, 1202–1211 (2002).
104. Schlenk, D., Lavado, R., Loyo-Rosales, J. E., Jones, W., Maryoung, L., Riar, N., Werner, I., & Sedlak, D. Reconstitution studies of pesticides and surfactants exploring the cause of estrogenic activity observed in surface waters of the San Francisco Bay delta. *Environ. Sci. Technol.* **46**, 9106–9111 (2012).
105. Llorca, M., Schirinzi, G., Martínez, M., Barceló, D. & Farré, M. Adsorption of perfluoroalkyl substances on microplastics under environmental conditions. *Environ. Pollut.* **235**, 680–691 (2018).
106. PIMA County Wastewater Reclamation. *PFAS in Biosolids: A Southern Arizona Case Study.* Pima County Wastewater Reclamation (2020).
107. Domingo, J. L. & Nadal, M. Human exposure to per- and polyfluoroalkyl substances (PFAS) through drinking water: A review of the recent scientific literature. Environ. Res. **177**, 108648 (2019).
108. Blunt, S. M. et al. Association between degradation of pharmaceuticals and endocrine-disrupting compounds and microbial communities along a treated wastewater effluent gradient in Lake Mead. *Sci. Total Environ.* **622–623**, 1640–1648 (2018).
109. Park, M. et al. Multivariate analyses for monitoring EDCs and PPCPs in a lake water. *Water Environ. Res.* **86**, 2233–2241 (2014).
110. Leiker, T. J., Abney, S. R., Goodbred, S. L., & Rosen, M. R. Identification of methyl triclosan and halogenated analogues in both male common carp (*Cyprinus carpio*) from Las Vegas Bay and semipermeable membrane devices from Las Vegas Wash, Nevada. *Sci. Total Environ.* **407**, 2102–2114 (2009).
111. Bai, X. & Acharya, K. Algae-mediated removal of selected pharmaceutical and personal care products (PPCPs) from Lake Mead water. *Sci. Total Environ.* **581–582**, 734–740 (2017).

2 The Circular Economy Concept and Its Application to SDG 6

Piotr F. Borowski

2.1 INTRODUCTION

Water is an invaluable and indispensable natural resource, playing a pivotal role in sustaining life on Earth. Hence, the adoption of the circular economy concept is progressively being employed to effectively address the challenges associated with the attainment of the Sixth Sustainable Development Goal (SDG 6). Encompassing more than 70% of the Earth's surface, water assumes a crucial function in upholding ecological equilibrium, life cycles, and human endeavors. Nevertheless, the global availability of freshwater, accounting for a mere 2.5% of the total water reservoir, presents intricate predicaments for societies, particularly due to its uneven distribution and inefficient management [1,2].

Only 0.6%–0.7% of freshwater serves as potable water, underscoring the necessity to prioritize sustainable management and efficient utilization of this precious asset [3,4].

Between 2015 and 2017, a decline in water security affected most global regions, sounding an alarm about the escalating issue of water scarcity. Recent research indicates that one-fourth of the world's population grapples with extreme water scarcity [5–7].

Water stress is an escalating challenge across diverse regions where water demand surpasses available supply. Fundamental water stress gauges the ratio of total water demand against renewable surface and groundwater resources. Populated regions enduring water scarcity confront diminishing freshwater sources due to factors like population expansion, climate fluctuations, and inefficient water management methodologies. This dilemma poses difficulties in fulfilling the needs of both local communities and industries, imparting extensive ecological, societal, and economic ramifications. Mitigation of water scarcity involves a blend of sustainable water consumption practices, enhanced infrastructure, and transboundary collaboration to guarantee equitable allocation and preservation. Water requisites encompass domestic, industrial, irrigation, and livestock uses. With heightened instances of intense rainfall in certain locales, affected communities face elevated drought and flooding hazards. As climate change renders some areas arid and others inundated, the inevitable flux in water supply and demand is illustrated in Figure 2.1.

DOI: 10.1201/9781003441007-4

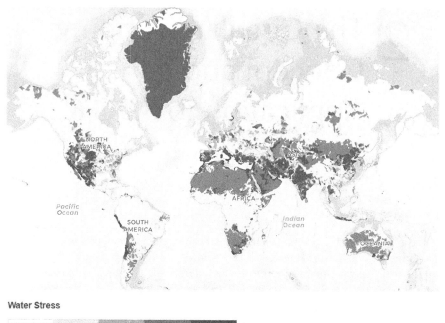

Water Stress

Low	Low-medium	Medium-high	High	Extremely high
(<10%)	(10-20%)	(20-40%)	(40-80%)	(>80%)

■ Arid and low water use
■ No data

FIGURE 2.1   World water stress [8].

To effectively mitigate and address water scarcity, a series of proactive measures and sustainable practices must be undertaken. These encompass water conservation and efficiency strategies, enhanced water management techniques, wastewater treatment, and reuse initiatives, implementation of water preservation practices at both individual and communal levels, enforcement of pertinent regulations, and systematic education on water resources' judicious use and safeguarding. The imperative for resolute action to ensure universal access to clean water and adequate sanitation becomes increasingly apparent, constituting the cornerstone of Sustainable Development Goal 6 (SDG 6). This objective guides the global community's endeavors to secure safe drinking water and provide sufficient sanitation, considering the ramifications of these factors for human health and ecological equilibrium.

Within this framework, the circular economy concept emerges as a potent instrument that can significantly contribute to the realization of SDG 6's objectives. Incorporating the principles of the circular economy is recommended for the achievement of SDG 6 [9,10].

Central to the circular economy philosophy is waste generation reduction, resource utilization optimization, and adverse impact minimization. This concept of circularity and self-sufficiency offers a practical framework for stakeholders seeking to actualize sustainable development goals. Its tenets, such as designing products for prolonged lifecycles, advocating recycling, and cultivating practices for material and resource reuse, exert a profound influence on the broader water management process and contribute to the sustainable stewardship of water resources [11].

Water occupies a unique position within the circular economy due to its unmatched role in the economic system. Consequently, it should be accorded treatment and valuation within the circular economy framework. Driven by the latest advancements in water science and technology, scientists, governments, water utilities, and industries are exploring more efficient models of water management. The circular economy concept has proven to be an effective framework for sustainable water management, with guiding principles like circular supply chains, value retention, waste minimization, and resource efficiency finding applicability in the water sector [12].

2.2 CIRCULAR ECONOMY ACTIONS TAKEN TO ACHIEVE SDG 6

The concept of a circular economy in the context of water-related issues has garnered significant attention, highlighting water's pivotal role in sustaining life and as a critical element of the global economy. Water, as the foundation of our existence and an essential component of life-sustaining ecosystem processes, has become the subject of intensive research and actions aimed at effectively managing its availability and utilization. Contemporary economies rely on numerous resources. Water, however, occupies a unique position as a raw material not only for sectors such as agriculture, industry, and energy but primarily as a substance without which life on Earth would be untenable. Given the mounting challenges associated with excessive consumption, pollution, and climate change, comprehending water's role as a finite resource becomes paramount for sustainable development. Actions related to water management within the framework of a circular economy encompass several key facets.

2.3 WATER CONSUMPTION REDUCTION

First, attention should be paid to water consumption reduction. The circular economy approach encourages water consumption minimization in production, agriculture, and daily activities. Water-saving technologies such as recirculation and rainwater harvesting systems allow this water to be used more efficiently.

2.3.1 WASTEWATER TREATMENT

Another significant aspect of the circular economy is advanced water waste treatment. In this context, great importance is attached to innovative, advanced technologies that enable effective processing and purification of used water to minimize its negative impact on the environment and, at the same time, recover valuable chemical substances. The circular economy concept of wastewater treatment goes beyond traditional treatment methods to focus on innovative approaches. Technologies such as advanced membrane processes, chemical and biological reactions, and hybrid treatment systems enable the selective removal of both organic and inorganic substances from used waters. The effect of the use of advanced wastewater processing is the possibility of recovering valuable chemical substances that can be reused in other applications. For example, recycling nutrients such as nitrogen and phosphorus allow them to be reused as fertilizers in agriculture, eliminating the need to extract these resources anew from natural sources. In addition, advanced methods of processing wastewater result in significant pollution reduction entering the environment. These processes allow for the effective removal of toxic substances, organic compounds, and pathogens, which in turn contributes to maintaining the quality of surface and groundwater. In a scientific context, advanced wastewater treatment requires an interdisciplinary approach, combining disciplines such as chemistry, biology, environmental engineering, and membrane technologies. Research into new technologies and treatment processes plays a key role in developing effective and sustainable water treatment methods, which is fundamental to achieving environmental and water resource goals as part of a circular economy strategy.

2.3.2 COOPERATION AND INNOVATION

In a broad sense, collaboration and innovation are also of paramount importance and pivotal components of transformation toward a circular economy that needs engagement and synergy across different sectors and stakeholders. Collaboration not only transcends the boundaries of individual domains but also unites the efforts of governments, industry, and civil society in achieving sustainable and efficient water resource management. Within the context of the circular economy, collaboration is imperative for a coordinated approach to water resource management.

The partnership between public and private sectors leads to technical knowledge and financing exchange, which is exceedingly significant in developing modern technologies geared toward more efficient water purification and treatment, such as the creation of closed-loop resource recovery systems and the establishment of sustainable sanitation infrastructure. Moreover, the engagement of civil society can aid in identifying local needs and issues, facilitating strategies and actions adaptation to real on-ground conditions. Innovation is pivotal to collaboration, as it is only through the introduction of innovative solutions that effectively respond to challenges associated with water management and sanitation issues can be achieved.

Partnerships can focus on developing novel technologies, such as advanced water purification systems, sewage recycling, or even modern desalination methods, to address drinking water shortages in regions with limited access to freshwater. It is worth emphasizing that collaboration holds potential not only for resolving local but also global challenges. Through knowledge and experiences exchange between different regions, more effective SDG achievement, including goal 6 related to clean water and sanitation, becomes attainable. Collaboration and innovation, thus, constitute the foundation of a future-oriented, sustainable water economy capable of addressing the mounting challenges concerning water resources, sanitation, and hygiene.

2.3.3 SUSTAINABLE AGRICULTURAL PRACTICES

Proper water resource utilization also extends to the realm of agriculture, asking for the implementation of sustainable agricultural practices. Transitioning toward agricultural models grounded in circular economy principles holds the potential to yield revolutionary changes in water usage efficiency in irrigation processes, water loss reduction, and sustainable soil use promotion. Within the framework of sustainable agricultural practices, the adoption of intelligent irrigation systems is a significant component that relies on precise measurements of soil moisture and plant water needs [13].

Through the application of intelligent systems based on the Internet of Things (IoT) and machine learning, water consumption can be precisely tailored to actual plant requirements, minimizing excessive water consumption, and reducing over-irrigation losses [14,15].

In the circular economy context, sustainable agriculture assumes a pivotal role. More efficient irrigation, employing erosion-limiting techniques, and strategies to curtail water losses during cultivation significantly contribute to water conservation in agriculture. Furthermore, strategies to reduce food waste, and promote local and seasonal produce, impact overall water consumption reduction in food production. Moreover, embracing circular economy-oriented agricultural models paves the way for leveraging renewable water sources, such as rainwater harvesting or treated wastewater reclamation for irrigation purposes. This approach not only enhances water availability in agriculture but also alleviates pressure on limited water resources, particularly in drought-prone areas [16].

Promoting agroecological techniques to prevent soil erosion, improve soil structure, and enhance water retention capacity is pivotal to sustainable agricultural practices. Encouraging perennial crops, implementing composting, and maintaining riparian vegetation around water bodies contribute to soil preservation and erosion minimization, resulting in more efficient use of available water resources [17].

Implementing sustainable agricultural practices constitutes a critical step toward striking a balance between food demand and the crucial protection of water resources [18]. Transitioning to such models not only enhances agricultural production efficiency but also fosters a sustainable approach to water utilization, with fundamental implications for SDGs. These practices optimize food production processes through intelligent water resource management. Consequently, farmers can achieve greater yield per water unit while concurrently mitigating inefficient irrigation and soil erosion losses. Such a holistic model not only mitigates water resource depletion but also eases pressure on aquatic ecosystems, providing a resilient foundation for future agricultural production.

By harmonizing food demand and water resource protection, sustainable agricultural practices significantly contribute to global sustainable development goals. As an agricultural community and food sector transform their methods toward greater sustainability, the opportunity emerges to shape a future in which food production and water resource preservation coexist harmoniously, fostering enduring and sustainable development [19].

2.4   EDUCATION AND RAISING AWARENESS

As part of promoting the circular economy, education, and awareness raising represent crucial actions in contemporary societies. The circular economy requires behavioral and approach changes toward resource management. Introducing the circular economy concept within the water context is linked with education and social awareness enhancement. Informing individuals about water-saving practices, recycling, and responsible resource utilization leads to behavioral changes and more effective water usage. Education and societal awareness are pivotal for the effective implementation of efficient water utilization and consumption reduction practices. Education plays a key role in advancing circular economy ideals and safeguarding water resources. Disseminating knowledge about the principles and benefits of an economic model based on sustainable resource use holds the potential to contribute to a long-term shift in societal mindset and attitudes [20].

Implementing educational programs in schools, higher education institutions, and local communities will solidify resource conservation habits fostering a responsible approach to water utilization. By educating younger generations, a foundation for a future society more oriented toward sustainable development can be established. In the realm of water resource preservation, education can focus on conveying information about the value of water, the water cycle, and the impact of anthropogenic activities on water quality/availability. This knowledge will influence behavior change, motivating individuals to conserve water, prevent pollution, and engage in actions to protect local water sources. Education involves not only acquiring information but also shaping attitudes and behavior [21]. Long-term engagement in societal education will lead to a mindset shift toward more sustainable and responsible resource utilization, ultimately contributing to the achievement of SDGs and safeguarding precious water resources [22].

The circular economy serves as a potent tool in achieving SDG 6 by focusing on sustainable water management and resource conservation. The transition from a linear economic model to a circular one holds the potential to enhance people's quality of life, protect aquatic ecosystems, and ensure safe drinking water for current and future generations. Through advanced technologies, education, and cross-sector collaboration, the circular economy emerges as a key instrument for attaining sustainable water resource utilization, thus contributing to SDG realization.

2.5   EXTREME WEATHER PHENOMENA

Climate change is a direct driver behind the escalation in frequency and magnitude of extreme weather events, posing a severe menace to both society and the environment [23,24]. It significantly impacts the dynamic water cycle, exerting a profound influence on the volume, frequency, and geographical distribution of precipitation. As the Earth undergoes alterations in global temperatures, mounting concentrations of greenhouse gases, and heightened weather volatility, abrupt modifications in water cycle processes have become commonplace. Air temperature increase, a characteristic consequence of these changes, exerts influence on surface water evaporation dynamics, accelerating evaporation rate and, consequently, contributing to heightened evaporation from soil surfaces and water reservoirs. This augment in water vapor in the atmosphere intensifies rainfall but not uniformly, as some areas may face heightened drought conditions due to escalated aridity. Climate change also disrupts atmospheric cycles and air mass movements causing shifts in conventional precipitation patterns. The ongoing climate alteration is disrupting established weather patterns, resulting in extreme weather events, erratic water availability, aggravated water scarcity, and the emergence of incidents that contaminate drinking water supplies. Areas that were historically characterized by higher moisture levels are witnessing rainfall reduction whereas regions previously subject to lesser precipitation are experiencing heightened rainfall intensity [25]. To effectively tackle these challenges, extensive measures are imperative, with the circular economy approach assuming a pivotal role. Implementing the circular economy model can markedly contribute to mitigating deleterious anthropogenic practices and attenuating the impacts of these extreme occurrences. This holistic strategy requires synchronized global efforts to address the multifaceted predicaments posed by climate change and its ramifications. Extreme weather events entail the peril of flooding, which can cause substantial material and human losses [26]. Paradoxically, however, they may not directly augment accessibility to safe drinking water because of pollution, infrastructure impairment, and limited groundwater retention capacity. Intensive precipitation, despite contributing to surface inundation, does not invariably elevate groundwater levels. Precipitated water can rapidly run off the surface, lacking the time required to infiltrate deeper layers. Heavy rainfall can lead to swift surges in river and lake levels, thereby amplifying flood risks. Excessive water quantities saturate the soil, provoking heightened surface runoff that can culminate in riverbed overflow and adjacent areas inundation. The nexus between intense precipitation and flood intensity shifts as well as spatial and seasonal water availability becomes more pronounced as events become more extreme [27].

To effectively manage these complex issues, developing strategies to both minimize flooding risk and ensure sustainable clean drinking water access even in the face of extreme weather conditions is paramount. One way to prevent extreme weather events and effectively minimize their negative impact on the environment and society is by implementing comprehensive strategies based on the circular economy principles. A circular economy is an approach that focuses on maximizing resource use, minimizing waste, and reducing emissions with the intention to reduce pressure on the natural environment and limit the effects of extreme weather events. Thanks to this approach, it is possible to effectively reduce the negative impacts of anthropogenic activities on the environment, including factors contributing to climate change. In the context of extreme weather phenomena, a circular economy brings a wide variety of benefits.

As part of the circular economy, greenhouse gas emissions reduction is the most important aspect. The introduction of modern production technologies, including industrial water consumption reduction and increasing energy efficiency can significantly reduce negative anthropogenic impacts on the climate. The implementation of such a strategy for reducing negative environmental impacts requires the active involvement of government, business, and society. Achieving a circular economy can bring multi-faceted benefits, including improving air and water quality, reducing extreme weather events risk, and increasing society's resilience to climate change. In the long term, investing in the circular economy is an investment in sustainability and the future of our planet.

2.6 RECYCLING IN THE WATER SECTOR

In the water and wastewater sector, a circular economy encompasses efficient water collection, purification, and reuse systems. This approach alleviates pressure on water resources and effectively manages water availability, even during extreme rainfall or drought periods. Circular economy strategies also extend to wastewater management. Through innovative technologies and processes, wastewater can be more effectively treated, rendering it suitable for safe reuse in various applications, including irrigation and industrial processes. The circular economy fosters innovative solutions in the realm of water waste management. Sustainable sanitation solutions play a pivotal role in optimizing water resource utilization, waste reduction, and minimizing adverse environmental impacts [28].

The principles of circular economy influence sanitation practices by promoting sustainable and resource-efficient technologies. Enhanced sanitation systems, such as composting toilets and decentralized wastewater treatment, align with the circular economy paradigm, converting waste into valuable resources. These solutions offer dual benefits, mitigating water pollution and enhancing soil fertility, thus contributing to the achievement of SDG 6 by enhancing clean water availability and sustainable sanitation conditions. In the long run, sustainable solutions may lead to reduced water supply and wastewater treatment costs, benefiting both households and businesses.

Waste materials can be processed to yield valuable products (e.g., biogas or compost), with applications in agriculture or horticulture [29,30]. Embracing water management practices within the circular economy framework, communities can curtail escalating freshwater source demand, ensuring access to clean water for essential needs while concurrently diminishing water pollution [31].

Circular economy advocates recycling and resource repurposing, including water, thereby reducing the need for fresh extraction and production based on new resources, minimizing waste generation. In the context of extreme weather, reducing waste quantities and enhancing waste management phenomena translates into lower risks of water and soil contamination during floods and other natural disasters. Water recycling in the circular economy pertains to the practice of treating and reusing wastewater, as well as optimizing water management processes to minimize water waste and enhance resource efficiency. Various water recycling techniques exist, some of which involve mechanical and biological purification, membrane filtration, coagulation, and flocculation processes, reverse osmosis desalination, and greywater and industrial water recycling. Water recycling entails municipal wastewater purification and disinfection to provide water suitable for industrial applications, such as using recycled water for filling lakes, ponds, and decorative fountains, irrigating parks, campgrounds, golf courses, highways, green strips, school sports fields, food cultivation, nursery, and nursery material, controlling dust on construction sites, and aiding street cleaning. Recycled water can also be utilized in certain industrial

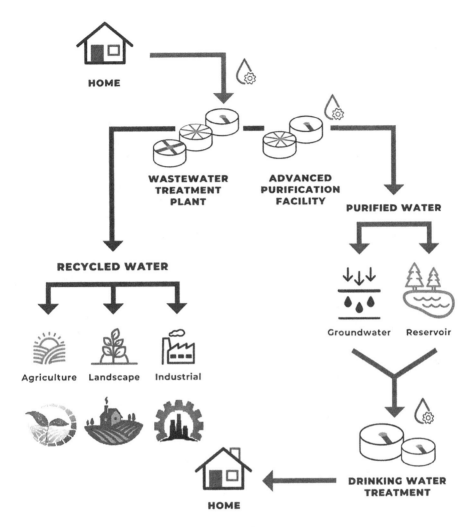

FIGURE 2.2 The scheme of water recycling [32].

processes. Moreover, advanced projects are increasingly focusing on purifying water for direct potable reuse (Figure 2.2).

This approach is pivotal in addressing water scarcity, promoting sustainable water usage, and bolstering water conservation and environmental objectives. Water recycling within the water sector encompasses diverse methods and strategies that contribute to a more comprehensive and sustainable approach to water management. Fundamental components of water recycling in this sector encompass both direct and indirect water reuse. The treatment and reuse of wastewater constitute a critical part of water recycling. Sophisticated treatment technologies have the capability to eliminate contaminants and pathogens from wastewater, rendering it fit for safe reuse without compromising human health. The treatment and reuse of wastewater play a significant role in ensuring sustainable stewardship of water resources, particularly in regions confronted with water scarcity and environmental pressures. Through the application of advanced treatment techniques (e.g., membrane filtration, activated carbon adsorption, UV disinfection), wastewater treatment facilities can achieve a high purification level, ensuring the resulting water adheres to rigorous quality standards for its intended reuse. The implementation of a wastewater reuse strategy requires a holistic approach encompassing regulatory frameworks, technological innovations, and public awareness. Robust regulations and guidelines must be established to ensure the secure and controlled utilization of purified water. Highlighting the economic, environmental, and social advantages of wastewater reuse can foster a favorable outlook on sustainable practices. Wastewater treatment and reuse present a promising resolution to water scarcity and environmental pressures. With the adoption of advanced treatment methodologies, wastewater treatment plants can transform wastewater into a valuable resource applicable for diverse purposes, alleviating the strain on limited freshwater sources. The integration of stringent regulations, innovative technologies, and community engagement is imperative in constructing a robust framework for effective wastewater reuse and the sustainable management of water resources.

There are two main options for water reuse: (i) Direct reuse of potable water. In certain regions, advanced water treatment processes yield high-quality purified water conforming to drinking water standards. This treated wastewater, termed "direct potable water reuse," can be reintroduced into the water distribution network following rigorous treatment and monitoring procedures [33]. This pertains to situations where drinking water is reclaimed directly from processes or systems where it was previously utilized and then repurposed for the same application. As a result, both water consumption and energy usage associated with processes related to the supply of new water are diminished [34]. For instance, drinking water employed for flushing toilets can be collected and reused for subsequent flushes, obviating the necessity for fresh water in this context. Certain office buildings, shopping centers, and hotels have systems for recycling drinking water for toilet flushing or irrigating plants. Another illustration involves advanced household drinking water recycling technologies that permit the collection, purification, and subsequent reuse of water for diverse domestic applications, such as dishwashing and laundry; (ii) Indirect potable water reuse. This is a more complex approach involving water recovery from processes or wastewater. Potable water containing minor levels of contaminants can undergo purification and disinfection processes to render it safe for use in industrial processes or other applications that do not require the highest quality water. Indirect potable water reuse entails introducing treated wastewater into natural water reservoirs such as rivers or groundwater basins, subjecting them to natural cleansing processes before extraction and subsequent purification for drinking water supply. This method often involves additional treatment stages to ensure the safety of recycled water.

Direct and indirect potable water reuse pertains to two distinct approaches for water recovery and reuse in industrial processes or other purposes. In both cases, the aim is to reduce fresh potable water consumption by maximizing the use of available resources and minimizing water waste. Indirect potable water reuse within recycling is particularly significant as it enables efficient water reuse in industrial processes and other applications, simultaneously alleviating the strain on water sources [35].

Various techniques mentioned above are also employed in recycling to ensure rational water utilization. One popular technique is rainwater harvesting. Collecting and storing rainwater for subsequent use, such as irrigation or toilet flushing, constitutes a form of water recycling. Rainwater that falls on building roofs can be collected, purified, and subjected to treatment processes to make it suitable for industrial purposes or irrigating green spaces. Rainwater harvesting reduces the demand for freshwater sources and can be particularly valuable in areas with limited access to traditional water sources.

Industrial water recycling is a process involving the collection, purification, and reuse of used or contaminated water across various industrial sectors. This activity aims to reduce wastewater volume, mitigate environmental burdens, and conserve natural resources. Industries can implement water recycling practices in their processes to minimize water consumption and decrease pollutant emissions. Techniques such as closed-loop systems and water recovery aid industries in maximizing water reuse within their operations. An example is the recirculating aquaculture systems (RAS), employed in fish production [36].

Another circular economy method in the water sector is graywater recycling. The term "graywater" in the context of recycling refers to water with a low contamination level. Such water can undergo purification processes and then be used for nonpotential purposes, such as garden irrigation, street cleaning, or industrial process cooling. In practice, this means reusing water that is not intended for consumption. Graywater that can undergo recycling processes includes water from sources such as showers, sinks, baths, and laundry [37]. After proper treatment, greywater can be reused for green areas irrigation or in toilets. The introduction of graywater recycling aims to reduce freshwater consumption in cases where high-quality drinking water is not required [38].

The main benefit of graywater recycling is the alleviation of sewage systems and wastewater treatment plants' load, although it requires additional investment [39]. Additionally, it allows for the conservation of freshwater traditionally used in processes where the highest water quality is not necessary. These actions align with the goals of sustainable water resource management and the reduction of negative impacts on the natural environment.

Promoting water recycling requires raising societal awareness regarding the benefits and safety of water derived from recycling. Educational campaigns have the potential to alter perceptions and behaviors, fostering acceptance of recycled water for various applications. Researchers highlight the imperative of employing pedagogical tools to achieve significant social transformation from the standpoint of sustainable development [37]. Legal instruments also hold importance. Hence, policies and regulations that support actions toward sustainable development and water recycling are critical. Governments and regulatory bodies play a pivotal role in promoting and regulating water recycling practices. The formulation and enforcement of guidelines, standards, and regulations about recycled water utilization ensures its safety and compliance with established water quality norms.

Water recycling within the water sector presents a multifaceted approach to addressing water scarcity, reducing pollution, and enhancing water resource sustainability. By integrating recycling practices into water management strategies, communities can better meet their water needs, concurrently contributing to broader environmental protection and sustainable development objectives.

2.7   URBANIZATION AND WATER PROBLEMS

Urbanization gives rise to water-related challenges. Therefore, a comprehensive understanding of water scarcity and potential solutions for global cities is urgently needed to foster a more sustainable and livable urban future. Dynamic urban growth poses intricate water resource challenges, exerting pressure on existing water supply and sanitation systems [40]. As urban populations surge, drinking water demand escalates, along with greater wastewater generation. This phenomenon urges rethinking and restructuring water management approaches within cities. Population increases and urbanization lead to a rise in water users and applications, rendering water resources scarcer and more contaminated [28]. Consequently, there is an urgent need for a comprehensive understanding of the complex water scarcity issue, which is becoming increasingly prevalent in developing urban areas [41].

To achieve a sustainable and future-proof urban environment, multiple facets must be considered, such as climate change, burgeoning populations, household and industrial water consumption, and threats to surface and groundwater quality. Implementation of integrated water management solutions, including the use of efficient water-saving technologies, development of renewable water source-based infrastructure, promotion of water recycling, and fostering public awareness about water conservation and protection, represents a pivotal step toward a future in which cities can adequately supply water to their residents [42].

Shaping a more sustainable and livable urban future requires collaboration across different sectors, including local authorities, scientists, water specialists, and the community [43,44]. Global cities should act as innovation hubs in water management, striving to formulate sustainable strategies that account for the unique challenges of each location. As urbanization continues to progress, it is imperative to recognize that water issues in cities are dynamic and multifaceted. Global collaboration to exchange best practices, develop new technologies and solutions, and enhance public awareness becomes paramount in constructing sustainable, resilient, and livable urban spaces capable of meeting future challenges [45,46]. The rise in urbanization leads to increased water consumption and heightened pollution levels. The introduction of circular economy-based initiatives calls for changes in social attitudes and behaviors regarding consumption and waste production. Educational and awareness campaigns associated with the circular economy model promote patterns of responsible consumption, highlighting the significance of resource conservation. As ecological awareness among individuals and communities grows, there is an increasing inclination toward adopting water-saving practices, reducing single-use plastics, and engaging in sustainable sanitation practices [47]. All these aspects constitute integral components of achieving the Sustainable Development Goal.

2.8 SDG 6 WATER AND ENERGY

In the context of ambitious objectives aligned with the attainment of SDGs, particularly guided by the directives delineated in SDG 6 that emphasizes sustainable stewardship of water resources and assurance of equitable access to adequate sanitation, intricate interconnections emerge between the dynamic framework of circular economy and evolving energy sector. As these two domains intertwine, auspicious prospects emerge for heightening resource management efficiency and optimizing energy utilization, thereby effectuating a curtailment of deleterious environmental impact. The integration of circular economy tenets into the energy sector holds the potential to engender synergistic outcomes spanning diverse economic sectors, accruing advantages in the realms of sustainable water management and energy provisioning. Enhanced water resources economization within energy-related processes, exemplified by cooling mechanisms in power plants, can engender a diminution in water requisition and concomitant mitigation of localized aquatic ecosystem perturbations. Equally encouraging are pioneering methodologies leveraging waste as a reservoir of energy. Instances encompassing thermal energy harnessing from wastewater treatment processes to fuel heating or cooling infrastructures underscore the symbiotic nexus between the circular economy and the energy sphere. Wastewater treatment procedures confer a substantial quantum of heat as a byproduct, conventionally relegated as thermal waste and dissipated into the environment. However, by harnessing this thermal energy for energy provisioning, systemic energy efficiency gains of substantial magnitude can be attained.

In practical terms, recuperated thermal energy sourced from wastewater treatment can be channeled to heating systems in residential, commercial, or institutional buildings, engendering a marked abatement in reliance on conventional energy sources for space heating [48]. This, in turn, culminates in diminished combustion of fossil fuels or reduced electricity consumption in heating processes, thereby furnishing a concomitant reduction in greenhouse gas emissions and an abatement in adverse environmental impact. Analogously, the thermal energy harnessed from wastewater treatment processes can be judicially deployed to temper structures or industrial systems. The distribution of this recuperated heat to cooling systems can usher mitigation in electricity consumption requisite for air conditioning, thus yielding favorable ramifications for energy equilibrium and amelioration of greenhouse gas emissions [49]. The transition from regarding heat generated from wastewater treatment processes as an expendable byproduct to leveraging it as a valuable energy asset constitutes a stride toward an enhanced sustainable energy paradigm. This transformation also offers a compelling exemplar of how the circular economy can generate value through judicious utilization of waste-derived energy streams. The adoption of this paradigm holds the potential to substantively curtail reliance on traditional energy sources, thus advancing the trajectory toward a sustainable energy transition. Moreover, the energy potential latent within the biomass derived from sanitary sewage treatment processes can be harnessed to yield biogas, a sustainable alternative to fossil fuels [50]. The realization of these aspirations mandates multidisciplinary cooperation, technological innovation, and public awareness, which collectively underpin a trajectory of sustainable and harmonious development, where the realms of water and energy mutually reinforce each other in the pursuit of a more sustainable future.

2.9 CONCLUSIONS

In summary, the implementation of a closed-loop economy approach in water resource management constitutes a fundamental stride toward achieving the sustainable and balanced utilization of invaluable natural resources. These endeavors wielded a profound impact on the realization of overarching objectives embedded within the Sustainable Development Agenda, particularly within the ambitious SDGs closely linked to water and sanitation (SDG 6).

Transitioning to a closed-loop economy model unveils avenues conducive to more efficient utilization of available water resources. Through the minimization of losses and waste, the water management paradigm assumes a more sustainable stance, contributing to the safeguarding of aquatic ecosystems and their biological diversity. Concurrently, the reduction of pollutants discharged into water bodies and the environment augments the human quality of life, health, and biodiversity conservation.

In embracing the closed-loop economy framework, we also propagate a responsible approach to consumption and resource utilization. Water conservation, recycling, and material recovery from waste metamorphose into daily practices, thus facilitating the sustainable development of society. We bequeath to our generation the responsibility for natural resources, ensuring their availability not just for the present, but also for generations to come.

Consequently, by focusing on the closed-loop economy paradigm within the realm of water resource management, we not only establish the foundations for enduring and sustainable development but also play a pivotal role in advancing global sustainable development objectives. Our actions in the water management domain align with the trajectory toward a harmonious equilibrium between progress and environmental preservation, thereby constituting a pivotal determinant shaping the future of our planet.

REFERENCES

1. See https://www.nationalgeographic.com/environment/article/freshwater-crisis#:~:text=While%20nearly%2070%20percent%20of%20the%20world%20is,much%20of%20it%20trapped%20in%20glaciers%20and%20snowfields (accessed 15.07.2023).
2. See https://www.nasa.gov/feature/when-it-comes-to-water-you-have-to-think-global (accessed 15.07.2023).
3. See https://www.worldatlas.com/articles/what-percentage-of-the-earth-s-water-is-drinkable.html (accessed 15.07.2023).
4. Mishra, R. K. (2023). Fresh water availability and its global challenge. *Brit. J. Multidiscipl. Adv. Stud.*, **4**(3), 1–78.
5. See https://www.wri.org/ (access 15.07.2023).
6. See https://www.euronews.com/green/2023/08/17/25-countries-now-face-extreme-water-stress-every-year-three-of-them-are-in-europe (accessed 15.07.2023).
7. See https://www.theguardian.com/environment/2023/aug/16/extreme-water-stress-faced-by-countries-home-to-quarter-of-world-population (accessed 15.07.2023).

8. See   https://www.wri.org/applications/aqueduct/water-risk-atlas/#/?advanced=false&base
    map=hydro&indicator=bws_cat&lat=43.273987071514895&lng=-
    408.37501287460327&mapMode=view&month=1&opacity=0.81&ponderation=DEF&pr
    edefined=false&projection=absolute&scenario=optimistic&scope=baseline&timeScale=
    annual&year=baseline&zoom=2 (accessed 15.07.2023).
9. de Oliveira, C. T., & Oliveira, G. G. A. (2023). What circular economy indicators really
    measure? An overview of circular economy principles and sustainable development
    goals. *Resour. Conserv. Recycl.*, **190**, 106850.
10. Valverde, J. M., & Avilés-Palacios, C. (2021). Circular economy as a catalyst for prog-
    ress towards the Sustainable Development Goals: A positive relationship between two
    self-sufficient variables. *Sustainability*, **13**(22), 12652.
11. Puntillo, P. (2023). Circular economy business models: Towards achieving sustainable
    development goals in the waste management sector-empirical evidence and theoretical
    implications. *Corp. Soc. Responsib. Environ. Manag.*, **30**(2), 941–954.
12. Morseletto, P., Mooren, C. E., & Munaretto, S. (2022). Circular economy of water:
    Definition, strategies and challenges. *Circ. Econ. Sustain.*, **2**(4), 1463–1477.
13. Oussama, G., Rami, A., Tarek, F., Alanazi, A. S., & Abid, M. (2022). Fast and intel-
    ligent irrigation system based on WSN. *Comput. Intell. Neurosci.* 5086290.
14. Borowski, P., Patuk, I., Hasegawa, H., Whitaker, A., Boiarskii, B., & Borodin, I. (2021).
    Current status and perspectives of rice farming in Sivakovka Village, Primorsky Krai,
    Russia. *AMA-Agric. Mech. Asia Africa Latin America*, **52**(4), 80–87.
15. Gujar, K. A., & Jagtap, S. (IEEE, 2022). Intelligent irrigation system for agriculture
    using IoT and machine learning. In A. Ambikapathy, K. Hazarika (eds.), 2022 2nd
    International Conference on Advance Computing and Innovative Technologies in
    Engineering (ICACITE) (pp. 86–90).
16. Singh, S., Yadav, R., Kathi, S., & Singh, A. N. (2022). Treatment of harvested rainwa-
    ter and reuse: Practices, prospects, and challenges. *Cost Effect. Technol. Solid Waste
    Wastew. Treat.*, 161–178.
17. Ward, N. D., Megonigal, J. P., Bond-Lamberty, B., Bailey, V. L., Butman, D., Canuel,
    E. A., ... & Windham-Myers, L. (2020). Representing the function and sensitivity of
    coastal interfaces in Earth system models. *Nat. Commun.*, **11**(1), 2458.
18. Borowski, P. F., & Patuk, I. (2021). Environmental, social and economic factors in sus-
    tainable development with food, energy and eco-space aspect security. *Present Environ.
    Sustain. Dev.*, **15**(1), 153–169.
19. Borowski, P. F. (2020). Nexus between water, energy, food and climate change as chal-
    lenges facing the modern global, European and Polish economy. *AIMS Geosci.*, **6**(4),
    397–421.
20. Bugallo-Rodríguez, A., & Vega-Marcote, P. (2020). Circular economy, sustainability
    and teacher training in a higher education institution. *Int. J. Sustain. High. Educ.*, **21**(7),
    1351–1366.
21. Dumitrescu, C. I., Moiceanu, G., Dobrescu, R. M., & Popescu, M. A. M. (2022).
    Analysis of UNESCO ESD priority areas' implementation in Romanian HEIs. *Int. J.
    Environ. Res. Public Health*, **19**(20), 13363.
22. Tiippana-Usvasalo, M., Pajunen, N., & Maria, H. (2023). The role of education in
    promoting circular economy. *Int. J. Sustain. Eng.*, **16**(1), 92–103.
23. Stott, P. (2016). How climate change affects extreme weather events. *Science*, **352**(6293),
    1517–1518.
24. Konisky, D. M., Hughes, L., & Kaylor, C. H. (2016). Extreme weather events and
    climate change concern. *Clim. Change*, **134**, 533–547.
25. Yoon, P. R., Lee, S. H., Choi, J. Y., Yoo, S. H., & Hur, S. O. (2022). Analysis of climate
    change impact on resource intensity and carbon emissions in protected farming systems
    using water–energy–food–carbon nexus. *Resour. Conserv. Recycl.*, **184**, 106394.

26. Gills, B., & Morgan, J. (2022). Global climate emergency: After COP24, climate science, urgency, and the threat to humanity. *Globalizations*, 17(6), 885–902. https://doi.org/10.1080/14747731.2019.1669915

27. Tabari, H. (2020). Climate change impact on flood and extreme precipitation increases with water availability. *Sci. Rep.*, **10**(1), 13768.

28. Tortajada, C. (2020). Contributions of recycled wastewater to clean water and sanitation sustainable development goals. *NPJ Clean Water*, **3**(1), 22.

29. Attwater, R., & Derry, C. (2017). Achieving resilience through water recycling in peri-urban agriculture. *Water*, **9**(3), 223.

30. Villamar, C. A., Vera-Puerto, I., Rivera, D., & De la Hoz, F. (2018). Reuse and recycling of livestock and municipal wastewater in Chilean agriculture: A preliminary assessment. *Water*, **10**(6), 817.

31. Mannina, G., Gulhan, H., & Ni, B. J. (2022). Water reuse from wastewater treatment: The transition towards circular economy in the water sector. *Bioresour. Technol.*, 363, 127951.

32. See https://www.sdcwa.org/your-water/local-water-supplies/water-recycling/.

33. Giakoumis, T., Vaghela, C., & Voulvoulis, N. (Elsevier, 2020). The role of water reuse in the circular economy. In Paola Verlicchi (ed.), *Advances in Chemical Pollution, Environmental Management and Protection* (Vol. 5, pp. 227–252).

34. Tow, E. W., Hartman, A. L., Jaworowski, A., Zucker, I., Kum, S., AzadiAghdam, M., ... & Warsinger, D. M. (2021). Modeling the energy consumption of potable water reuse schemes. *Water Res.* **13**(X), 100126.

35. Hooper, J., Funk, D., Bell, K., Noibi, M., Vickstrom, K., Schulz, C., ... & Huang, C. H. (2020). Pilot testing of direct and indirect potable water reuse using multi-stage ozone-biofiltration without reverse osmosis. *Water Res.*, **169**, 115178.

36. Borowski, P. F., & Zalewski, W. (2016). Innowacje w akwakulturze zapewniające zdrową produkcję ryb. *Inżynieria Przetwórstwa Spożywczego*, **2**, 15–18.

37. Kordana-Obuch, S., Starzec, M., Wojtoń, M., & Słyś, D. (2023). Greywater as a future sustainable energy and water source: Bibliometric mapping of current knowledge and strategies. *Energies*, **16**(2), 93.

38. Javadinejad, S., Dara, R., Hamed, M. H., Saeed, M. A. H., & Jafary, F. (2020). Analysis of gray water recycling by reuse of industrial waste water for agricultural and irrigation purposes. *J. Geogr. Res.*, **3**(2), 20–24.

39. Stec, A., & Kordana, S. (2015). Analysis of profitability of rainwater harvesting, gray water recycling and drain water heat recovery systems. *Resour., Conser. Recycl.*, *105*, 84–94.

40. He, C., Liu, Z., Wu, J., Pan, X., Fang, Z., Li, J., & Bryan, B. A. (2021). Future global urban water scarcity and potential solutions. *Nat. Commun.*, **12**(1), 4667.

41. Grison, C., Koop, S., Eisenreich, S., Hofman, J., Chang, I. S., Wu, J., ... & van Leeuwen, K. (2023). Integrated water resources management in cities in the world: Global challenges. *Water Resour. Manag.*, **37**(6–7), 2787–2803.

42. Blasi, S., Ganzaroli, A., & De Noni, I. (2022). Smartening sustainable development in cities: Strengthening the theoretical linkage between smart cities and SDGs. *Sustain. Cities Soc.*, **80**, 103793.

43. Chapagain, K., Aboelnga, H. T., Babel, M. S., Ribbe, L., Shinde, V. R., Sharma, D., & Dang, N.M. (2022). Urban water security: A comparative assessment and policy analysis of five cities in diverse developing countries of Asia. *Environ. Dev.*, **43**, 100713.

44. Stoker, P., Albrecht, T., Follingstad, G., & Carlson, E. (2022). Integrating land use planning and water management in US cities: A literature review. *JAWRA J. Am. Water Resour. Assoc.*, **58**(3), 321–335.

45. Wang, Y., Zhang, Y., Sun, W., & Zhu, L. (2022). The impact of new urbanization and industrial structural changes on regional water stress based on water footprints. *Sustain. Cities Soc.*, **79**, 103686.

46. Wu, W., & Lin, Y. (2022). The impact of rapid urbanization on residential energy consumption in China. *PLoS One*, **17**(7), e0270226.

47. Kumar, S., Singh, E., Mishra, R., Kumar, A., & Caucci, S. (2021). Utilization of plastic wastes for sustainable environmental management: A review. *ChemSusChem*, **14**(19), 3985–4006.

48. Nagpal, H., Spriet, J., Murali, M. K., & McNabola, A. (2021). Heat recovery from wastewater – a review of available resource. *Water*, **13**(9), 1274.

49. Cecconet, D., Raček, J., Callegari, A., & Hlavínek, P. (2019). Energy recovery from wastewater: A study on heating and cooling of a multipurpose building with sewage-reclaimed heat energy. *Sustainability*, **12**(1), 116.

50. Manikandan, S., Vickram, S., Sirohi, R., Subbaiya, R., Krishnan, R. Y., Karmegam, N., ... & Awasthi, M. K. (2023). Critical review of biochemical pathways to transformation of waste and biomass into bioenergy. *Bioresour. Technol.*, 372, 128679.

3 Life Cycle Assessment
A Tool for Evaluation of Circular Economy Applications

Gabriela E. Moeller-Chavez and
Catalina Ferat-Toscano

3.1 INTRODUCTION

Nowadays, ensuring enough water for use across all sectors and water security while respecting environmental flows in a sustainable manner represents a worldwide key challenge [1,2].

The circular economy (CE) approach has been proposed as an effective framework for sustainable water management, considering principles such as a closed-loop supply chain, value retention of materials and products, waste minimization, and resource efficiency that fit water sector concepts. CE is well suited to support efficient water supply management, intersects with multiple spatial scales (from local to global), governance levels (micro, meso, and macro), and implementation forms (from supply to discharge, treatment, and recycling or reuse), and can also be applied in all economic sectors. In a broad sense, CE is applied to water uses (for example, supply, sanitation, irrigation, and industry) along with sustainability criteria to achieve water security.

CE for water and water security also represent ways to achieve the Sustainable Development Goals (SDGs), in particular, SDG 6 "Clean Water and Sanitation", SDG 9 "Industry, Innovation and Infrastructure", and SDG 12 "Responsible Consumption and Production". More indirectly, they could help to achieve practically all other SDGs [2,3]. Water is considered as a special case in CE because it is simultaneously a resource, a product, and a service, depending on the specific case, with no equivalent in the economic system. This uniqueness demands a clear CE definition for water. In this chapter, the definition used was adopted from Morseletto [3] and Morseletto, Mooren, and Munaretto [4], who describe circular economy for water (CEW) as "an economic framework for reducing, preserving, and optimizing water use through waste avoidance, efficient use, and quality retention while ensuring environmental protection and conservation". As mentioned earlier, water plays different roles depending on the context. It is vital to the sustainment of life, and it is a critical input resource for the world economy (industrial processes, service delivery, food, and energy production), but it is also a finite resource needing quantity and quality preservation.

DOI: 10.1201/9781003441007-5

3.1.1 THE NATURAL WATER CYCLE AND THE EFFECT OF ANTHROPOGENIC ACTIVITIES

Water needs to be managed sustainably, respecting its natural cycle [1] in all processes where it is used, consumed, or treated. The water cycle not only is essential for all Earth system processes, interactions, and feedback but also is subject to anthropogenic manipulation at a global scale. Water security and sustainability depend on various water stores and fluxes that are being modified by global changes [5].

The natural water cycle is the continuous movement of water around the planet through evaporation, transpiration, condensation, precipitation, run-off, infiltration, and percolation [6]. Water is used by humans to satisfy all their needs. After being used, generated wastewater requires to be treated before being returned to the environment with, at least, the same quality as it had before it was taken. The water cycle is significantly altered by human activities, and CE is an interesting alternative to preserve water for future generations. Figure 3.1 shows a schematic view of the natural and anthropogenic water cycles. The colors green and brown in Figure 3.1 represent the stages where anthropogenic activities alter the water cycle to satisfy human needs, survival, and progress.

The water cycle involves a broad set of core functions in terrestrial and aquatic ecosystems, the atmosphere, and/or in direct interaction with human society that are crucial for biophysical stability, determinant for water resilience [9], and active across different segments. While these water functions can overlap with ecosystem services provided by water, water key functions are related not just to their benefits but also to their ability to generate ecosystem resilience. Water functions along the anthropogenic water cycle are described in Table 3.1. Some of these functions have been identified with specific benefits, including regulatory, productive, moisture feedback, supply, carrier/chemical load, state, productive, and regulatory/control. Because of the interrelation of these functions, several gaps exist, and coordinated activities are needed to achieve and maintain equilibrium between human activities and minimal environmental alteration to control actions for resilience to climate change.

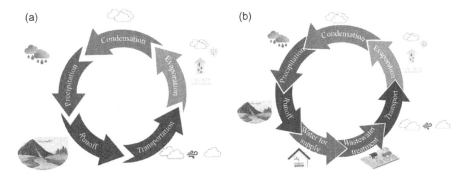

FIGURE 3.1 Schematic representation of (a) the natural [7] and (b) anthropogenic water [8] cycles.

TABLE 3.1

The Anthropogenic Water Cycle and its Functions (Adapted from Falkenmark, Wang-Erlandsson and Rockström [9])

Function	Description
1. Regulatory[a]	Includes air and soil moisture, water in living matter, evaporation, and transpiration to regulate energy balance and climate at a local or global scale through carbon sequestration, cloud formation, albedo regulation, and latent heat release.
2. Productive[a]	Includes evaporation to sustain food, bioenergy production, and biomass growth.
3. Moisture feedback[a]	Includes recycled evaporation to regulate the water cycle over land.
4. Water for supply[b]	Includes water withdrawn for supply.
5. Carrier/chemical load[b]	Includes river and base flows to carry pollutants.
6. State[b]	Includes water volumes (e.g., lakes, wetlands, groundwater, reservoirs, and rivers) to maintain the aquatic state.
7. Productive[b]	Includes water extraction for agriculture and to allow and sustain aquatic biomass growth.
8. Regulatory/control[b]	Includes river and base flows to regulate aquatic ecosystems, the Earth's energy balance, and climate through albedo regulation and carbon storage. Groundwater and glaciers regulate sea levels and geological processes such as subsidence.

[a] Green functions: Green water flows include evaporation and transpiration. Green water stocks include soil moisture.

[b] Blue functions: Blue water flows include river and groundwater flows. Blue water stocks include rivers, lakes, aquifers, and wetlands.

3.2 LIFE CYCLE ASSESSMENT PRINCIPLES

Used as a methodology to evaluate and quantify potential environmental impacts of a product or service throughout its life cycle [10], a life cycle assessment (LCA) identifies and measures inputs (e.g., raw materials and consumed energy), outputs (e.g., pollutant emissions, products, and by-products), and their associated environmental impacts [11,12]. Typically, the stages included in a product's life cycle are raw materials extraction and processing, production, transportation and distribution, use, and end-of-life treatment (e.g., reuse, recycling, recovery, and final disposal) [13].

LCA follows two international standards: ISO 14040 (Principles and Framework) and ISO 14044 (Requirements and Guidelines) [14]. According to the framework set by ISO 14040, LCA consists of four interdependent phases [15], shown in Figure 3.2.

The interdependence among different LCA phases is what makes this approach iterative, which means that as the study advances, some adjustments will have to take place. For example, data availability is often unknown at the beginning of an LCA study, so changes will have to be made accordingly [10,16].

Life Cycle Assessment Framework

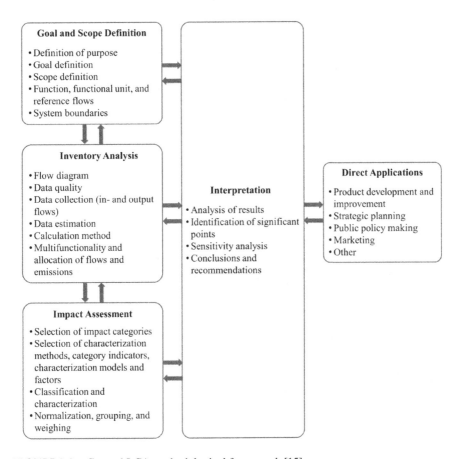

Goal and Scope Definition

- Definition of purpose
- Goal definition
- Scope definition
- Function, functional unit, and reference flows
- System boundaries

Inventory Analysis

- Flow diagram
- Data quality
- Data collection (in- and output flows)
- Data estimation
- Calculation method
- Multifunctionality and allocation of flows and emissions

Interpretation

- Analysis of results
- Identification of significant points
- Sensitivity analysis
- Conclusions and recommendations

Direct Applications

- Product development and improvement
- Strategic planning
- Public policy making
- Marketing
- Other

Impact Assessment

- Selection of impact categories
- Selection of characterization methods, category indicators, characterization models and factors
- Classification and characterization
- Normalization, grouping, and weighing

FIGURE 3.2   General LCA methodological framework [15].

3.2.1   LCA Goal and Scope Definition

The LCA goal states the reason for carrying out the study as well as the intended application of results, and target audience. The scope definition provides a product system description, its functions, and system boundaries [17]. Functional unit, reference flow, cut-off and allocation rules, impact assessment methodology type, impact categories, data requirements, limitations, critical review type, and final results type and format are other issues that need to be addressed in this phase [17,18].

Depending on the study, two types of Life Cycle Inventory (LCI) modeling approaches can be distinguished: attributional (also known as "accounting" or "descriptive approach") and consequential (or "change-oriented approach") [19]. Attributional LCA (ALCA) describes potential environmental impacts for a specific product or system over its entire life cycle [20], using average data [21]. Consequential LCA (CLAC) describes environmental burdens resulting from a particular decision [19] using marginal data [21]. According to Bergman, Gu, Page-Dumroese, and Anderson [20], ALCA is a snapshot of a moment in time whereas CLCA is used to evaluate market decisions. In LCA studies, a functional unit (FU) qualitatively and quantitatively describes the function or functions provided by a product for a set time [18,20]. The main purpose of an FU is to provide a reference to which all inputs and outputs are related [17]. To ensure comparability among LCA studies, all the compared product systems need to have the same FU [10,22].

According to Müller et al. [10], when comparing products with identical chemical structure and composition, mass shall be used as a basis for comparison. When products with different chemical structures and compositions are compared, an FU should be defined in terms of the products' technical performance. Once the FU is defined, the next step is determining the reference flow (i.e., the product amount needed to fulfill the FU), considering the product's performance [23].

The system boundary identifies the processes and life cycle stages required to fulfill the FU [10]. The analysis covers from raw materials extraction to the product's final disposal (i.e., cradle to grave) or from raw materials extraction to the factory gate (i.e., cradle to gate). It may focus only on manufacturing processes (i.e., gate to gate); include raw materials extraction, production, use, and recycling (i.e., cradle to cradle), or consider product use and end-of-life phases once it leaves the factory (i.e., gate to grave) [16,17].

In theory, the system boundaries should include all the processes and flows attributable to the investigated system. However, not all of them may be relevant, and cut-off criteria must be determined [18,20]. Cut-off criteria will indicate which material and energy flows should be included according to the relative contribution of each environmental impact category considered [18,24]. To simplify LCA modeling and enable completion within a reasonable time, auxiliary flows and those smaller than 1% are excluded [20,24]. The allocation rules attribute shares of the total environmental impact to the different products of a system [17].

3.2.2 Life Cycle Inventory Analysis

In a life cycle inventory (LCI) analysis, data are collected and calculations are performed to quantify relevant inputs (e.g., raw materials, energy, and water) and outputs (e.g., emissions and waste) within the system boundaries [18,25]. To facilitate this process, it is recommended to divide the system into several interconnected subsystems [18]. Data collection is a time-consuming and complex task. Primary data come from direct measurements, and secondary data can be retrieved from the literature and commercial databases, such as Ecoinvent or GaBi [17,18].

3.2.3 Life Cycle Impact Assessment

The life cycle impact assessment (LCIA) aims at using inventory analysis results to evaluate the potential environmental impact significance. In this step, inventory data relates to specific environmental impact categories and their respective category indicators, for example, global warming potential (GWP) as an indicator for climate change [25].

The LCIA phase consists of the following mandatory steps [20]:

- Impact categories, indicators, and characterization models selection.
- Classification: assignment of individual inventory inputs and outputs to impact categories.
- Characterization: LCI flow conversion to common units within each impact category so that the LCI flows can be aggregated into category indicator outputs.

Additionally, the following optional steps may also be considered [17,18]:

- Normalization: impact assessment results are multiplied by normalization factors to compare the quantified impact of a certain flow to a reference value.
- Grouping: impact categories are assigned into sets to better allow results interpretation for specific areas of concern.
- Weighting: normalized results are multiplied by weighting factors reflecting the perceived relative importance of the various impact categories.

There are different methods to perform LCIA, which classify and characterize environmental impact using two main approaches: midpoint (or problem-oriented) and endpoint (or damage-oriented) [17]. The former relates to environmental problems, such as climate change, acidification, and ecotoxicity. The latter is related to human health, natural environment, and natural resources [18]. Midpoint-oriented LCIA methods include CML 2001, EDP, TRACI, and USETox. Examples of endpoint-oriented LCIA methodologies include EcoIndicator 99, Impact 2002+, and ReCiPe [17].

3.2.4 INTERPRETATION

In this final step, results are analyzed to draw conclusions and offer recommendations [10]. Life cycle interpretation of an LCA comprises three main elements: (i) identification of significant problems (i.e., environmental hotspots), based on the LCI and LCIA outcomes; (ii) outcome evaluation, considering completeness, sensitivity, and consistency checks; and (iii) conclusions and recommendations [20].

LCA is a useful methodology to develop new products/services or improve existing ones. In order to apply it successfully, it is necessary to have a clear and deep knowledge of the product system in question, determine the study goal and scope, be familiar with LCA methods, identify and collect relevant data to make up the life cycle inventory, perform an impact assessment using the proper LCA method, and interpret the results to see the potential environmental impacts and offer recommendations. LCA allows producers or relevant stakeholders for strategic planning and proper decision-making, always aiming for continued improvement to minimize impacts on natural resources and human health.

Using this methodology to eliminate or minimize environmental impacts remains a major challenge, despite its evident advantages. Coordination between LCA practitioners and policymakers is needed to create synergy and comply with commonly established goals when dealing with water-related scenarios.

3.3   WATER SUSTAINABILITY USING CE AND LCA

3.3.1   CURRENT WATER SITUATION

According to the UN World Water Development Report 2023, two billion people cannot access safe drinking water [26]. Water shortages affect between two and three billion people for at least one month every year [26]. In some situations, lack of water last longer, as Cape Town, South Africa residents have witnessed recently. Between 2015 and 2018, this city's population—close to 4.6 million people—had to live through a 1-in-400-year drought that brought Cape Town to the verge of disaster [27]. The current global water situation is the consequence of human negligence and behavior, as well as inefficient and unsustainable management of water resources [28]. With the world's population expected to reach 9.7 billion by 2050 [29], appropriately managing finite natural resources has become extremely important [30]. Failing to change how water is perceived and used will have severe consequences for humanity [28].

3.3.2   CAN CE SUPPORT ACHIEVING WATER SUSTAINABILITY?

Since the start of the industrial age, economic systems have followed a traditional linear model for resource consumption based on the "take, make, and dispose" production theory [28,31]. According to this model, resources are perceived as limitless, accessible, and cheap with no negative consequences [28]. Therefore, goods are designed to be thrown away once they reach the end of their useful life or even before that [31]. This results in the generation of large quantities of managed and mismanaged waste and a rapid depletion of available resources [31,32]. In contrast, the CE model focuses on eliminating the waste concept by promoting the sustainable use of natural resources and minimizing environmental impacts [32]. The idea behind CE is that any material at the end of its life cycle will act as input for another life cycle, keeping components and materials within the market at the highest quality possible for as long as possible [32]. With this idea in mind, products must be designed for reuse and remanufacturing, reducing energy costs, and natural resource depletion [30].

Regarding water management, the CE approach seeks to shift from a linear economy model of extraction, use, and discharge—which is no longer viable [33]—to one where wastewater is purified, regenerated, and re-integrated back into the supply cycle as a renewed resource [28].

Several CE strategies that can be applied to water management have been identified in the past [4,3335]. Table 3.2 shows some examples of nine CE strategies divided into three categories [4].

TABLE 3.2

CE Strategies for Water Management [4]

Category	CE strategy	Definition
Decrease (using less o no water at all)	Rethink	Reconfiguring and reconceptualizing water use to favor circular utilization.
	Avoid	Preventing water use
	Reduce	Using less water than usual scenario
Optimizing (intensified/ efficient water use)	Replace	Substituting water with other substances
	Reuse	Using water again without treatment
	Recycle	Using water again after treatment
Retaining (keeping water, materials, molecules, and energy)	Cascading	Consecutive water uses for different purposes
	Store	Transfer used water to a reservoir
	Recover	Retrieving valuable materials (e.g., organic matter, chemical elements, biochemical compounds) and energy generation.

3.3.3   WHY IS LCA A VALUABLE TOOL TO SUPPORT CE?

According to Mannan et al. [31], the need for more quantitative measures to assess environmental performance is the major reason for delayed CE implementation. The authors suggest that current mass-based metrics serve to evaluate the circularity of any process or product but cannot assess environmental perspectives. The material circularity indicator (MCI) illustrates this point clearly. MCI measures the "degree of circularity" of materials composing products [34]. However, it cannot indicate which CE strategy is the best environmental option [31]. Haupt et al. [36] noted that to implement the CE approach, circularity and environmental performance should be successfully evaluated together, as circularity does not always equate to environmental sustainability. This view is supported by other studies [37], where authors suggest circularity by itself does not ensure social, economic, and environmental performance.

There is an urgent need to consider a product's social, economic, and environmental impacts along its lifespan (from cradle to grave) to appropriately assess CE effectiveness [30]. LCA helps with this task by quantifying environmental impacts and assessing the benefits of CE strategies in terms of supply and production, sustainable consumption, and waste management [38]. LCA supports CE in different ways by analyzing CE strategy's pros and cons and identifying better alternatives for the entire life cycle. Because it is an iterative process, LCA continuously reformulates the goals and objectives to improve circularity scenarios [31].

The CE concept seeks to develop community well-being and minimize ecological damage, while LCA is a trustworthy method to evaluate environmental impacts. Combining both concepts provides a comprehensive approach that strengthens available CE alternatives. Furthermore, it allows product manufacturers to compare CE strategies with environmental evaluations to achieve a balanced environmental and ecological system for new products [31].

3.3.4 WATER QUANTITY AND QUALITY

Despite Earth's water abundance—71% of Earth's surface is water-covered—96.5% of Earth's water is salty [35]. With only a small amount of freshwater available to sustain human, plant, and animal life, its distribution is space and time irregular all over the world, and it is under pressure due to human activity and economic development [39,40]. Water availability dictates where people can live and determines their life quality. Water sufficiency strongly correlates with the rate and degree of society development within a specific region [41] and is undoubtedly recognized as essential for human survival and well-being, playing a dominant role in many economic sectors.

Accelerated urbanization and expanding municipal water supply and sanitation systems have also contributed to the rising water demand [42], projected to reach 4,350 billion m^3 by 2040 and likely to vary based on region and sector [43]. Trends in water use and needs largely depend on urbanization, population growth, and living standards. Also, a vast number of people worldwide still lack access to proper drinking water (in quantity and quality), and an even larger population has no access to improved sanitation services [42].

According to Molle and Mollinga [44], water demands are classified into five water use categories: (i) *Drinking water,* the most vital and inelastic use of water, is considered a human right. Depending on climate, humans need between 1 and 5 L of water daily; (ii) *Domestic water*, including other basic domestic uses such as cooking, hygiene, and laundry; (iii) *Food security needs*, related to water use to grow food for self-consumption or other activities on which their subsistence depends critically. The most common example is smallholders and peasants who irrigate their fields and depend on this agricultural production for their food and subsistence; (iv) *Economic production*, related to water for goods production by people economically dependent on these goods, but whose basic domestic and food needs are not drastically affected by water shortages that might constrain their production process; and (v) *Environmental needs*, humans are affected when the amount, quality, and timing of water flows are not ensured. Water scarcity causes biodiversity loss, pollution's health impacts, aesthetic degradation, and other negative impacts.

According to the intended specific use, water quantity (i.e., how much water is needed) and quality (i.e., what it is needed for) need to comply with specific characteristics. Certain chemical and biological parameters are selected depending on their significance and concentrations in water. According to the World Economic Forum Global Risk Reports 2016–2021 [45], lack of water availability is also intrinsically linked to water quality because water pollution may prevent specific uses. Increased discharges of untreated sewage, combined with agricultural runoff and unproperly treated wastewater, have resulted in the degradation of water quality around the world [42], threatening human health and ecosystems and constraining sustainable economic development.

In different countries, wastewater is released directly into the environment without adequate treatment, with significantly detrimental impacts [42,46]. For example, it has been estimated that 80% of all infections and over one-third of demises in developing countries are caused by polluted drinking water [47]. Surface water is a carrier of chemical and microbiological pollutants. Pathogenic microorganisms in human waste or sewage from surrounding localities can cause gastrointestinal diseases [47], making microbiologically contaminated water extremely dangerous for drinking, swimming, or bathing. Releasing raw sewage in water bodies significantly disturbs oxygen balance because of the organic matter and other chemical pollutants load in sewage wastewater, threatening aquatic flora and fauna. One example is the excess of nitrogen and phosphorus causing eutrophication in diverse water bodies. Water quality monitoring has been carried out to evaluate the suitability of a water resource for a particular usage. Water suitability is assessed concerning acceptable concentrations of water quality variables, which are defined by guidelines, standards, or maximum permissible concentrations [47].

With regard to scarcity, it is crucial to define three essential concepts that are interrelated: scarcity, water stress, and water risk. Water scarcity refers to the volumetric abundance, or lack thereof, of water supply. It is typically calculated as a ratio of human water consumption to available water supply for a given area. Water scarcity is physical, objective data that can be measured consistently across regions and over time [48]. Water scarcity can be related to water demand and/or infrastructure, and the water amount that can be physically accessed varies as supply and demand change [49]. About four billion people, representing nearly two-thirds of the global population, experience severe water scarcity during at least one month of the year [49,50]. Water stress refers to the ability, or lack thereof, to meet human and ecological demands for water. Compared to scarcity, water stress is a broader concept. It considers several physical aspects related to water resources, including water scarcity, quality, environmental flows, and accessibility. It is directly related to water quality and environmental flows that reflect volumetric availability and accessibility [48]. Water risk is the probability of experiencing a deleterious water-related event. Because different sectors and organizations will experience water risk differently, it is defined and interpreted differently even when the same degree of water scarcity or water stress is experienced. Water-related conditions (e.g., water scarcity, pollution, governance, infrastructure, and climate change) simultaneously create risk for many sectors and organizations [48].

Several methods have been proposed to measure water scarcity. One of the most used was developed by Falkenmark based on water availability per capita per year within the country or region. The threshold used is 1,700 m^3/per capita/per year, and countries or regions with availability below this threshold are considered to experience water stress [41,51,52]. Water scarcity is an increasing problem on every continent, with poorer communities most badly affected. To build resilience against climate change and to serve a growing population, an integrated and inclusive approach must be taken to managing this finite resource. This integrated approach must be built from particular to general, and CEW is an essential and helpful tool to achieve these goals. Otherwise, achieving SDGs 6, 9, and 12 will be compromised [53]. Water reuse, regardless of the intent, will alleviate water scarcity by applying wastewater treatment technologies to remove pollutants of concern and achieving CEW. At the same time, developing new water treatment technologies is imperative and has great potential in reducing the gap between availability, demand, and quality.

3.3.5 REUSE AND RECYCLE: TECHNOLOGICAL TOOLS TO ACHIEVE WATER SUSTAINABILITY

Availability, scarcity, and water management are top global risks, according to the World Economic Forum Global Risk Report [54]. The global shortfall in water supply by 2030 is estimated to be as high as 40%, a trend that will continue if we do not change how we manage water [42,55]. To reduce the use of resources and favor the practical application of the waste hierarchy, the European Community [56] designed a scheme for waste management, including water and wastewater aspects (see Figure 3.3). This scheme establishes a priority order for prevention, preparation for reuse, recycling, energy recovery, and, finally, disposal [57].

Wastewater reuse (also known as water recycling or water reclamation) reclaims water from various sources and then treats and reuses it for different purposes (e.g., agriculture groundwater replenishment, industrial uses, potable water supplies, and environmental restoration). Water reuse provides alternatives to conventional water supplies and enhances water security, sustainability, and resilience [58,59]. The United States Environmental Protection Agency (USEPA), in collaboration with the United States Agency for International Development (USAID), issued a document whose primary purpose was to facilitate further water reuse development by serving as a reference for water reuse practices [59].

Before carrying out any water reclamation and reuse activities, it is necessary to pursue sustainable water resource management. Promoting wastewater reuse as an alternative water supply can satisfy different needs depending on water quality requirements, mitigate water scarcity, and help satisfy water supply demands, specifically in arid and semi-arid regions. Worldwide, in different countries, several governments have implemented water reuse projects [60–65]. Existing facilities need to be re-engineered to satisfy water demands, in the context of water scarcity, to improve the supply. New wastewater treatment facilities must be designed to comply with more rigorous water quality requirements, including removing conventional (e.g., nutrients, suspended solids, and pathogens) and emerging pollutants as well as implementing best practices for facilities management [66].

Water reuse will transform the linear human water cycle (i.e., abstract, treat, distribute, consume, collect, treat, and discharge) into a circular flow by closing the loop [46]. The motivation to implement reuse programs mainly addresses urbanization and water supply scarcity, achieving efficient use, and environmental and public

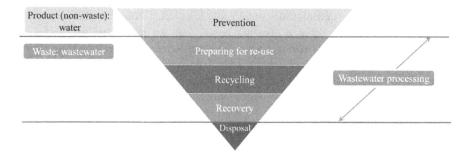

FIGURE 3.3   Water and wastewater aspects in the EU waste management hierarchy [57].

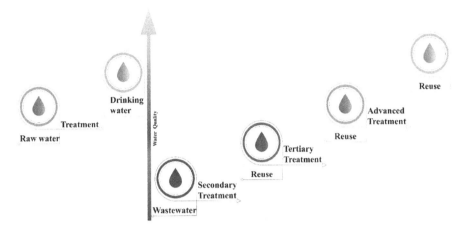

FIGURE 3.4 Available treatment technologies to achieve any desired water quality level [59,63,64].

health protection [59]. Nevertheless, reclaimed water typically poses more significant financial, technical, and institutional challenges than traditional sources. A range of treatment options are available, so any level of water quality can be achieved depending upon the expected use when wastewater is treated properly to meet fit-for-purpose specifications [66] (see Figure 3.4).

Two important concepts to introduce are treatment levels and treatment trains. The former brings together water quality and applications, so process integration and unit operations achieve the treated wastewater quality level that the specific reuse demands. The latter is defined as the combination of unitary processes employed to achieve a specific treatment objective to obtain the desired water quality for the purpose for which it is intended [61]. There are different treatment combinations and options that allow to achieve the desired quality for the different reuse options, including direct and indirect potable reuse. These combinations can be customized to satisfy the required quality according to the objectives of specific reuse [61]. Various treatment options, including engineered and managed natural treatment processes, mitigate microbial and chemical contaminants in reclaimed water, facilitating the achievement of specific water quality objectives (see Figure 3.5) [58].

3.3.6 CIRCULARITY IN DRINKING AND WASTEWATER TREATMENT

The main advantage of CEW and wastewater processing is resource recovery and reuse, transforming drinking water delivery and sanitation into self-sustaining and value-adding systems [67]. The CE concept was introduced in 1990 [68,69], and its application in the water sector to combat water scarcity was proposed later on [46,70,71]. Several opportunities exist to manage water and wastewater facilities from the CE perspective. For example, reclaiming wastewater for potable or non-potable purposes [72], where nutrients are recovered from wastewater and sludge to then be used for agricultural purposes. To integrate the CE concept into the treatment processes in water and wastewater facilities in their daily work, it is essential to plan

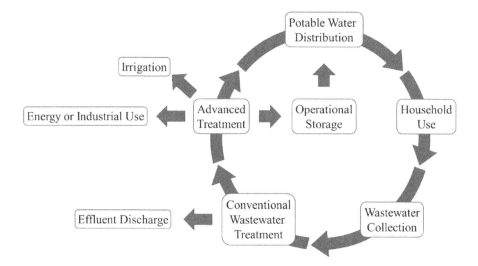

FIGURE 3.5   Water flows to close the loop (with potential applications of direct and indirect water reuse) [46].

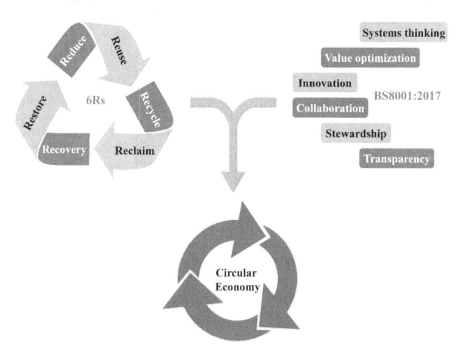

FIGURE 3.6   Confluence of 6Rs and 6BS principles on CE [71].

and put into action the 6Rs concepts (i.e., reduce, reuse, recycle, reclaim, recover, and restore) coupled with the British Standard for circularity (i.e., systems thinking, value optimization, innovation, collaboration, stewardship, and transparency) [71,73,74] (see Figure 3.6).

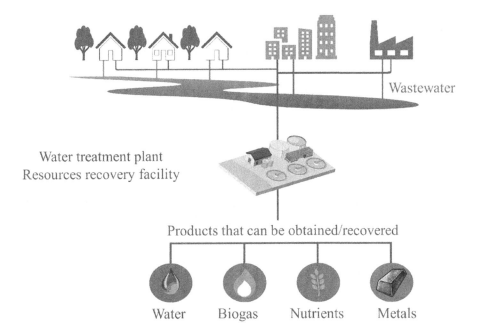

FIGURE 3.7   Products obtained from a water treatment plant [67].

When CE is applied to water treatment plants, water resource usage, and generated waste will be guaranteed to put less pressure on the effects of water scarcity. Once CE is incorporated into water treatment plants, it creates a participatory environment that links plants and various municipalities and industries seeking sustainability (see Figure 3.7) and revalorizes useful products or resource recovery [75–78]. Table 3.3 shows some examples of valuable resources recuperated from water and wastewater treatment facilities [75,79].

As shown in Table 3.3, different opportunities for using waste from water treatment plants are identified. For example, potential for biogas generation through organic sludge or using non-hazardous inert sludge as raw material for construction. In addition, inorganic substances recovery concentrated in advanced treatments is also possible, but this usually involves a significant extra expense, so it must be analyzed carefully. Future research related to treatment plant design should not only aim at optimizing efficiency and treatment costs but also seek the best use of generated wastes. Changing the paradigm in which contaminated water is perceived as a problem towards an opportunity for sustainable development remains a long way but, fortunately, we have begun.

3.4   CONCLUDING REMARKS

Humankind needs to be aware that the water cycle is a natural function of the planet and that human beings depend on this precious resource to survive, develop, and grow. However, throughout the years people have managed water resources by

TABLE 3.3

Valuable Resources Recuperated from Water and Wastewater Treatment Facilities [79]

Treatment stages	Processes	Waste	Exploitation strategy
Primary	Physic	Inert sludge	• Providing it is non-hazardous, sludge can be used for construction • Recirculate recovered effluent
Secondary	Physicochemical	Inert sludge	
Tertiary	Biological	Aerobic: Inert sludge	• Use for construction • Recirculate recovered effluent
		Anaerobic: Organic loaded-sludge	• Use as fertilizer and in biogas production
Advanced	Reverse osmosis	Inorganic substances	• Substance recovery through advanced methods
	Ion exchange	Mineral-containing resins	• Resin recovery through demineralization, hot softening, or reverse osmosis plants

practicing a linear economy principle (i.e., extract, use, and discharge). Fortunately, this way of thinking and acting has been gradually changing to a CE mindset, and water management topics have started to move in this direction. People working on the CEW concept have been dedicating their efforts to establishing and improving guidelines and criteria to develop activities that better resource management.

As water functions overlap with ecosystem services provided by water, the key to water functions lies in providing the equilibrium for resilience. All the water functions are interrelated. There are still several gaps to fill and coordinated activities that are needed to maintain equilibrium for the success of human activities with minimal environmental alteration and resilience to climate change.

The LCA methodology was reviewed to show its usefulness as a tool to improve water resource management, by determining potential environmental impacts, particularly in activities aimed at achieving water sustainability and hydric security, following ISO Standards (14040 and 14044). Unfortunately, using this methodology to eliminate or minimize environmental impacts remains a major challenge. LCA practitioners and policymakers need to work in tandem to develop or improve environmentally friendly products and services. This joint effort should be made in a coordinated and synergistic manner to comply with commonly established goals.

The CE strategies for water management are described and emphasized to be applied in every activity related to water management.

Also, water scarcity, water risk, and water stress concepts were introduced to raise awareness about the problems related to water availability and management and to highlight the need for water reuse and recycling as a way to achieve Sustainable

Development Goals, in particular SDGs 6, 9, and 12. The waste management hierarchy was also underlined to change the widely held perception of waste production as a problem by considering waste as a resource or raw material for further use.

As an example of circularity applied to water and wastewater treatment facilities, treatment options for reuse were mentioned to preserve water, emphasizing resource recovery.

There is still a long way to go, but we are on the way.

REFERENCES

1. Dolan, F. et al. Evaluating the economic impact of water scarcity in a changing world. *Nat. Commun.* 12, 1915 (2021). https://doi.org/10.1038/s41467-021-22194-0
2. Martínez-Austria, P. F., Díaz-Delgado, C. & Moeller-Chavez, G. Seguridad hídrica en México: diagnóstico general y desafíos principales. *Ingeniería del agua* 23(2), 107–121 (2019). https://doi.org/10.4995/ia.2019.10502
3. Morseletto, P. Targets for a circular economy. *Resour., Conserv. Recycl.* 153, 104553 (2020). https://doi.org/10.1016/j.resconrec.2019.104553
4. Morseletto, P., Mooren, C. E. & Munaretto, S. Circular economy of water: definition, strategies and challenges. *Circ. Econ. Sust.* 2, 1463–1477 (2022). https://doi.org/10.1007/s43615-022-00165-x
5. Gleeson, T. et al. Illuminating water cycle modifications and Earth system resilience in the Anthropocene. *Water Resour. Res.* 56(4), e2019WR024957 (2020). https://doi.org/10.1029/2019WR024957
6. USGS. The Water Cycle. (2022). https://www.usgs.gov/special-topics/water-science-school/science/water-cycle
7. Principado de Asturias – Consorcio de Aguas. (n.d.). Ciclo natural del agua. https://consorcioaa.com/divulgacion/ciclo-del-uso-del-agua/
8. Principado de Asturias – Consorcio de Aguas. (n.d.). Ciclo del uso del agua en el abastecimiento y el saneamiento. https://consorcioaa.com/divulgacion/ciclo-del-uso-del-agua/
9. Falkenmark, M., Wang-Erlandsson, L. & Rockström J. Understanding of water resilience in the Anthropocene. *J. Hydrol.* 2(X), 100009 (2019). https://doi.org/10.1016/j.hydroa.2018.100009
10. Müller, L. J. et al. A guideline for life cycle assessment of carbon capture and utilization. *Front. Energy Res.* 8, (2020). https://doi.org/10.3389/fenrg.2020.00015
11. Lunardi, M. M., Alvarez-Gaitan, J. P., Bilbao, J. I. & Corkish, R. Comparative life cycle assessment of end-of-life silicon solar photovoltaic modules. *Appl. Sci.* 8(8), 1396 (2018). https://doi.org/10.3390/app8081396
12. Laca, A., Herrero, M., & Díaz, M. Life cycle assessment in biotechnology. In *Comprehensive Biotechnology* (second edition) (ed. Moo-Young, M.) 839–851. https://doi.org/10.1016/B978-0-08-088504-9.00140-9 (Elsevier, 2011).
13. Lazarevic, D., Liljenström, C. & Finnveden, G. *Silicon-Based Nanomaterials in a Life-Cycle Perspective, Including a Case Study on Self-Cleaning Coatings* (KTH-Royal Institute of Technology, Stockholm, Sweden, 2013).
14. Trompeta, A-F, Koklioti, M. A., Perivoliotis, D. K., Lynch, I. & Charitidis, C. A. Towards a holistic environmental impact assessment of carbon nanotube growth through chemical vapor deposition. *J. Clean. Prod.* 129, 384–394 (2016). https://doi.org/10.1016/j.jclepro.2016.04.044
15. Guinée, J. B. (Ed.). *Handbook on Life Cycle Assessment. Operational Guide to the ISO Standards* (Kluwer Academic Publishers, 2002).

16. Golsteijn, L. Life Cycle Assessment (LCA) explained. (2022). https://pre-sustainability.com/articles/life-cycle-assessment-lca-basics/
17. Fokaides, P. A. & Christoforou, E. Life cycle sustainability assessment of biofuels. In *Handbook of Biofuels Production: Processes and Technologies* (second edition) (eds. Luque, R., Sze Ki Lin, C., Wilson, K. & Clark, J.) 41–60. https://doi.org/10.1016/B978-0-08-100455-5.00003-5 (Elsevier, 2016).
18. REACHnano Consortium. Guidance on Available Methods for Risk Assessment of Nanomaterials. (REACHnano Consortium, 2015). https://invassat.gva.es/documents/16 1660384/162311778/01+Guidance+on+available+methods+for+risk+assesment+of+nan omaterials/8cae41ad-d38a-42f7-90f3-9549a9c13fa0
19. Valdivia, S. UNEP/SETAC Life Cycle Initiative. (2011). https://wedocs.unep.org/handle/20.500.11822/32747
20. Bergman, R. D., Gu, H., Page-Dumroese, D. S. & Anderson, N. M. Life cycle analysis of biochar. In *Biochar: A Regional Supply Chain Approach in View of Climate Change Mitigation* (eds. Bruckman, V. J., Apaydın Varol, E., Uzun, B. B. & Liu, J.) 46–69. https://doi.org/10.1017/9781316337974 (Cambridge University Press, 2016).
21. Ekvall, T. Attributional and consequential life cycle assessment. In Sustainability Assessment at the 21st *Century* (eds. Bastante-Ceca, M. J., Fuentes-Bargues, J. L., Hufnagel, L., Mihai, F-C. & Iatu, C.) https://dx.doi.org/10.5772/intechopen.89202 (IntechOpen, 2020).
22. Weidema, B. Wenzel, H., Petersen, C. & Hansen, K. The product, functional unit and reference flows in LCA. *Environ.News* **70**, 1–46, (2004).
23. International Organization for Standardization [ISO]. Environmental Management – Life Cycle Assessment – Illustrative Examples on How to Apply ISO 14044 to Goal and Scope Definition and Inventory Analysis *(Reference number* ISO/TR 14049:20129(E)). (2012).
24. Agarski, B., Vukelic, D., Ilic Micunovic, M., and Budak, I. (2020). Screening of cut-off and allocation rules in environmental product declarations *(Paper Presented at the Scientific Conference "Environmental Labelling in Circular Economy", Virtual on-Line Conference*, November 30th, 2020). https://www.vstecb.cz/wp-content/uploads/2021/06/ecolabelling2020_boris_agarski.pdf
25. Klöpffer, W. Publishing scientific articles with special reference to LCA and related topics (6 pp). *Int. J. Life Cycle Assess.* **12**(2), 71–76 (2007). https://doi.org/10.1065/lca2007.01.306
26. UNESCO. *Imminent Risk of a Global Water Crisis, Warns the UN World Water Development Report 2023* (2023). https://www.unesco.org/en/articles/imminent-risk-global-water-crisis-warns-un-world-water-development-report-2023#:~:text=Between%20two%20and%20three%20billion,UN%20World%20Water%20Development%20Report.
27. Hill-Lewis, G. Cape Town: Lessons from managing water scarcity. Brookings (2023). https://www.brookings.edu/articles/cape-town-lessons-from-managing-water-scarcity/
28. Urrea Vivas, M. Avoid water stress by utilizing a circular economy model. *Harvard Advanced Leadership Initiative Social Impact Review* (2021). https://www.sir.advancedleadership.harvard.edu/articles/avoid-water-stress-by-utilizing-a-circular-economy-model
29. United Nations. World population to reach 8 billion this year, as growth rate slows. *UN News. Global Perspective Human Stories* (2022). https://news.un.org/en/story/2022/07/1122272
30. Broadbent, C. Steel's recyclability: demonstrating the benefits of recycling steel to achieve a circular economy. *Int. J. Life Cycle Assess.* **21**(11), 1658–1665 (2016). https://doi.org/10.1007/s11367-016-1081-1

31. Mannan, M. & Al-Ghamdi, S.G. Complementing circular economy with life cycle assessment: Deeper understanding of economic, social, and environmental sustainability. In *Circular Economy and Sustainability (Volume 1: Management and Policy)* (eds. Stefanakis, A. & Nikolaou, I.) 145–160. https://doi.org/10.1016/C2019-0-00505-5 (Elsevier, 2021).

32. Ingemarsdotter, E. & Dumont, M. Why the circular economy and LCA make each other stronger. *PRé* (2022). https://pre-sustainability.com/articles/the-circular-economy-and-lca-make-each-other-stronger/

33. Jazbec M., Mukheibir, P. & Turner, A. *Transitioning the Water Industry with the Circular Economy* (Prepared for the Water Services Association of Australia). (Institute for Sustainable Futures, University of Technology Sydney, 2020). https://www.wsaa.asn.au/sites/default/files/publication/download/Transitioning%20the%20water%20industry%20with%20the%20circular%20economy%20FINAL%2012102020.pdf

34. Rocchi, L., Paolotti, L., Cortina, C., Fagioli, F.F., & Boggia, A. Measuring circularity: an application of modified Material Circularity Indicator to agricultural systems. *Agric. Food Econ.* 9, 9 (2021). https://doi.org/10.1186/s40100-021-00182-8

35. USGS. *The Distribution of Water on, in, and Above the Earth.* (2019). https://www.usgs.gov/media/images/distribution-water-and-above-earth

36. Haupt, M. & Hellweg, S. Measuring the environmental sustainability of a circular economy. *Environ. Sustainab– Indicators* 1–2, 100005 (2019). https://doi.org/10.1016/j.indic.2019.100005

37. Carpenter, A. Role of Life Cycle Assessment in the Circular Economy – ASTM Workshop: Fostering a Circular Economy for Manufacturing Materials. (NREL, 2022). https://www.nrel.gov/docs/fy22osti/82677.pdf

38. Teixeira Pontes, A. & Maia Angelo, A. C. Use of life cycle assessment in the context of circular economy: A literature review. *Syst. Manag.* 14, 424–434 (2019). https://doi.org/10.20985/1980-5160.2019.v14n4.1576

39. UNICEF. Water Scarcity: Addressing the Growing Lack of Available Water to Meet Children's Needs. (n.d.). https://www.unicef.org/wash/water-scarcity

40. UNESCO. Water Scarcity and Quality. (2021). https://en.unesco.org/themes/water-security/hydrology/water-scarcity-and-quality

41. Nepomilueva, D. Water Scarcity Indexes: Water Availability to Satisfy Human Needs. (Helsinki Metropolia University of Applied Sciences, 2017).

42. United Nations World Water Assessment Programme. Wastewater: The Untapped Resource – The United Nations World Water Development Report 2017 (2017). https://wedocs.unep.org/20.500.11822/20448

43. Tiseo, I. Global Water Withdrawal and Consumption 2014–2040. (2023). https://www.statista.com/statistics/216527/global-demand-for-water/

44. Molle, F. & Mollinga, P. Water poverty indicators: conceptual problems and policy issues. *Water Policy* 5(5–6), 529–544 (2013). https://doi.org/10.2166/wp.2003.0034

45. World Economic Forum. The Global Risks Report 2022, 17th Edition (2022). https://www.weforum.org/publications/global-risks-report-2022/

46. Voulvoulis, N. Water reuse from a circular economy perspective and potential risks from an unregulated approach. *Curr. Opin. Environ. Sci. Health* 2, 32–45 (2018). https://doi.org/10.1016/j.coesh.2018.01.005

47. World Health Organization (WHO). *International Standards for Drinking Water*, 2nd Edition. (World Health Organization, 1971).

48. Schulte, P. Defining Water Scarcity, Water Stress, and Water Risk. https://pacinst.org/water-definitions/#:~:text=Water%20scarcity%20is%20a%20physical,across%20regions%20and%20over%20time.&text=%E2%80%9CWater%20stress%E2%80%9D%20refers%20to%20the,and%20ecological%20demand%20for%20water (Pacific Institute, 2014).

49. United Nations. Water Scarcity. (n.d.). https://www.unwater.org/water-facts/water-scarcity

50. Mekonnen, M. M. & Hoekstra, A. Y. Four billion people facing severe water scarcity. *Sci. Adv.* **2**(2), e1500323 (2016). https://doi.org/10.1126/sciadv.1500323

51. Falkenmark, M., Lundqvist, J. & Widstrand, C. Macro-scale water scarcity requires micro-scale approaches. *Nat. Resour. Forum* **13**(4), 258–267 (1989). https://doi.org/10.1111/j.1477-8947.1989.tb00348.x

52. Luo, T., Young, R. & Reig, P. *Aqueduct Projected Water Stress Country Rankings.* (World Resources Institute, 2015). https://www.wri.org/data/aqueduct-projected-water-stress-country-rankings

53. United Nations. Transforming Our World: The 2030 Agenda for Sustainable Development (A/RES/70/1). (2015).

54. World Economic Forum (WEF). The Global Risks Report 2018, 13th Edition (2018). https://www3.weforum.org/docs/WEF_GRR18_Report.pdf

55. World Economic Forum (WEF). The Global Water Initiative. (2017). https://www3.weforum.org/docs/Environment_Team/00042829_Water_initiative.pdf

56. European Parliament, Council of the European Union. *Directive 2008/98/EC of the European Parliament and of the Council of 19* November 2008 *on Waste and Repealing Certain Directives.* (2008).

57. Smol, M., Adam, C. & Preisner, M. Circular economy model framework in the European water and wastewater sector. *J. Mater. Cycles Waste Manag.* **22**, 682–697 (2020). https://doi.org/10.1007/s10163-019-00960-z

58. United States Environmental Protection Agency [EPA]. Uses for Recycled Water. (2023). https://www.epa.gov/waterreuse/basic-information-about-water-reuse#uses

59. United States Environmental Protection Agency [EPA]. Guidelines for Water Reuse 2012 (AR-1530 EPA/600/R-12/618). (2012). https://www.epa.gov/sites/default/files/2019-08/documents/2012-guidelines-water-reuse.pdf

60. Roccaro, P. & Verlicchi, P. Wastewater and reuse. *Curr. Opin. Environ. Sci. Health* **2**, 61–63 (2018). https://doi.org/10.1016/j.coesh.2018.03.008

61. Moeller, G., Guillén, R. A., Treviño, L. G. & Lizama, C. Reúso de aguas residuales tratadas como fuente directa e indirecta de agua potable. In *Uso seguro de Aguas Residuales para Reúso* (Eds. Tello Espinoza, P., Mijailova, P. & Chamy, R.) 93–107 (AIDIS, UNESCO & PHI-LAC; 2016). https://climatesmartwater.org/wp-content/uploads/sites/2/2018/09/AIDIS-Uso_seguro_del_agua_26_sep.pdf

62. Lee, H. & Tan, T. P. Singapore's experience with reclaimed water: NEWater. *Int. J. Water Resour. Dev.* **32**(4), 611–621 (2016). https://doi.org/10.1080/07900627.2015.1120188

63. Leverenz, H. L., Tchobanoglous, G. & Asano, T. Direct potable reuse: a future imperative. *J. Water Reuse Desalin.* **1**(1), 2–10 (2011). https://doi.org/10.2166/wrd.2011.000

64. Ghimire, U., Sarpong, G. & Gnaneswar Gude, V. Transitioning wastewater treatment plants toward circular economy and energy sustainability. *ACS Omega* **6**(18), 11794–11803 (2021). https://doi.org/10.1021/acsomega.0c05827

65. Mehmeti, A. & Canaj, K. Environmental assessment of wastewater treatment and reuse for irrigation: a mini-review of LCA studies. *Resources* **11**(10), 94 (2022). https://doi.org/10.3390/resources11100094

66. Corominas, L. et al. The application of life cycle assessment (LCA) to wastewater treatment: A best practice guide and critical review. *Water Res.* **184**, 116058 (2020). https://doi.org/10.1016/j.watres.2020.116058

67. Rodríguez, D. J., Serrano, H. A., Delgado Martin, A., Nolasco, D. & Saltiel, G. *From Waste to Resource: Shifting Paradigms for Smarter Wastewater Interventions in Latin America and the Caribbean.* (World Bank Group, 2020). https://documents.worldbank.org/en/publication/documents-reports/documentdetail/161611584134018929/from-waste-to-resource-shifting-paradigms-for-smarter-wastewater-interventions-in-latin-america-and-the-caribbean

68. Pearce, D. W. & Turner R. K. *Economics of Natural Resources and the Environment.* (Harvester Wheatsheaf, 1990).
69. Ellen MacArthur Foundation. Towards the Circular Economy *Vol. 1:* An Economic and Business Rationale for an Accelerated Transition. (2013). https://www.ellenmacarthur-foundation.org/towards-the-circular-economy-vol-1-an-economic-and-business-ratio-nale-for-an
70. International Water Association [IWA]. Water Utility Pathways in a Circular Economy. (2016). https://iwa-network.org/wp-content/uploads/2016/07/IWA_Circular_Economy_screen.pdf
71. Kakwani, N. S. & Kalbar, P. P. Review of circular economy in urban water sector: challenges and opportunities in India. *J. Environ. Manag.* **271**, 111010 (2020). https://doi.org/10.1016/j.jenvman.2020.111010
72. Angelakis, A. N., Asano, T., Bahri, A., Jimenez, B. E. & Tchobanoglous, G. Water reuse: from ancient to modern times and the future. *Front. Environ. Sci.* **6**, 26 (2018). https://doi.org/10.3389/fenvs.2018.00026
73. Niero, M. & Schimdt Rivera, X. C. The role of life cycle sustainability assessment in the implementation of circular economy principles in organizations. *Proc. CIRP* **69**, 793–798 (2018). https://doi.org/10.1016/j.procir.2017.11.022
74. British Standard Institution. BS 8001: 2017 Framework for Implementing the Principles of the Circular Economy in Organisations – Guide. (BSI Standards Publication, 2017). https://www.bsigroup.com/en-SG/About-BSI/sustainability/understanding-bs-8001-principles-of-the-circular-economy-in-organizations-training/#:~:text=The%20guidance%20standard%20BS%208001,the%20principles%20of%20the%20circular
75. Moeller Chávez, G. E. & Guerrero García Rojas, H. El ciclo del agua, la economía circular y la influencia de su alteración antropogénica. *Influencia Politécnica* 30–33 (2023). https://issuu.com/upemor/docs/revista_influencia_politecnica_junio2023
76. Daigger G. T. Evolving urban water and residuals management paradigms: water reclamation and reuse, decentralization, and resource recovery. *Water Environ. Res.* **81**(8), 809–823 (2009). https://doi.org/10.2175/106143009X425898
77. Gherghel, A., Teodosiu, C. & De Gisi, S. A review on wastewater sludge valorisation and its challenges in the context of circular economy. *J. Clean. Prod.* **228**, 244–263 (2019). https://doi.org/10.1016/j.jclepro.2019.04.240
78. Smol, M. Circular economy in wastewater treatment plant–water, energy and raw materials recovery. *Energies* **16**(9), 3911 (2023). https://doi.org/10.3390/en16093911
79. Alvizuri Tintaya, P. A., Astete Dalence, V., Torregrosa López, J. I., Lo Iacono Ferreria, V. G. & Lora García, J. Circular economy and its incorporation in water treatment plants. In Amin Saidoun (ed.), 25th *International Congress on Project Management and Engineering,* Alcoi, July 6–9, 2021. (2021).

Part II

Case Studies

4 Biogenic Nanomaterials for Water Detoxification and Disinfection

Thyerre Santana da Costa, Marcia Regina Assalin, Maelson Cardoso Lacerda, Nelson Durán, and Ljubica Tasic

4.1 INTRODUCTION

Earth has various types of water, primarily classified into two main categories: saltwater (also known as seawater) and freshwater. Saltwater includes about 97.5% of Earth's total water [1]. It is found in oceans, seas, and some saltwater lakes. Seawater contains dissolved salts and minerals, primarily sodium chloride, which gives it a characteristic salty taste [1,2]. The salinity of seawater varies depending on the location and other factors. Freshwater makes up the remaining 2.5% of Earth's total water [3,4]. This type of water contains a much lower concentration of salt compared to seawater, making it suitable for drinking and other anthropogenic uses. Freshwater is found in lakes, rivers, streams, groundwater, and ice caps/glaciers. Most of Earth's freshwater used to be stored in ice caps and glaciers and can be inaccessible as well as underground aquifers. Freshwater is a critical resource for human activities, including drinking, agriculture, and industry. Therefore, its availability, management, and quality are vital for sustainable growth.

Water quality is altered by the presence of various types and sources of impurities. In general, it is possible to distinguish between polluted water and contaminated water. Contaminated water is whose quality is degraded by the presence of undesirable substances or impurities of natural origin (e.g., sediment, nutrients, decaying organic matter) that affect primarily the environment and aquatic ecosystems (e.g., water bodies eutrophication, aquatic habitats degradation, biodiversity reduction) but are not necessarily harmful to human health [5–11]. On the other hand, polluted water has its quality altered by the presence of impurities of anthropogenic origin harmful to human health or ecosystems (e.g., pathogenic microorganisms, toxic chemicals, heavy metals, and other hazardous substances). In this case, water purification and treatment measures are required to make it safe for anthropogenic activities.

Freshwater and seawater are negatively affected by toxic substances and microorganisms. Contaminated water has a serious impact on the marine environment as oil spills are a growing problem as they enter the food chain and are toxic to marine life and humans [27]. Water treatment processes play a critical role in reducing water

DOI: 10.1201/9781003441007-7

TABLE 4.1

Comparison Factors between Freshwater and Seawater

Comparison Factors	Freshwater	Seawater
Composition	Low dissolved salts and minerals, presence of pollutants and contaminants, lower salinity [28]	Higher salinity level (NaCl); Presence of dissolved minerals (e.g., Mg, Ca, P) [29]
Desalination	Desalination may not be required [30]	Desalination is required, generally by reverse osmosis, distillation, or electrodialysis [31]
Pretreatment	Pretreatment involves debris removal and settling. Other processes (e.g., coagulation, flocculation, filtration) may be needed [32]	Requires extensive pretreatment including screening, pre-filtration, and chemical conditioning to prevent desalination fouling [33]
Energy requirements	Typically requires less energy [34]	High energy required for desalination [35]
Environmental impact	Low environmental impact (low energy-intensive processes, utilizes existing natural water sources) [36]	High environmental impacts (high energy consumption, brine discharge, potential harm to marine ecosystems) [37]

contaminants and improving water quality. Treatment of freshwater and seawater differs significantly due to the different compositions and challenges associated with each type of water [28]. The primary difference between freshwater and seawater lies in the need for desalination for seawater treatment. Freshwater treatment focuses on removing pollutants and contaminants, while seawater treatment involves additional steps to desalinate before it can be used for various purposes. Key considerations for both freshwater and saltwater treatment are presented in Table 4.1.

Water pollution has negative consequences for the environment, human health, and sustainable development. The presence of toxins in water often has adverse effects on aquatic life [5,6]. Examples of toxic matter in water include heavy metals (e.g., lead, mercury, arsenic, cadmium) [7], organic compounds (e.g., pesticides, dyes, plasticizers, polychlorinated biphenyls, dioxins, polycyclic aromatic hydrocarbons, emerging pollutants) [8] and pathogens like bacteria (e.g., *Escherichia coli*, *Salmonella*, and *Vibrio cholerae*), viruses (e.g., hepatitis A, norovirus, and rotavirus), and protozoa (e.g., *Giardia* and *Cryptosporidium*) that cause waterborne diseases [10–14]. These pollutants come from various sources, including industrial discharges, agricultural wastes, urban stormwater, and improper waste disposal [15–20].

Water pollution results in changes to the characteristics of this resource, leading to a loss of quality and portability. The following are some negative effects of contaminated water:

- Harm to aquatic life: contaminated water leads to the death of aquatic plants and animals. It disrupts the ecological balance of water ecosystems and can result in reduced biodiversity [21].

- Impact on agriculture: contaminated irrigation water affects crops and agricultural productivity [22].
- Economic implications: increased water treatment costs decrease property values near polluted waters and lost tourism revenue [23].
- Human health risks: consuming or contact with contaminated water leads to waterborne diseases, affecting millions of people worldwide, especially in regions with inadequate clean water access and proper sanitation [5,6].

Water pollution control efforts are critical to ensuring the availability of clean, safe water for human consumption and the environment. These efforts involve a multi-faceted approach including policies (e.g., water quality standards and allowable pollutants levels in water bodies), technologies (e.g., conventional, and innovative technologies), international cooperation, agreements and conventions (e.g., United Nations Sustainable Development Goals—SDGs and regional agreements), public awareness and education [13,14,24–26]. Overall, safeguarding water resources and minimizing pollution is vital for maintaining aquatic ecosystem health and ensuring safe drinking water for people around the world.

Freshwater and seawater are negatively affected by toxic substances and micro-organisms. Contaminated water has a serious impact on the marine environment as oil spills are a growing problem as they enter the food chain and are toxic to marine life and humans [27]. Water treatment processes play a critical role in reducing water contaminants and improving water quality. Treatment of freshwater and seawater differs significantly due to the different compositions and challenges associated with each type of water [28]. The primary difference between freshwater and seawater lies in the necessity of desalination for seawater treatment. Freshwater treatment focuses on removing pollutants and contaminants, while seawater treatment involves additional steps to desalinate the water before it can be used for various purposes.

4.2 BIOGENIC NANOMATERIALS

The Sustainable Development Goals (SDGs), also known as Global Goals, are a set of 17 interlinked goals proposed by the United Nations in 2015 as part of the 2030 Agenda for Sustainable Development [38]. One of these targets is dedicated to the successful management of wastewater (SDG 6, Clean Water and Sanitation) to ensure water availability and sustainable management and sanitation for all. Green synthesis has become a popular technique to achieve SDG targets by 2030, describing biological techniques used to create nanomaterials [39,40]. The concept of "green synthesis" encompasses methods, techniques, and processes for producing nanoparticles with minimal energy consumption and avoiding harmful byproducts through regulation, management, remediation, and cleanup, increasing their environmental friendliness [38–46].

BNMs have shown advantageous properties in various applications [40–44], including water treatment [44–47]. BNMs are nanoparticles or nanomaterials produced by living organisms or derived from biological sources such as bacteria, fungi, biomolecules, and plant extracts– commonly water extracts. In addition to the diverse sources of biological extracts, biogenic nanomaterials (BNMs) exhibit various morphologies

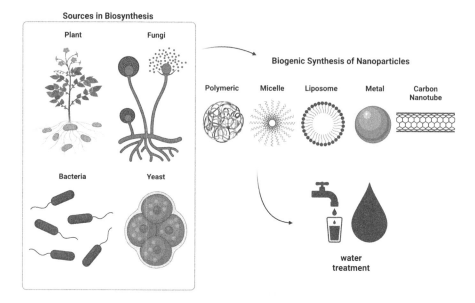

FIGURE 4.1   Sources of biological extracts for biosynthesis of nanoparticles with applications in water treatment.

including polymeric, micelle, liposome, metallic, and nanotube structures [48,50] (Figure 4.1). They possess unique properties and have garnered significant interest in research and technology development, due to their eco-friendly and sustainable nature [50–52]. Besides that, BNMs show high regeneration efficiency and could be reused several times for the removal of environmental pollutants [53]. In summary, BNMs are emerging as an environmentally friendly, cost-effective, less energy-intensive, and superior alternative to physicochemical methods for the synthesis of nanomaterials in terms of cost, human health, and environmental impact [54,55].

Biogenic nanoparticles can be used to remediate water by reducing or eliminating hazardous contaminants found in water resources. BNMs have been utilized for the removal of heavy metals, degradation of inorganic and organic contaminants, as well as disinfection.

Some of the most important applications of BNMs for water treatment include:

- Nanoparticles for removal of heavy metals: magnetic, metal, and metal oxide nanoparticles have oxidative and adsorptive properties. These nanoparticles have been used for the removal of various heavy metals and their ions from contaminated water using both microbial nanoparticles (e.g., Pd, Se, CdS, Fe_2O_3) and plant-mediated synthesis nanoparticles (e.g., Fe_2O_3, Fe_3O_4, Ag, CuO) [56–62]. Reduced graphene oxide has also been investigated for its ability to remove heavy metals (e.g., Pb(II)) from water [63,64].
- Nanoparticles for disinfection: metal nanoparticles (e.g., Ag and Au nanoparticles), metal oxide nanoparticles (e.g., CuO and ZnO), graphene, and carbon nanotubes (CNTs) have been studied for water disinfection because their

antimicrobial properties, although this property depends on nanoparticle's size and shape [65]. They can be used for water disinfection to control harmful microorganisms' growth, including bacteria and viruses [66,67].

*   Nanomaterials for water filtration: some BNMs possess unique porous structures or surface properties that can be harnessed for water filtration applications. For instance, cellulose-based nanomaterials have shown potential for removing organic pollutants and nanoparticles from water [68–70]. Metallic biogenic nanoparticles, such as silver nanoparticles, have been used to provide membrane substrate modification and improve both antifouling and antibacterial properties [45].
*   Nanoparticles for remediation of organic pollutants: BNMs have the potential to break down organic pollutants (e.g., dyes, pesticides, and pharmaceuticals) through biodegradation or catalytic processes [8,71]. BNMs in combination with other materials (nanocomposites) enhance water treatment capabilities. For example, graphene oxide nanocomposites with BNMs have shown improved adsorption properties for organic contaminants [72–77].

Pavan and collaborators [46] conducted a study that underscores a crucial aspect of employing nanomaterials in water treatment. Traditional nanomaterials synthesis methods often involve using hazardous and/or volatile chemicals, leading to secondary pollution. Consequently, biogenic alternatives have emerged as promising solutions. There are variations in the sources of biosynthesis of nanoparticles with potential for water treatment. To illustrate some examples, Table 4.2 presents different biological extract sources used and the affinity of the nanoparticles, with a focus on water treatment.

Among studies on the application of BNMs for water treatment, González-Pedroza and collaborators [88] have showcased impressive results in the biosynthesis of silver nanoparticles using *Cuphea procumbens* extract. In this study, authors aimed to assess the photocatalytic potential of AgNPs in the degradation of specific dyes (Congo red and malachite green), found among water pollutants and notorious for their high acute and chronic toxicity. Degradation achieved was remarkable (86.61% for Congo red and 82.11% for malachite green) suggesting that biologically synthesized nanoparticles from *C. procumbens* exhibit exceptional promise for a range of biomedical and environmental applications.

Water treatment using nanoparticles employs various approaches, including adsorption, photocatalysis, and coagulation/flocculation. For adsorption, BNMs with specific adsorptive properties are introduced into the water, effectively capturing undesirable contaminants (e.g., metal ions or organic pollutants) thereby making the water cleaner. Photocatalysis involves the use of photocatalytic nanoparticles, typically based on semiconductor materials which, when exposed to solar or artificial radiation, generate reactive oxygen species capable of degrading persistent organic pollutants. Coagulation/flocculation employs nanoparticles acting as coagulants to agglomerate suspended particles in water, facilitating their removal by settling [89]. These approaches offer innovative and effective solutions to enhance the quality of drinking water and treat it in a more sustainable manner (Figure 4.2).

TABLE 4.2

Nanoparticles for Water Treatment: Diverse Particle Types and Applications

Biogenic Nanoparticle	Source	Pollutants Removed	References
Silver nanoparticles (AgNPs)	*Linum usitatissimum*	Industrial wastewater treatment	[78]
Biogenic zinc oxide nanoparticles (ZnONPs)	*Bacillus cereus*	Textile dyes	[79]
ZnONPs	*Citrus limon*	Bacteria	[80]
Silver-iron oxide nanoparticles($Ag\text{-}Fe_2O_3NPs$)	*Kulekhara* leaves	Crystal violet, malachite green	[81]
Palladium nanoparticles (PdNPs)	*Lantana trifolia* seeds	Crystal violet, methyl orange	[82]
AgNPs	*Coleus vettiveroids*	Textile dyes	[83]
Boron-Fe and Boron-NiFe nanoparticles	*Terminalia bellirica*	Cr(VI), methylene blue	[84]
Copper nanoparticles (CuNPs)	*Escherichia* sp.	Azo dye degradation	[85]
Pd nanoparticles (PdNPs)	*Erigeron canadensis* L.	Cr(VI) reduction	[86]
Copper oxide nanoparticles (CuO NPs)	*Psidium guajava*	Degradation of dyes	[87]

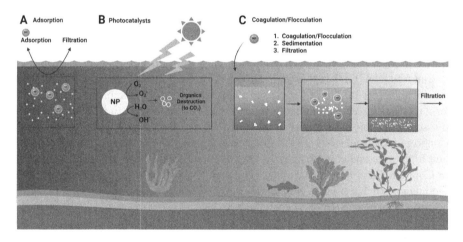

FIGURE 4.2   Water treatment using nanoparticles (NP), through the techniques of (A) adsorption, (B) photocatalysis, and (C) coagulation and flocculation.

Carbon, single-walled carbon nanotubes, multi-walled carbon nanotubes, covalent organic frameworks, metal and metal oxide-based nanoparticles, graphene-based nanoparticles, their oxides, and reduced graphene oxide are examples of BNMs used as nanomaterials for water treatment [90]. This text will focus on graphene-based nanoparticles, carbon nanotubes, metal, and metal oxide nanoparticles.

4.2.1 GRAPHENE AND CARBON NANOTUBES

Graphene and carbon nanotubes (CNTs) are both members of the carbon nanomaterial family, showing considerable potential for various water treatment applications because of their exceptional properties [91,92]. Among the wide range of their applications, recent studies mainly focus on:

- Adsorption of contaminants: graphene and CNTs possess high surface areas and unique surface chemistry, making them excellent adsorbents for a wide range of contaminants in water [93]. They can effectively adsorb heavy metals, organic pollutants, dyes, and other harmful substances [94,95]. The adsorption process involves physical or chemical interactions between the nanomaterials' surfaces and the pollutants, leading to their removal from the water [96].
- Membrane filtration: graphene and CNTs enhance membrane filtration processes [97,98]. These nanomaterials can be incorporated into polymer membranes to create nanocomposite membranes with improved permeability, selectivity, and fouling resistance [98,99]. This advancement has the potential to improve water purification and desalination processes efficiency.
- Disinfection: graphene and CNTs show antimicrobial properties and are utilized in water disinfection applications [100–102]. By incorporating these nanomaterials into water treatment systems or coatings, they help inhibit bacteria and other microorganisms' growth, contributing to improved water quality [103,104].
- Catalysis: CNTs and graphene-based materials act as catalysts for advanced oxidation processes (AOPs) [105,106]. AOPs involve generating highly reactive chemical species that break down organic pollutants in water [107,108]. The high surface area and unique electronic properties of graphene and CNTs make them suitable candidates for catalytic applications in water treatment [109].
- Sensors for water quality monitoring: graphene and CNTs have been integrated into sensors for real-time water quality monitoring [110,111]. These nanomaterial-based sensors can detect and quantify various contaminants, including heavy metals and organic compounds, in water samples with high sensitivity and specificity [112,113].
- Electrochemical water treatment: graphene and CNTs are used in electrochemical water treatment systems (e.g., electrochemical oxidation and electrocoagulation). These nanomaterials enhance the efficiency of electrochemical processes, leading to pollutant removal from water through oxidation or coagulation [114–116].

Several studies have focused on reporting applications of these nanomaterials in water treatment. For example, Nasrollahzadeh et al. [117] published a comprehensive review that considered recent advanced water treatment methods using materials involving magnetic activated carbon, CNTs, graphene (G), graphene oxide (GO), quantum dots (QDs), carbon nanorods, carbon nano-onions, and reduced graphene oxide (RGO). They concluded that nanomaterials have adsorptive and photocatalytic properties that make them highly effective in removing heavy metal ions and organic dyes, as well as degrading pollutants through photocatalysis. Their high specific surface area and higher adsorption and decomposition kinetics make them suitable for wastewater treatment, even when the pollutants are present in very low concentrations. Another study [118] highlighted advanced carbon-based nanomaterials and methods used to eliminate contaminants and ionic metals in aqueous media. These nanomaterials act as new nano-sorbents for the treatment of wastewater, drinking water, and groundwater. The study also highlights recent trends and challenges related to sustainable nanomaterials derived from carbon QD and G, as well as their devices for wastewater treatment and purification. It highlighted that greener nanotechnology and applications of sustainable nanomaterials can be exploited for wastewater treatment. Another study [119] highlighted recent advances in new-generation nanomaterials—CNTs, G, zeolites, and aquaporin (AQP)—and their applications in water desalination. The roles of these nanomaterials in different desalination processes are critically discussed. The study also briefly addresses the potential impact of these nanomaterials on human health and the ecosystem, as well as their possible interference with water treatment processes.

It is worth mentioning that while G and CNTs offer exciting possibilities for water treatment, some challenges still need to be addressed, such as large-scale production, cost-effectiveness, and potential environmental impacts. Additionally, ensuring the safe disposal or removal of nanomaterials after their use is essential to avoid any unintended consequences on aquatic ecosystems. As research in nanotechnology continues to advance, we can expect more innovations and practical applications of graphene and carbon nanotubes for water treatment, further contributing to addressing global water challenges and improving water quality.

4.2.2 Metal and Metal-Oxide Nanoparticles

Metal (MNPs), metal-oxide (MNOs), and their composites have been studied and utilized in various water treatment applications because of their high catalytic activity, large specific surface area, high physical/chemical, thermal stability, substantial chemical reactivity, strong ability to transfer electrons, and potential for removing toxic organic/inorganic contaminants [111]. These nanoparticles have shown promising results in both adsorption and catalytic processes for water treatment [120,121].

MNPs such as iron nanoparticles (FeNPs) and silver nanoparticles (AgNPs) possess high surface areas and can be further functionalized with inorganic material (e.g., silica, metal, nonmetal, metal oxides) to have specific properties for adsorbing various contaminants. They are commonly used for the removal of heavy metals, organic pollutants, dyes, and even some microorganisms from water [120]. On the other side, MNOs like iron oxide nanoparticles (Fe_2O_3, Fe_3O_4), copper oxide nanoparticles (CuO), zinc oxide nanoparticles (ZnO), cobalt oxide nanoparticles (CoO), titanium dioxide nanoparticles (TiO_2) also possess high surface areas and strong adsorption capabilities [122]. They are effective adsorbents for heavy metals, organic compounds, and other contaminants due to their surface reactivity [37,123]. Iron oxide nanoparticles (e.g., Fe_2O_3, Fe_3O_4), have been used for the removal of heavy metals from water through adsorption or co-precipitation processes. They can effectively capture and remove metal ions and metalloids (e.g., lead, cadmium, and arsenic) from water [124].

Additionally, MNPs possess high catalytic ability because of their high density of catalytically active sites and surface area [125]. MNPs can serve as catalysts in advanced oxidation processes (AOPs) for the degradation of organic pollutants in water [126]. For example, palladium nanoparticles (PdNPs) and platinum nanoparticles (PtNPs) have been used for catalytic removal of organic compounds through the Fenton process or photocatalysis [127,128]. Metal-oxide nanoparticles (MNOs) such as TiO_2 are widely used in the photocatalytic degradation of organic pollutants in the presence of UV radiation generating reactive oxygen species that break down organic contaminants into harmless by-products [129–131]. Several other types of metal and metal oxide nanoparticles have demonstrated strong antimicrobial and antibacterial properties and can effectively inhibit the growth of bacteria, viruses, and other microorganisms, making them suitable candidates for water disinfection [132,133]. Certain MNPs, like silver nanoparticles (AgNPs), can be used in water treatment to inhibit the growth of bacteria and other microorganisms, reducing the risk of waterborne diseases, and they have been incorporated into water treatment systems for disinfection purposes [134–136]. AgNPs can be added directly to water or used as coatings on surfaces to inhibit the growth of bacteria and viruses. When in contact with microorganisms, AgNPs release Ag(I) ions that damage microbial cell membranes and interfere with their metabolic processes, leading to disinfection [137]. Copper nanoparticles (CuNPs) also exhibit strong antimicrobial activity and have been studied for water disinfection applications [138]. Like AgNPs, CuNPs can disrupt microbial cell membranes and cause oxidative stress, leading to microbial inactivation.

Some MNOs, such as copper oxide nanoparticles (CuO NPs), also exhibit antimicrobial activity and can be employed to disinfect water. Zinc oxide nanoparticles (ZnO NPs) possess antimicrobial properties and can be used for water disinfection applications [139]. Their mode of action involves the generation of reactive oxygen species, leading to microbial inactivation. Some BNMs, such as titanium dioxide nanoparticles (TiO_2), have photocatalytic properties when exposed to UV light, these nanoparticles generate reactive oxygen species, such as hydroxyl radicals, which are highly effective in disinfecting water by damaging and killing microorganisms [106].

4.3   BNM IMPLEMENTATION FOR WATER TREATMENT: FEASIBILITY CHALLENGES

It is worth noting that while BNMs offer promising opportunities for water treatment, their implementation and practical application in large-scale systems may require further research and development. Challenges include scaling up production, understanding potential ecological impacts [140], and ensuring the long-term stability and safety of BNMs [141]. As research in nanotechnology and BNMs progresses, there may be more advancements and innovative applications in water treatment and environmental remediation. The large-scale production of BNMs for water treatment is still in the early stages of development. While BNMs show promising potential for water treatment applications, there are still some challenges to be addressed before large-scale production becomes widespread. We can high-light three of the existing challenges:

- Production efficiency: scaling up BNM production while maintaining its quality and efficacy is a challenge. The production processes need to be optimized to ensure cost-effectiveness and reproducibility [142–144].
- Regulation and safety: large-scale BNM production will require adherence to strict regulatory standards to ensure their safety for both human health and the environment. Thorough risk assessments and safety measures are essential [145,146].
- Characterization and quality control: large-scale BNM production demands rigorous characterization and quality control to ensure that the nanoparticles meet the desired specifications and are free from impurities [147–150].

While these challenges exist, research and development in nanotechnology and water treatment are ongoing. As technology advances and more research is conducted, large-scale BNM production for water treatment could become more feasible in the future. Collaboration between researchers, industries, and regulatory bodies will play a vital role in overcoming these challenges and promoting the responsible and sustainable use of BNMs for water treatment applications.

4.4   FINAL REMARKS AND PERSPECTIVES

The field of BNMs is an active area of research and innovation. Scientists and engineers are exploring various novel applications and advancements related to BNMs. The removal of BNMs after water treatment is an essential step to prevent their unintended release into the environment and ensure the safety of water users. If BNMs are potentially hazardous, their proper disposal is essential to prevent unintended consequences on ecosystems and human health. Responsible waste management practices and adherence to regulations and guidelines are necessary to ensure the safe handling and disposal of BNMs after water treatment. The specific method for BNM removal depends on their type and the water treatment process. Among the most used methods for removal of BNMs after water treatment, filtration, sedimentation, coagulation, and flocculation are found. Other processes generally used for this purpose are adsorption, ion exchange, electrodialysis chemical precipitation, and advanced oxidation processes.

As research in nanotechnology and water treatment continues, more specific and effective methods for BNM removal from treated water may be developed, ensuring both efficient water treatment and environmentally responsible practices.

4.4.1   REFLECTION ON BNMS

Safety use and disposal in the environment and human health. The safety, use, and disposal of BNMs in the environment and their potential effects on human health are areas of active research and concern. One concern is the potential for human exposure to BNMs during their production, use, and disposal. Inhalation, skin contact, and ingestion are routes through which BNMs can enter the human body. The toxicological properties of BNMs can vary depending on their composition, size, shape, and surface chemistry. Some BNMs may have adverse effects on human health, such as respiratory issues, skin irritation, or systemic toxicity. The release of BNMs into the environment raises questions about their potential impact on ecosystems and wildlife. It is crucial to assess their potential for bioaccumulation and toxicity to aquatic and terrestrial organisms. The use of BNMs in various applications, including water treatment, must be subject to regulations and guidelines to ensure safety. Regulatory bodies assess the potential risks associated with BNMs and set standards for their safe use. Proper engineering controls and personal protective equipment should be employed during the handling, production, and application of BNMs to minimize human exposure. The disposal of BNMs is a critical aspect to prevent unintended environmental release. Proper waste management and disposal practices should be followed to minimize potential harm. In some cases, BNMs can be recycled or reused, reducing waste and potential environmental impacts. Long-term studies are necessary to evaluate the chronic effects of BNMs on human health and the environment. Researchers and industries working with BNMs should adopt principles of responsible research and innovation, considering ethical, environmental, and social considerations. Open and transparent communication about the risks and benefits of BNMs is vital to fostering public trust and understanding. BNMs offer exciting opportunities in various fields, including water treatment. Nevertheless, their safety and potential environmental and health impacts need to be thoroughly evaluated and managed. Collaborative efforts among researchers, regulatory bodies, industries, and the public are essential to ensure the responsible and safe development and application of BNMs.

The water treatment industry typically relies on well-established and proven technologies for large-scale water treatment processes. These technologies, such as conventional filtration, chemical coagulation, sedimentation, and disinfection, have been in use for years and have demonstrated reliability and efficiency in purifying water for various applications. However, research and interest in BNMs for water treatment is steadily growing, and there is a possibility that some industries may have started exploring the potential applications of BNMs in their day-to-day routines. The technology landscape and industry practices are continually evolving, and innovations may have emerged or gained traction in the time since. Finally, it is essential to keep in mind that the field of nanotechnology is dynamic, and ongoing research and development may have led to advancements in the use of BNMs in water treatment.

ACKNOWLEDGMENTS

The authors would like to thank FAPESP (#2023/02338-0, #2023/03760-8, and #INCTBio 2014/50867-3) and CNPq for their financial support.

REFERENCES

1. Adebayo, S. B., Cui, M., Williams, T. J., Martin, E. & Johannesson, K. H. Evolution of the rare earth element and εNd compositions of Gulf of Mexico seawater during interaction with Mississippi River sediment. *Geochim. Cosmochim. Acta.* 335, 231–242. https://doi.org/10.1016/j.gca.2022.08.024 (2022).

2. Lebrato, M. et al. Global variability in seawater Mg:Ca and Sr:Ca ratios in the modern ocean. *Proc. Natl. Acad. Sci. U.S.A.* 117(36) 22281-22292. https://10.1073/pnas.1918943117 (2020).

3. Stephens, G. L. et al. Earth's water reservoirs in a changing climate. *Proc. Math. Phys. Eng. Sci.* 476, 20190458. https://doi.org/10.1098/rspa.2019.0458 (2020).

4. Livanov, D. Fresh Water on the Earth. In (D. Livanov, Editor) *The Physics of Planet Earth and Its Natural Wonders.* https://doi.org/10.1007/978-3-031-33426-9_6 (2023).

5. Qadri, R. & Faiq, M. A. Freshwater pollution: Effects on aquatic life and human health In Qadri H., Bhat R., Mehmood M., Dar G. (eds). *Fresh Water Pollution Dynamics and Remediation.* https://doi.org/10.1007/978-981-13-8277-2_2 (2020), Springer, Singapore.

6. Singh, J., Yadav, P., Pal, A. K.&Mishra, V. Water pollutants: Origin and status. *In: Pooja D., Kumar P., Singh P., Patil S. (Eds) Sensors in water pollutants monitoring: Role of material. Advanced Funcional materials and sensors, Springer, Singapore.* https://doi.org/10.1007/978-981-15-0671-0_2 (Springer, Cham, 2020).

7. Kumar, V. et al. Global evaluation of heavy metal content in surface water bodies: A meta-analysis using heavy metal pollution indices and multivariate statistical analyses. *Chemosphere.* 236, 124364. https://doi.org/10.1016/j.chemosphere.2019.124364 (2019).

8. Lu, F.&Astruc, D. Nanocatalysts and other nanomaterials for water remediation from organic pollutants. *Coord. Chem. Rev.* 408, 213180 https://doi.org/10.1016/j.ccr.2020.213180 (2020).

9. Madhav, S. et al. Water pollutants: Sources and impact on the environment and human health. In: Pooja D., Kumar P., Singh P., Patil S. (eds) *Sensors in Water Pollutants Monitoring: Role of Material. Advanced Functional Materials and Sensors.* https://doi.org/10.1007/978-981-15-0671-0_4 (Springer, Singapore, 2020).

10. Alegbeleye, O. O.&Sant'Ana, A. S. Manure-borne pathogens as an important source of water contamination: An update on the dynamics of pathogen survival/transport as well as practical risk mitigation strategies. *Int. J. Hyg. Environ. Health.* 277, 113524 https://doi.org/10.1016/j.ijheh.2020.113524 (2020).

11. Dai, C. et al. Review on the contamination and remediation of polycyclic aromatic hydrocarbons (PAHs) in coastal soil and sediments. *Environ. Res.* 205, 112423 (2022) https://doi.org/10.1016/j.envres.2021.112423

12. Banks, D. et al. Selected advanced water treatment technologies for perfluoroalkyl and polyfluoroalkyl substances: A review. *Sep. Purif. Technol.* 231, 115929 https://doi.org/10.1016/j.seppur.2019.115929 (2020) 115929.

13. Jarin, M., Dou, Z., Gao, H., Chen, Y. & Xie, X. Salinity exchange between seawater/brackish water and domestic wastewater through electrodialysis for potable water. *Front. Environ. Sci. Eng.* 17, 16 https://doi.org/10.1007/s11783-023-1616-1 (2023).

14. Wang, K., Close, H. G., Tuller-Ross, B. & Chen, H. Global average potassium isotope composition of modern seawater. *ACS Earth Space Chem.* 4(7), 1010-1017 https://doi.org/10.1021/acsearthspacechem.0c (2020).

15. Dhakal, N. et al. Is desalination a solution to freshwater scarcity in developing countries? *Membranes.* 12(4), 381 https://doi.org/10.3390/membranes12040381 (2022).

16. Lin, S. et al. Seawater desalination technology and engineering in China: A review. *Desalination.* 498, 114728 https://doi.org/10.1016/j.desal.2020.114728 (2021).

17. Liu, Y. et al. Pretreatment method for the analysis of phosphate oxygen isotope (δ18OP) of different phosphorus fractions in freshwater sediments. *Sci. Total Environ.* 685, 229-238 https://doi.org/10.1016/j.scitotenv.2019.05.238 (2019).
18. Kavitha, J., Rajalakshmi, M., Phani, A. R. & Padaki, M. Pretreatment processes for seawater reverse osmosis desalination systems – A review. *J. Water Process. Eng.* 32, 100926 https://doi.org/10.1016/j.jwpe.2019.100926 (2019).
19. Mir, N. & Bicer, Y. Integration of electrodialysis with renewable energy sources for sustainable freshwater production: A review. *J. Environ. Manage.* 289, 112496 https://doi.org/10.1016/j.jenvman.2021.112496 (2021).
20. Wei, X., Binger, Z. M., Achilli, A., Sanders, K. T. & Childress, A. E. A modeling framework to evaluate blending of seawater and treated wastewater streams for synergistic desalination and potable reuse. *Water Res.* 170, 115282 https://doi.org/10.1016/j.watres.2019.115282 (2020).
21. Panagopoulos, A. & Haralambous, K. J. Environmental impacts of desalination and brine treatment – Challenges and mitigation measures. *Mar. Pollut. Bull.* 161, 111773 https://doi.org/10.1016/j.marpolbul.2020.111773 (2020).
22. Alhaj, M. & Al-Ghamdi, S. G. Integrating concentrated solar power with seawater desalination technologies: A multi-regional environmental assessment. *Environ. Res.* 14(7), 074014 https://doi.org/10.1088/1748-9326/ab1d74 (2019).
23. Ezugbe, E. O. & Rathilal, S. Membrane technologies in wastewater treatment: A review. *Membranes.* 10(5), 89 https://doi.org/10.3390/membranes10050089 (2020).
24. Inamdar, A. K. et al. A review on environmental applications of metal oxide nanoparticles through wastewater treatment. *Mater Today Proc.* In press https://doi.org/10.1016/j.matpr.2023.05.527 (2023).
25. Bwire, G. et al. The quality of drinking and domestic water from the surface water sources (lakes, rivers, irrigation canals and ponds) and springs in cholera prone communities of Uganda: an analysis of vital physicochemical parameters. *BMC Public Health.* 20, 1–18 https://doi.org/10.1186/s12889-020-09186-3 (2020).
26. Mushtaq, N., Singh, D. V., Bhat, R. A., Dervash, M. A. & Hameed, O. Bin. Freshwater Contamination: Sources and Hazards to Aquatic Biota. Fresh Water Pollution Dynamics and Remediation. 27–50 https://doi.org/10.1007/978-981-13-8277-2_3 (2020).
27. Banks, D. et al. Selected advanced water treatment technologies for perfluoroalkyl and polyfluoroalkyl substances: A review. *Sep. Purif. Technol.* 231 https://doi.org/10.1016/j.seppur.2019.115929 (2020) 115929.
28. Jarin, M., Dou, Z., Gao, H., Chen, Y. & Xie, X. Salinity exchange between seawater/brackish water and domestic wastewater through electrodialysis for potable water. *Front. Environ. Sci. Eng.* 17, 16 https://doi.org/10.1007/s11783-023-1616-1 (2023).
29. Wang, K., Close, H. G., Tuller-Ross, B. & Chen, H. Global average potassium isotope composition of modern seawater. *ACS Earth Space Chem.* Preprint at https://doi.org/10.1021/acsearthspacechem.0c00047 (2020).
30. Dhakal, N. et al. Is desalination a solution to freshwater scarcity in developing countries? *Membranes.* Preprint at https://doi.org/10.3390/membranes12040381 (2022).
31. Lin, S. et al. Seawater desalination technology and engineering in China: A review. *Desalination.* Preprint at https://doi.org/10.1016/j.desal.2020.114728 (2021).
32. Liu, Y. et al. Pretreatment method for the analysis of phosphate oxygen isotope (δ18OP) of different phosphorus fractions in freshwater sediments. *Sci. Total Environ.* Preprint at https://doi.org/10.1016/j.scitotenv.2019.05.238 (2019).
33. Kavitha, J., Rajalakshmi, M., Phani, A. R. & Padaki, M. Pretreatment processes for seawater reverse osmosis desalination systems - – A review. *J. Water Process. Eng.* Preprint at https://doi.org/10.1016/j.jwpe.2019.100926 (2019).

34. Mir, N. & Bicer, Y. Integration of electrodialysis with renewable energy sources for sustainable freshwater production: A review. *J. Environ. Manage.* Preprint at https://doi.org/10.1016/j.jenvman.2021.112496 (2021).

35. Wei, X., Binger, Z. M., Achilli, A., Sanders, K. T. & Childress, A. E. A modeling framework to evaluate blending of seawater and treated wastewater streams for synergistic desalination and potable reuse. *Water Res.* Preprint at https://doi.org/10.1016/j.watres.2019.115282 (2020).

36. Panagopoulos, A. & Haralambous, K. J. Environmental impacts of desalination and brine treatment – Challenges and mitigation measures. *Mar. Pollut. Bull.* Preprint at https://doi.org/10.1016/j.marpolbul.2020.111773 (2020).

37. Alhaj, M. & Al-Ghamdi, S. G. Integrating concentrated solar power with seawater desalination technologies: A multi-regional environmental assessment. *Environ. Res.* Preprint at https://doi.org/10.1088/1748-9326/ab1d74 (2019).

38. Alprol, A. E., Mansour, A. T., Abdelwahab, A. M. & Ashour, M. Advances in green synthesis of metal oxide nanoparticles by marine algae for wastewater treatment by adsorption and photocatalysis techniques. *Catalysts.* Preprint at https://doi.org/10.3390/catal13050888 (2023).

39. Mansour, A. T. et al. The using of nanoparticles of microalgae in remediation of toxic dye from industrial wastewater: Kinetic and isotherm studies. *Materials.* Preprint at https://doi.org/10.3390/ma15113922 (2022).

40. Mabrouk, M. M. et al. Nanoparticles of Arthrospira platensis improves growth, antioxidative and immunological responses of Nile tilapia (Oreochromis niloticus) and its resistance to Aeromonas hydrophila. *Aquacult. Res.* Preprint at https://doi.org/10.1111/are.15558 (2022).

41. El-Deeb, N. M., Khattab, S. M., Abu-Youssef, M. A. & Badr, A. M. A. Green synthesis of novel stable biogenic gold nanoparticles for breast cancer therapeutics via the induction of extrinsic and intrinsic pathways. *Sci. Rep.* Preprint at https://doi.org/10.1038/s41598-022-15648-y (2022).

42. Ibrahim, S. et al. Optimization for biogenic microbial synthesis of silver nanoparticles through response surface methodology, characterization, their antimicrobial, antioxidant, and catalytic potential. *Sci. Rep.* Preprint at https://doi.org/10.1038/s41598-020-80805-0 (2021).

43. Patil, S. & Chandrasekaran, R. Biogenic nanoparticles: A comprehensive perspective in synthesis, characterization, application and its challenges. *J. Genet. Eng. Biotechnol.* Preprint at https://doi.org/10.1186/s43141-020-00081-3 (2020).

44. Mughal, B., Zaidi, S. Z. J., Zhang, X. & Hassan, S. U. Biogenic nanoparticles: Synthesis, characterisation and applications. *Appl. Sci.* Preprint at https://doi.org/10.3390/app11062598 (2021).

45. Wu, X., Fang, F., Zhang, B., Wu, J. & Zhang, K. Biogenic silver nanoparticles-modified forward osmosis membranes with mitigated internal concentration polarization and enhanced antibacterial properties. *NPJ Clean Water.* Preprint at https://doi.org/10.1038/s41545-022-00190-1 (2022).

46. Gautam, P. K., Singh, A., Misra, K., Sahoo, A. K. & Samanta, S. K. Synthesis and applications of biogenic nanomaterials in drinking and wastewater treatment. *J. Environ. Manage.* Preprint at https://doi.org/10.1016/j.jenvman.2018.10.104 (2019).

47. Bandala, E. R., Stanisic, D. & Tasic, L. Biogenic nanomaterials for photocatalytic degradation and water disinfection: A review. *Water Sci. Technol.* Preprint at https://doi.org/10.1039/d0ew00705f (2020).

48. Gu, X. et al. Preparation and antibacterial properties of gold nanoparticles: A review. *Environ. Chem. Lett.* Preprint at https://doi.org/10.1007/s10311-020-01071-0 (2021).

49. Rheder, D. T. et al. Synthesis of biogenic silver nanoparticles using Althaea officinalis as reducing agent: Evaluation of toxicity and ecotoxicity. *Sci. Rep.* Preprint at https://doi.org/10.1038/s41598-018-30317-9 (2018).

50. Truong, L. B. et al. Biogenic metal nanomaterials to combat antimicrobial resistance. *Emerging Nanomaterials and Nano-Based Drug Delivery Approaches to Combat Antimicrobial Resistance.* Preprint at https://doi.org/10.1016/B978-0-323-90792-7.00011-7 (2022).

51. Alsamhary, K. I. Eco-friendly synthesis of silver nanoparticles by *Bacillus subtilis* and their antibacterial activity. *Saudi J. Biol. Sci.* Preprint at https://doi.org/10.1016/j.sjbs.2020.04.026 (2020).

52. Soni, V. et al. Sustainable and green trends in using plant extracts for the synthesis of biogenic metal nanoparticles toward environmental and pharmaceutical advances: A review. *Environ. Res.* Preprint at https://doi.org/10.1016/j.envres.2021.111622 (2021).

53. Kumari, S., Tyagi, M. & Jagadevan, S. Mechanistic removal of environmental contaminants using biogenic nano-materials. *Int. J. Dent.* Preprint at https://doi.org/10.1007/s13762-019-02468-3 (2019).

54. Laouini, S. E. et al. Green synthesized of Ag/Ag2O nanoparticles using aqueous leaves extracts of Phoenix dactylifera l. And their azo dye photodegradation. *Membranes.* Preprint at https://doi.org/10.3390/membranes11070468 (2021).

55. Hidangmayum, A., Debnath, A., Guru, A. et al. Mechanistic and recent updates in nanobioremediation for developing green technology to alleviate agricultural contaminants. *Int. J. Environ. Sci. Technol.* Preprint at https://doi.org/10.1007/s13762-022-04560-7 (2023).

56. Nguyen, V. P., Le Trung, H., Nguyen, T. H., Hoang, D. & Tran, T. H. Synthesis of biogenic silver nanoparticles with eco-friendly processes using *Ganoderma lucidum* extract and evaluation of their theranostic applications. *J. Nanomater.* Preprint at https://doi.org/10.1155/2021/6135920 (2021).

57. Goutam, S. P. & Saxena, G. Biogenic nanoparticles for removal of heavy metals and organic pollutants from water and wastewater: Advances, challenges, and future prospects. *Bioremediation for Environmental Sustainability: Toxicity, Mechanisms of Contaminants Degradation, Detoxification and Challenges.* Preprint at https://doi.org/10.1016/B978-0-12-820524-2.00025-0 (2020).

58. Bouafia, A. et al. Removal of hydrocarbons and heavy metals from petroleum water by modern green nanotechnology methods. *Sci. Rep.* Preprint at https://doi.org/10.1038/s41598-023-32938-1 (2023).

59. Qasem, N. A. A., Mohammed, R. H. & Lawal, D. U. Removal of heavy metal ions from wastewater: A comprehensive and critical review. *Clean Water.* Preprint at https://doi.org/10.1038/s41545-021-00127-0 (2021).

60. Kumar, M., Nandi, M. & Pakshirajan, K. Recent advances in heavy metal recovery from wastewater by biogenic sulfide precipitation. *J. Environ. Manage.* Preprint at https://doi.org/10.1016/j.jenvman.2020.111555 (2021).

61. Sunanda, S., Misra, M. & Ghosh Sachan, S. Nanobioremediation of heavy metals: Perspectives and challenges. *J. Basic Microbiol.* Preprint at https://doi.org/10.1002/jobm.202100384 (2022).

62. Arjaghi, S. K., Alasl, M. K., Sajjadi, N., Fataei, E. & Rajaei, G. E. Retraction note to: Green synthesis of iron oxide nanoparticles by RS lichen extract and its application in removing heavy metals of lead and cadmium. *Biol. Trace Elem. Res.* Preprint at https://doi.org/10.1007/s12011-021-02650-0 (2021).

63. Weng, X., Wu, J., Ma, L., Owens, G. & Chen, Z. Impact of synthesis conditions on Pb(II) removal efficiency from aqueous solution by green tea extract reduced graphene oxide. *Chem. Eng. J.* Preprint at https://doi.org/10.1016/j.cej.2018.11.089 (2019).

64. Ahmad, S. et al. Algal extracts based biogenic synthesis of reduced graphene oxides (rGO) with enhanced heavy metals adsorption capability. *J. Ind. Eng. Chem.* Preprint at https://doi.org/10.1016/j.jiec.2018.12.009 (2019).

65. Khan, F. et al. Prospects of algae-based green synthesis of nanoparticles for environmental applications. *Chemosphere.* Preprint at https://doi.org/10.1016/j.chemosphere. 2022.133571 (2022).

66. Rikta, S. Y. Application of nanoparticles for disinfection and microbial control of water and wastewater. *Nanotechnology in Water and Wastewater Treatment: Theory and Applications.* Preprint at https://doi.org/10.1016/B978-0-12-813902-8.00009-5 (2019).

67. Al-Issai, L., Elshorbagy, W., Maraqa, M. A., Hamouda, M. & Soliman, A. M. Use of nanoparticles for the disinfection of desalinated water. *Water.* Preprint at https://doi.org/10.3390/w11030559 (2019).

68. Saleem, H. & Zaidi, S. J. Developments in the application of nanomaterials for water treatment and their impact on the environment. *Nanomaterials.* Preprint at https://doi.org/10.3390/nano10091764 (2020).

69. Yaqoob, A. A., Parveen, T., Umar, K. & Ibrahim, M. N. M. Role of nanomaterials in the treatment of wastewater: A review. *Water.* Preprint at https://doi.org/10.3390/w12020495 (2020).

70. Abdelhamid, H. N. & Mathew, A. P. Cellulose-based nanomaterials advance biomedicine: A review. *Int. J. Mol.* Preprint at https://doi.org/10.3390/ijms23105405 (2022).

71. Choi, Y. & Lee, S. Y. Biosynthesis of inorganic nanomaterials using microbial cells and bacteriophages. *Nat. Rev. Chem.* Preprint at https://doi.org/10.1038/s41570-020-00221-w (2020).

72. Alfryyan, N., Kordy, M. G. M., Abdel-Gabbar, M., Soliman, H. A. & Shaban, M. Characterization of the biosynthesized intracellular and extracellular plasmonic silver nanoparticles using Bacillus cereus and their catalytic reduction of methylene blue. *Sci. Rep.* Preprint at https://doi.org/10.1038/s41598-022-16029-1 (2022).

73. Liu, X. et al. Biosynthesis of silver nanoparticles with antimicrobial and anticancer properties using two novel yeasts. *Sci. Rep.* Preprint at https://doi.org/10.1038/s41598-021-95262-6 (2021).

74. Priya, N., Kaur, K. & Sidhu, A. K. Green synthesis: An eco-friendly route for the synthesis of iron oxide nanoparticles. *Front. Nanotechnol.* Preprint at https://doi.org/10.3389/fnano.2021.655062 (2021).

75. Hnamte, M. & Pulikkal, A. K. Clay-polymer nanocomposites for water and wastewater treatment: A comprehensive review. *Chemosphere.* Preprint at https://doi.org/10.1016/j.chemosphere.2022.135869 (2022).

76. Basnet, P. & Chatterjee, S. Nanocomposites of ZnO for water remediation. *Environ. Eng.* Preprint at https://doi.org/10.1002/9781119555346.ch7 (2019).

77. Folawewo, A. D. & Bala, M. D. Nanocomposite zinc oxide-based photocatalysts: Recent developments in their use for the treatment of dye-polluted wastewater. *Water.* Preprint at https://doi.org/10.3390/w14233899 (2022).

78. Karimi, F. et al. One-step synthesized biogenic nanoparticles using Linum usitatissimum: Application of sun-light photocatalytic, biological activity and electrochemical H2O2 sensor. *Environ. Res.* Preprint at https://doi.org/10.1016/j.envres.2022.114757 (2023).

79. Abbas, A. et al. Immobilized biogenic zinc oxide nanoparticles as photocatalysts for degradation of methylene blue dye and treatment of textile effluents. *Int. J. Dent.* Preprint at https://doi.org/10.1007/s13762-021-03872-4 (2022).

80. Singh, K., Nancy, Singh, G. & Singh, J. Sustainable synthesis of biogenic ZnO NPs for mitigation of emerging pollutants and pathogens. *Environ. Res.* Preprint at https://doi.org/10.1016/j.envres.2022.114952 (2023).

81. Kolya, H. & Kang, C. W. Biogenic synthesis of silver-iron oxide nanoparticles using Kulekhara leaves extract for removing crystal violet and malachite green dyes from water. *Sustainability*. Preprint at https://doi.org/10.3390/su142315800 (2022).

82. Banu, R., Bhagavanth Reddy, G., Ayodhya, D., Ramakrishna, D. & Kotu, G. M. Biogenic Pd-nanoparticles from Lantana trifolia seeds extract: Synthesis, characterization, and catalytic reduction of textile dyes. *Results Chem*. Preprint at https://doi.org/10.1016/j.rechem.2022.100737 (2023).

83. Ajay, S., Panicker, J. S., Manjumol, K. A. & Subramanian, P. P. Photocatalytic activity of biogenic silver nanoparticles synthesized using *Coleus vettiveroids*. *Inorg. Chem. Commun*. Preprint at https://doi.org/10.1016/j.inoche.2022.109926(2022).

84. Gunarani, G. I., Raman, A. B., Dilip Kumar, J., Natarajan, S. & Jegadeesan, G. B. Biogenic synthesis of Fe and NiFe nanoparticles using Terminalia bellirica extracts for water treatment applications. *Mater. Lett*. Preprint at https://doi.org/10.1016/j.matlet.2019.03.104 (2019).

85. Noman, M. et al. Use of biogenic copper nanoparticles synthesized from a native Escherichia sp. as photocatalysts for azo dye degradation and treatment of textile effluents. *Environ. Pollut*. Preprint at https://doi.org/10.1016/j.envpol.2019.113514 (2020).

86. Tripathi, R. M. & Chung, S. J. Reclamation of hexavalent chromium using catalytic activity of highly recyclable biogenic Pd(0) nanoparticles. *Sci. Rep*. Preprint at https://doi.org/10.1038/s41598-020-57548-z (2020).

87. Singh, J., Kumar, V., Kim, K. H. & Rawat, M. Biogenic synthesis of copper oxide nanoparticles using plant extract and its prodigious potential for photocatalytic degradation of dyes. *Environ. Res*. Preprint at https://doi.org/10.1016/j.envres.2019.108569 (2019).

88. González-Pedroza, M. G. et al. Biogeneration of silver nanoparticles from Cuphea procumbens for biomedical and environmental applications. *Sci. Rep*. Preprint at https://doi.org/10.1038/s41598-022-26818-3 (2023).

89. Kefeni, K. K. & Mamba, B. B. Photocatalytic application of spinel ferrite nanoparticles and nanocomposites in wastewater treatment: Review. *SM&T*. Preprint at https://doi.org/10.1016/j.susmat.2019.e00140 (2020).

90. Jabbar, K. Q., Barzinjy, A. A. & Hamad, S. M. Iron oxide nanoparticles: Preparation methods, functions, adsorption and coagulation/flocculation in wastewater treatment. *Environ. Nanotechnol. Monit. Manag*. Preprint at https://doi.org/10.1016/j.enmm.2022.100661 (2022).

91. Yang, Y. et al. Super-adsorptive and photo-regenerable carbon nanotube based membrane for highly efficient water purification. *J. Memb. Sci*. Preprint at https://doi.org/10.1016/j.memsci.2020.119000 (2021).

92. Chen, Z. et al. Hierarchical poly(vinylidene fluoride)/active carbon composite membrane with self-confining functional carbon nanotube layer for intractable wastewater remediation. *J. Memb. Sci*. Preprint at https://doi.org/10.1016/j.memsci.2020.118041 (2020).

93. Asghar, F. et al. Fabrication and prospective applications of graphene oxide-modified nanocomposites for wastewater remediation. *RSC Adv*. Preprint at https://doi.org/10.1039/D2RA00271J (2022).

94. Wang, X. et al. Fabrication of graphene oxide/polydopamine adsorptive membrane by stepwise in-situ growth for removal of rhodamine B from water. *Desalination*. Preprint at https://doi.org/10.1016/j.desal.2021.115220 (2021).

95. Khan, F. S. A. et al. A comprehensive review on micropollutants removal using carbon nanotubes-based adsorbents and membranes. *J. Environ. Chem. Eng*. Preprint at https://doi.org/10.1016/j.jece.2021.106647 (2021).

96. Paul, S., Bhoumick, M. C., Roy, S. & Mitra, S. Carbon nanotube enhanced membrane filtration for trace level dewatering of hydrocarbons. *Sep. Purif. Technol*. Preprint at https://doi.org/10.1016/j.seppur.2022.121047 (2022).

97. Giwa, A., Ahmed, M. & Hasan, S. W. Polymers for membrane filtration in water purification. *Sep. Purif.* Preprint at https://doi.org/10.1007/978-3-030-00743-0_8 (2019).
98. Li, J. et al. Fabrication and characterization of carbon nanotubes-based porous composite forward osmosis membrane: Flux performance, separation mechanism, and potential application. *J. Memb. Sci.* Preprint at https://doi.org/10.1016/j.memsci.2020.118050 (2020).
99. Saleemi, M. A., Kong, Y. L., Yong, P. V. C. & Wong, E. H. An overview of antimicrobial properties of carbon nanotubes-based nanocomposites. *Adv. Pharm. Bull.* Preprint at https://doi.org/10.34172/apb.2022.049 (2022).
100. Azizi-Lalabadi, M., Hashemi, H., Feng, J. & Jafari, S. M. Carbon nanomaterials against pathogens; the antimicrobial activity of carbon nanotubes, graphene/graphene oxide, fullerenes, and their nanocomposites. *Adv. Colloid Interface Sci.* Preprint at https://doi.org/10.1016/j.cis.2020.102250 (2020).
101. Kassem, A., Ayoub, G. M. & Malaeb, L. Antibacterial activity of chitosan nanocomposites and carbon nanotubes: A review. *Sci. Total Environ.* Preprint at https://doi.org/10.1016/j.scitotenv.2019.02.446 (2019).
102. Baek, S., Joo, S. H., Su, C. & Toborek, M. Antibacterial effects of graphene- and carbon-nanotube-based nanohybrids on *Escherichia coli*: Implications for treating multidrug-resistant bacteria. *J. Environ. Manage.* Preprint at https://doi.org/10.1016/j.jenvman.2019.06.077 (2019).
103. Fatima, N. et al. Recent developments for antimicrobial applications of graphene-based polymeric composites: A review. *J. Ind. Eng. Chem.* Preprint at https://doi.org/10.1016/j.jiec.2021.04.050 (2021).
104. Zhang, M. et al. MXene-like carbon sheet/carbon nanotubes derived from metal–organic frameworks for efficient removal of tetracycline by non-radical dominated advanced oxidation processes. *Sep. Purif. Technol.* Preprint at https://doi.org/10.1016/j.seppur.2022.121851 (2022).
105. Masood, Z. et al. Application of nanocatalysts in advanced oxidation processes for wastewater purification: Challenges and future prospects. *Catalysts.* Preprint at https://doi.org/10.3390/catal12070741 (2022).
106. Liu, X. et al. A review on percarbonate-based advanced oxidation processes for remediation of organic compounds in water. *Environ. Res.* Preprint at https://doi.org/10.1016/j.envres.2021.111371 (2021).
107. Pedrosa, M., Figueiredo, J. L. & Silva, A. M. T. Graphene-based catalytic membranes for water treatment – A review. *J. Environ. Chem. Eng.* Preprint at https://doi.org/10.1016/j.jece.2020.104930 (2021).
108. Navalón, S., Ong, W. J. & Duan, X. Sustainable catalytic processes driven by graphene-based materials. *Processes.* Preprint at https://doi.org/10.3390/PR8060672 (2020).
109. He, Q. et al. Research on the construction of portable electrochemical sensors for environmental compounds quality monitoring. *Mater. Today. Adv.* Preprint at https://doi.org/10.1016/j.mtadv.2022.100340 (2023).
110. Nasture, A. M., Ionete, E. I., Lungu, F. A., Spiridon, S. I. & Patularu, L. G. Water quality carbon nanotube-based sensors technological barriers and late research trends: A bibliometric analysis. *Chemosensors.* Preprint at https://doi.org/10.3390/chemosensors10050161 (2022).
111. Venkateswara Raju, C. et al. Emerging insights into the use of carbon-based nanomaterials for the electrochemical detection of heavy metal ions. *Coord. Chem. Rev.* Preprint at https://doi.org/10.1016/j.ccr.2022.214920 (2023).
112. Chawla, S., Rai, P., Garain, T., Uday, S. & Hussain, C. M. Green carbon materials for the analysis of environmental pollutants. *Trends Environ. Anal. Química.* Preprint at https://doi.org/10.1016/j.teac.2022.e00156 (2022).

113. Rodríguez-Narváez, O. M. et al. Electrochemical oxidation technology to treat textile wastewaters. *Curr. Opin. Electrochem.* Preprint at https://doi.org/10.1016/j.coelec. 2021.100806 (2021).

114. Salazar-Banda, G. R., Santos, G. de O. S., Duarte Gonzaga, I. M., Dória, A. R. & Barrios Eguiluz, K. I. Developments in electrode materials for wastewater treatment. *Curr. Opin. Electrochem.* Preprint at https://doi.org/10.1016/j.coelec.2020.100663 (2021).

115. Fan, X., Wei, G.&Quan, X. Carbon nanomaterial-based membranes for water and wastewater treatment under electrochemical assistance. *Environ. Sci. Nano* Preprint at https://doi.org/10.1039/d2en00545j (2022).

116. Chenab, K. K., Sohrabi, B., Jafari, A. & Ramakrishna, S. Water treatment: Functional nanomaterials and applications from adsorption to photodegradation. *Mater. Today Chem.* Preprint at https://doi.org/10.1016/j.mtchem.2020.100262 (2020).

117. Nasrollahzadeh, M., Sajjadi, M., Iravani, S. & Varma, R. S. Carbon-based sustainable nanomaterials for water treatment: State-of-art and future perspectives. *Chemosphere.* Preprint at https://doi.org/10.1016/j.chemosphere.2020.128005 (2021).

118. Teow, Y. H. & Mohammad, A. W. New generation nanomaterials for water desalination: A review. *Desalination.* Preprint at https://doi.org/10.1016/j.desal.2017.11.041 (2019).

119. Nasrollahzadeh, M., Sajjadi, M., Iravani, S. & Varma, R. S. Green-synthesized nano-catalysts and nanomaterials for water treatment: Current challenges and future perspectives. *J. Hazard. Mater.* Preprint at https://doi.org/10.1016/j.jhazmat.2020.123401 (2021).

120. Nguyen, N. T. T. et al. Recent advances on botanical biosynthesis of nanoparticles for catalytic, water treatment and agricultural applications: A review. *Sci. Total Environ.* Preprint at https://doi.org/10.1016/j.scitotenv.2022.154160 (2022).

121. Alsaiari, N. S., Osman, H., Amari, A. & Tahoon, M. A. The synthesis of metal–organic-framework-based ternary nanocomposite for the adsorption of organic dyes from aqueous solutions. *Magnetochemistry.* Preprint at https://doi.org/10.3390/magnetochemistry 8100133 (2022).

122. Inamdar, A. K., Rajenimbalkar, R. S., Thabet, A. E., Shelke, S. B. & Inamdar, S. N. Environmental applications of flame synthesized CuO nanoparticles through removal of Congo Red dye. *Mater. Today Proc.* Preprint at https://doi:10.1016/j.matpr.2023.03.698 (2023)

123. Gebre, S. H. & Sendeku, M. G. New frontiers in the biosynthesis of metal oxide nanoparticles and their environmental applications: An overview. *SN Appl. Sci.* Preprint at https://doi.org/10.1007/s42452-019-0931-4 (2019).

124. Du, Z., Zhang, Y., Xu, A., Pan, S. & Zhang, Y. Biogenic metal nanoparticles with microbes and their applications in water treatment: A review. *Environ. Sci. Pollut. Res.* Preprint at https://doi.org/10.1007/s11356-021-17042-z (2022).

125. Mahmoud, A. E. D., Al-Qahtani, K. M., Alflaij, S. O., Al-Qahtani, S. F. & Alsamhan, F. A. Green copper oxide nanoparticles for lead, nickel, and cadmium removal from contaminated water. *Sci. Rep.* Preprint at https://doi.org/10.1038/s41598-021-91093-7 (2021).

126. Velidandi, A., Sarvepalli, M., Pabbathi, N. P. P. & Baadhe, R. R. Biogenic synthesis of novel platinum-palladium bimetallic nanoparticles from aqueous Annona muricata leaf extract for catalytic activity. *Biotechnology* Preprint at https://doi.org/10.1007/s13205-021-02935-0 (2021).

127. Rehman, K. U. et al. Optimization of platinum nanoparticles (PtNPs) synthesis by acid phosphatase mediated eco-benign combined with photocatalytic and bioactivity assessments. *Nanomaterials.* Preprint at https://doi.org/10.3390/nano12071079 (2022).

128. Islam, M. T. et al. Development of photocatalytic paint based on TiO2 and photopolymer resin for the degradation of organic pollutants in water. *Sci. Total Environ.* Preprint at https://doi.org/10.1016/j.scitotenv.2019.135406 (2020).

129. Bibi, I. et al. Green synthesis of iron oxide nanoparticles using pomegranate seeds extract and photocatalytic activity evaluation for the degradation of textile dye. *J. Mater. Res. Technol.* Preprint at https://doi.org/10.1016/j.jmrt.2019.10.006 (2019).
130. Tilahun Bekele, E., Gonfa, B. A. & Sabir, F. K. Use of different natural products to control growth of titanium oxide nanoparticles in green solvent emulsion, characterization, and their photocatalytic application. *Bioinorg. Chem. Appl.* Preprint at https://doi.org/10.1155/2021/6626313 (2021).
131. Saikia, J., Gogoi, A. & Baruah, S. Nanotechnology for water remediation. Environmental Chemistry for a Sustainable World. Preprint at https://doi.org/10.1007/978-3-319-98708-8_7 (Springer, Cham, 2019).
132. Ogunsona, E. O., Muthuraj, R., Ojogbo, E., Valerio, O. & Mekonnen, T. H. Engineered nanomaterials for antimicrobial applications: A review. *Appl. Mater. Today.* Preprint at https://doi.org/10.1016/j.apmt.2019.100473 (2020).
133. Ojha, A. Nanomaterials for removal of waterborne pathogens: Opportunities and challenges. *Waterbor. Pathogens.* Preprint at https://doi.org/10.1016/B978-0-12-818783-8.00019-0 (2020).
134. Kumar, A. et al. Biogenic metallic nanoparticles: Biomedical, analytical, food preservation, and applications in other consumable products. *Front. Nanotechnol.* Preprint at https://doi.org/10.3389/fnano.2023.1175149 (2023).
135. Wan, H. et al. Bioreduction and stabilization of nanosilver using Chrysanthemum phytochemicals for antibacterial and wastewater treatment. *Chem. Select.* Preprint at https://doi.org/10.1002/slct.202200649 (2022).
136. Bhardwaj, A. K., Sundaram, S., Yadav, K. K. & Srivastav, A. L. An overview of silver nano-particles as promising materials for water disinfection. *Environ. Technol. Innov.* Preprint at https://doi.org/10.1016/j.eti.2021.101721 (2021).
137. Manjula, N. G., Sarma, G., Shilpa, B. M. & Suresh Kumar, K. Environmental applications of green engineered copper nanoparticles. *Phytonanotechnology.* Preprint at https://doi:10.1007/978-981-19-4811-4_12(2022)
138. Panchal, P. et al. Biogenic mediated Ag/ZnO nanocomposites for photocatalytic and antibacterial activities towards disinfection of water. *J. Colloid Interface Sci.* Preprint at https://doi.org/10.1016/j.jcis.2019.12.079 (2020).
139. Nzilu, D.M., Madivoli, E.S., Makhanu, D.S. et al. Green synthesis of copper oxide nanoparticles and its efficiency in degradation of rifampicin antibiotic. *Sci. Rep.* Preprint at https://doi.org/10.1038/s41598-023-41119-z (2023).
140. Esposito, M. C. et al. The era of nanomaterials: A safe solution or a risk for marine environmental pollution? *Biomolecules.* Preprint at https://doi.org/10.3390/biom11030441 (2021).
141. Ma, M. et al. A facile preparation of super long-term stable lignin nanoparticles from black liquor. *ChemSusChem.* Preprint at https://doi.org/10.1002/cssc.201902287 (2019).
142. Ahmed, S. F. et al. Green approaches in synthesising nanomaterials for environmental nanobioremediation: Technological advancements, applications, benefits and challenges. *Environ. Res.* Preprint at https://doi.org/10.1016/j.envres.2021.111967 (2022).
143. Ashrafi, G., Nasrollahzadeh, M., Jaleh, B., Sajjadi, M. & Ghafuri, H. Biowaste- and nature-derived (nano)materials: Biosynthesis, stability and environmental applications. *Adv. Colloid Interface Sci.* Preprint at https://doi.org/10.1016/j.cis.2022.102599 (2022).
144. Jacinto, M. J., Silva, V. C., Valladão, D. M. S. & Souto, R. S. Biosynthesis of magnetic iron oxide nanoparticles: A review. *Biotechnol. Lett.* Preprint at https://doi.org/10.1007/s10529-020-03047-0 (2021).
145. Singh, K. K., Singh, A. & Rai, S. A study on nanomaterials for water purification. *Mater. Today: Proc.* Preprint at https://doi.org/10.1016/j.matpr.2021.07.116 (2021).
146. Zhang, P. et al. Nanomaterial transformation in the soil-plant system: Implications for food safety and application in agriculture. *Small.* Preprint at https://doi.org/10.1002/smll.202000705 (2020).

147. Lee, S. K., Jo, M. S., Kim, H. P., Kim, J. C. & Yu, I. J. Quality assurance for nanomaterial inhalation toxicity testing. *Inhal. Toxicol.* Preprint at https://doi.org/10.1080/08958378.2021.1926602 (2021).
148. Pallotta, A. et al. Quality control of gold nanoparticles as pharmaceutical ingredients. *Int. J. Pharm.* Preprint at https://doi.org/10.1016/j.ijpharm.2019.118583569 (2019).
149. Durán, N., Fávaro, W., Alborés, S., da Costa, T. & Tasic, L. Biogenic silver nanoparticles capped with proteins: Timed knowledge and perspectives. *J. Braz. Chem. Soc.* Preprint at https://doi:10.21577/0103-5053.20230062 (2023)
150. Selvaraj, M., Hai, A., Banat, F. & Haija, M. A. Application and prospects of carbon nanostructured materials in water treatment: A review. *J. Water Process. Eng.* Preprint at https://doi.org/10.1016/j.jwpe.2019.100996 (2020).

5 Metallurgical Slag Applications for Photocatalytic Degradation of Emerging Pollutants

Tania-Ariadna García-Mejía,
Karen Valencia-García, José-Alberto Macías-Vargas,
Claudia Montoya-Bautista, and
Rosa-María Ramírez-Zamora

5.1 INTRODUCTION

Since mankind discovered and began to use metals for various applications, very significant amounts of metallurgical slag (MS), both ferrous and non-ferrous, have been generated. Ferrous slags due to their chemical composition and physical properties are generally classified as non-hazardous materials, so they have found many applications for valorization in various industries. This is not the case with non-ferrous slags, due to the content of transition metals, mainly such as iron, which gives it reactivity or photocatalytic properties, which is why in some countries such as Mexico they are classified as hazardous materials. However, in recent decades this property has attracted the attention of scientists who develop photocatalysts (PCs) for water or air treatment to decrease the high production costs of these materials. In addition, the direct use of MS as photocatalysts or indirectly as a source of metals to produce CP represents a niche opportunity to reduce the production of waste, also contributing to minimizing the use of resources and the environmental impact that this represents. This, framed by the concepts of circular economy, can help mitigate and adapt to climate change, which is key to achieving the United Nations Sustainable Development Goals and the Paris Agreement on climate change.

Based on the above, this book chapter presents an analysis of the most relevant information related to the physicochemical characteristics that confer photocatalytic properties to ferrous and non-ferrous slags. Likewise, works on the application of these materials for the treatment of emerging contaminants (EC) in water are described. Finally, challenges and perspectives are proposed on research niche opportunities to improve the performance of MS as CPs for environmental applications.

DOI: 10.1201/9781003441007-8

5.2 METALLURGICAL SLAGS CLASSIFICATION AND PRODUCTION PROCESSES

5.2.1 MS CLASSIFICATION

Slag is generated as a byproduct during extractive pyrometallurgical operations and is considered a relatively undesirable material playing an essential role in the extraction process [1]. In general, slags are divided into two main types: ferrous slags and nonferrous slags. The ferrous slag term describes non-metallic oxides generated while manufacturing iron and steel (e.g., iron slag and steel slag, respectively). Ferronickel, titaniferrous, and stainless-steel slag are also considered within this classification [2]. Steelmaking integrated process consists of ironmaking in the blast furnace (BF) and steelmaking in the basic oxygen furnace (BOF) or electric arc furnace (EAF) [3]. Subsequently, the molten steel undergoes a secondary refining process in a ladle furnace (LF) (refining slag) to obtain different steel qualities or send it to continuous casting [4].

Nonferrous slags are generated during the processing of nonferrous metal ores to produce base and precious metals. The term nonferrous refers to ores where iron is not the main alloy source. They are widely used for their high conductivity, malleability, corrosion resistance, and high strength/weight ratio [5]. Some examples of non-ferrous slags are copper, nickel, lead, zinc, tin, manganese, silicomanganese, and phosphorus slags [6].

5.2.2 FERROUS AND NONFERROUS SLAGS PRODUCTION PROCESSES

5.2.2.1 Ferrous Slags

More than 1,900 million tons of raw steel were generated around the world in 2022 (World Steel Association) with demand increasing representing 95% in the total metal processing industry. Steel manufacturing involves two main processes, one of them using iron ore as raw material, which includes the BF and BOF processes and the direct reduction and electric arc furnace (EAF) processes. BF and BOF processes use a combination of scrap steel and iron ore [7]. Scrap steel is used as a substitute for iron ore, especially in EAF, because of its low-cost generating that it is used in one-third of global steel production [4]. In the primary process, iron is produced in the BF by reducing iron-containing ores, producing about 200–400 kg of slag for each ton of melted material [8]. About 65% of global steel production depends on BF [7]. Annually, 160–240 Mt BOFS are produced with slag production variating between 0.1 and 0.2 tons per ton produced steel representing 15%–20% of the final crude steel volume [9]. BOF is currently the dominant steelmaking manufacturing technology, generating approximately 60% of global crude steel production and generating a considerable amount of slag dumped in landfills [10]. After molten steel is extracted from BOF or EAF, it undergoes further refinement to obtain a specific chemical composition [4]. Slag is generated by adding different fluxes during ladle refining, so the chemical composition and properties of these slags are different compared to BOF slags [1].

TABLE 5.1
World Mine Production of Some Non-Ferrous Metals in 2022 (Tons)

Year	Copper	Nickel	Zinc	Tin	Lead	Aluminum
2022[a]	22,000	3,300,000	13,000	310,000	4,500	69,900
Location[b]	Chile	Indonesia	China	China	China	China
	Peru	Philippines	Australia	Indonesia	USA	India
	China	Russia	Peru	Congo	Australia	Brazil

[a] [13].
[b] [11].

5.2.2.2 Nonferrous Slags

The nonferrous metals and alloys industry is developing rapidly, with an estimated production of 15.5 billion tons in the last five years and generating 4.95 billion tons of nonferrous slag yearly [11]. Table 5.1 shows the mining production of some non-ferrous metals and the countries where their corresponding slags are generated. Non-ferrous slags are mainly classified into heavy, light, and precious metal slag. More than one process is required to extract the metal from the mineral with metal properties largely determining the extraction method. For metals with melting points lower than 1,500°C (e.g., titanium, zirconium, or tantalum), pyrometallurgical casting processes are used. Mineral's chemical reactivity and content are also important aspects to consider. For example, high concentrations (>25%) are required for the pyrometallurgical process, but other processes are used in the case of low-grade metals (gold, uranium), such as leaching, for metal recovery [12].

5.3 PHYSICAL, CHEMICAL, AND OPTICAL PROPERTIES AND PHOTOCATALYTIC ACTIVITY OF MS

MS' chemical and physical characteristics depend on the extractive process, material processing, processing stages, amount and type of added flux, furnace used, and cooling conditions [14].

5.3.1 PHYSICAL PROPERTIES OF FERROUS AND NONFERROUS SLAGS

Physical characteristics are closely related to: (i) raw materials characteristics, (ii) operating practices used, and (iii) slag treatments [15]. The cooling rate and method also influence the physical characteristics of molten slag, determining their textural properties, density, hardness, size, and crystallinity [14]. Textural properties of metallurgical slags determine their photocatalytic performance as well as their participation in other pollutant removal processes, such as the heterogeneous Fenton reaction (HFR). These properties are described next [16].

Physical properties of ferrous slag. The morphology of three different ferrous slag types: BOF, EAF, and LF refining is shown in Figure 5.1 [17].

FIGURE 5.1   (a) Basic oxygen furnace slag (BOFS), (b) Electric arc furnace slag (EAFS) and (c) Ladle furnace refining slag (LFS).

BOFS physical characteristics and composition vary widely probably because of the variable nature of mineral and scrap sources used. Slow cooling generates crystalline slag with particle size <300 mm while fast cooling generates glassy slag with particle size <300 mm [7]. BOFS particle density ranges from 2.5 to 3.6 g/cm^3 [18]. BOFS is rock-like and dark (Figure 5.1a) because of its high iron content, with an angular surface and cavernous inside. It has low crushing value (i.e., high hardness) and low specific surface area (e.g., 0.4–0.5 m^2/g) [18]. EAFS is a rough, porous, blackish grey, rounded aggregate with small inclusions of metallic iron particles (Figure 5.1b), average particle size ~30–45 μm, surface area 0.7–4.8 m^2/g [19], and density ranging 3.2–3.8 g/cm^3 [20]. The rate and cooling method of LFS significantly affect its properties, when air-cooled. LFS is crystalline and vesicular. Expanded slags are porous crystalline and glassy material, granulated slags are vitreous and crystalline granules with generally smooth texture and rounded shape ranging 1–20 mm [14,21], specific surface area 0.2–0.3 m^2/g (Figure 5.1c), and density between 2.6 and 2.8 g/cm^3 [22].

Physical properties of nonferrous slag. Air-cooled copper slag possesses a dull black color and glassy appearance while granulated copper slag is more vesicular and porous. In general, copper slag specific gravity varies with its iron content within the 1.76–1.92 g/cm^3 range. Particle size ranges from 6 to 25 mm and specific surface area from 0.67 to 1.37 m^2/g [23]. Nickel slag is made up of glassy phase, small particles, and, to a lesser extent, a crystalline phase that may be present in larger particles, which are visible to the naked eye. Broken particles looking like glass and particles with honeycomb-shaped surfaces can be also observed. Granulated nickel slag is essentially a black, angular, glassy slag "sand" with most particles within the ~2 mm size range, a specific surface area of ~0.4 m^2/g, and a density of 3.16–3.31 g/cm^3 [24]. Lead and lead-zinc slags are black or red, with glassy, sharp, angular (cubic) particles. Granulated lead and lead-zinc slags tend to be porous with particle size 0.1–4 mm, and density 2.5–3.6 g/cm^3 [25]. Granulated zinc slag is a glassy, dark material similar to lead and lead-zinc slag. It tends to be porous with particle size within 30–110 mm, a surface area ranging 0.18–0.73 m^2/g, and density between 2.5 and 3.6 g/cm^3 (Xin et al., 2020). Granulated tin slag is a glassy, black material with a particle size larger than 10 mm and an apparent density of 3.3–3.6 g/cm^3.

5.3.2 CHEMICAL PROPERTIES OF FERROUS AND NONFERROUS SLAGS

The chemical composition of steel slag is variable, depending on the process used and the materials added, characteristics of ore, fuel, fluxes, and furnace conditions. Solidification of slag results in the crystallization of a variety of specific minerals, amorphous matter, or both, depending on its chemical composition and cooling conditions. Therefore, differences in furnace charge chemistry and cooling rates result in a variety of slag types, with different compositions of major and trace chemical elements and mineralogy [14].

Chemical properties of Ferrous slag. Differences in furnace charge chemistry and cooling rates result in a variety of slag types, with different major and trace element compositions and primary mineralogy. Predominant elements in ferrous slags include Ca, Si, Al, and Mg besides other elements present in minor or minimal quantities including K, Na, and Ti [14]. BOFS tends to develop crystalline microstructures, even under rapid cooling, reported to be suitable for alkaline activation. BOFS may include CaO (40%–60%), Fe_2O_3 (20%–30%), SiO_2 (10%–20%), Al_2O_3 (1%–6%), MgO (2%–10%), MnO, TiO_2, V_2O_5, ZrO_2, Cr_2O_3, P_2O_5, NiO [18] providing the ability to increase pH and alkalinity. Several minerals can be found in BOFS, including larnite, Ca_2SiO_4, and different types of CaMg oxides (Naidu et al., 2020). BOFS generally possesses a relatively high basicity (ratio of alkaline oxides to acid oxides), present in different mineral phases, including tricalcium silicate (C_3S), dicalcium silicate (C_2S), ferrite (C_2F), MgO, CaO and RO phase (CaO, MgO, MnO and FeO solid solution) [17]. EAFS composition varies over a wider range compared to BOFS. EAFS contains lower FeO content but higher Cr content being its main chemical components are iron (FeO, Fe_2O_3), calcium (CaO), silica (SiO_2), magnesia (MgO), and alumina (Al_2O_3) oxides. Minor components include chromium, manganese, and phosphorus oxides. Mineral phases identified for EAFS include merwinite ($3CaO \cdot MgO \cdot 2SiO_2$), wüstite (FeO), olivine, C_2S, and C_3S [17].

The chemical and mineralogical composition of LFS varies with many factors of the refining process. The main oxides present in LFS are CaO, SiO_2, MgO, and Al_2O_3 with CaO content (44.5%–58.4%) higher than BOFS and EAFS, while iron-containing constituents are much lower. To a lesser extent, iron, manganese, and titanium oxides can be found along with calcium sulfides and fluorides. Some LFS may also contain heavy metals (e.g., chromium, zinc, or lead) traces with a total content of less than 1% [22].

Chemical properties of nonferrous slags. Nonferrous slags have a wide range of compositions depending on the smelter feed and processing. Base metal slags from smelting sulfide ores are generally dominated by Fe and Si with minor Al and Ca amounts. Iron is usually higher and Ca lowers in nonferrous base metal slags compared to ferrous slags [14]. Nonferrous slag contains 30%–55% FeO, 30%–45% SiO_2, and less than 10% CaO, although variations occur depending on the system where the slag was produced [26].

Copper slag is a combination of molten oxides, consisting primarily of Fe (28%), SiO_2 (30%) from the flux, small amounts of impurities (e.g., Al_2O_3 (3%), CaO (9%), and MgO) from the concentrate. Oxides commonly found in copper slag include ferrous (FeO), ferric (Fe_2O_3), silica (SiO_2), alumina (Al_2O_3), calcium (CaO), and magnesia (MgO) oxides. It is often called fayalite slag since fayalite (Fe_2SiO_4) precipitates from the slag upon cooling [27]. Nickel slag is usually rich in iron and silica, but with high MgO levels which increases its melting point. The typical chemical composition of Ni slag is 0.2% Ni, 5.5% Fe, 38% MgO, and 53% SiO_2. It can contain up to 50%–55% of silica and high magnesium content. Compared with copper slag, it contains less iron, lime, and alumina [26].

Lead slag mineralogical composition includes ferrous phases such as hematite (Fe_2O_3), pyrrhotite (FeS), magnetite (Fe_3O_4), and minor amounts of fayalite (Fe_2SiO_4). Lead has been found as galena (PbS), anglesite ($PbSO_4$), and litharge (PbO). Also [28], along with trace elements including Sr, As, Cu, Cl, Sb, Co, Ba, Ni, Zr, Cr, Ce, V, Cd, U, Nd, Mo, Sn, Y, W, Li, Th, Se, In, Rb, Ag, Cs, Bi, Te, and Tl ranging between 0.18 and 2,470 mg/kg. The chemical composition of zinc slag includes compounds such as Fe (25%–52%), SiO_2 (19%–40%), Al_2O_3 (2%–10%), CaO (15%–23%), MgO (0.5%–5%), S (1.1%), PbO (0%–2%), and ZnO (0%–5%). Minor variations of some components, however, can have significant effects. For example, a high level of Al_2O_3 results in hercynite ($FeO\cdot Al_2O_3$) precipitation, raising the melting point [28]. Tin slag contains lime, iron, silica, and naturally occurring radioactive materials such as uranium, thorium, and/or their decay products. Its normal composition is 35% SiO_2, 30% CaO, 15% FeO, and 20% SnO_2 very small amount of SnO, ZnO, and Cd [29].

5.3.3 Optical Properties of Metallurgical Slags

MS optical properties are influenced by composition, crystallinity, and impurities contributing to valence and conduction band energy levels and affecting the band gap value [30], the minimum energy required for electrons transition from the valence band to the conduction band in semiconductors [31]. When photons with energy higher than the band gap are absorbed, electrons are excited to the conduction band, creating electron–hole pairs. Photocatalysis relies on electron–hole pairs generation, which then produce valuable species by redox reactions, like hydroxyl radicals, for pollutant degradation [32]. Understanding the band gap enables strategic changes in material properties to suit specific applications.

BFS photocatalyst. A unique cementitious composite, known as $CaWO_4$/ alkali-activated BFS-based cementitious composite (ASCC), was created through the impregnation of ammonium meta tungstate into alkali-activated BFS-based cementitious material. The composite was employed in the photocatalytic decomposition of water to produce hydrogen [33]. BSF containing calcium silicate and gehlenite ($Ca_2Al_2SiO_7$) showed low crystallinity but, after treatment with 1%–5% $CaWO_4$, a new phase containing calcium tungstate was generated exhibiting photoluminescence excitation bands at 469 nm (violet–blue) suggesting this material inhibits electron–hole recombination. The material was tested for photocatalytic hydrogen production showing the highest hydrogen production at 50 mmol/L after 6 h using UV radiation [33]. ZnO-modified ground granulated BFS (ZnO-GGBFS) has been reported as a highly porous geopolymer with effective photocatalyst capability using visible and/or UV radiation using methylene blue (MB) proxy of organic pollutant (40 mg/L). The experiments were carried out using solar radiation at room temperature in a glass conical flask finding MB discoloration as high as 95% within 1 h [34].

Steel slag (converter slag) photocatalyst. Recently, a converter slag consisting of 40% CaO, SiO_2, 20% Fe_2O_3, MgO, and MnO was used to create iron-doped hydroxyapatite (HAp) and employed for xanthate degradation and chromium reduction [35]. The authors found that pure HAp showed limited xanthate photodegradation efficiency, but various Fe(III)-doped HAp samples (ranging from 4% to 28%

doping) exhibited significantly enhanced photocatalytic activity, with degradation efficiencies reaching up to 94.1%.

Copper slagphotocatalyst. Copper slag was found with remarkable outcomes in the photocatalytic degradation of alcohols, accompanied by hydrogen production [36]. The copper slag band gap was found at 2.7 eV [37] linking photocatalytic activity to magnetite and characterizing copper slag as an n-type semiconductor with a role as a hole donor. Total hydrogen production, amounting to 2 μmol, was achieved after 6 h of continued illumination, positioning the material as a highly promising candidate for photocatalysis processes.

5.4 EC DEGRADATION USING MS

MS has been modified to improve their properties (e.g., specific area, porosity, optical properties) and thus their activity, safety, and stability [38] through physical, chemical, and thermochemical processes [39]. Acid treatment of MS uses sulfuric (H_2SO_4), hydrochloric (HCl), or nitric acid (HNO_3) to increase specific surface area and pore diameter, generating active sites, and improving activity. Furthermore, MS can be transformed into core–shell structure materials through chemical coating or nanoparticle deposition. Additionally, since some MS contain metallic elements (e.g., Fe, Ca, Al), they can be used to activate/modify biochar, a carbon-rich solid material obtained through pyrolysis of biomass in an oxygen-limited environment, useful for wastewater treatment [38]. Furthermore, blending steel slag with one or more materials followed by high-temperature calcination has been used to create composite-modified electrode slag particles, resulting in organic materials with enhanced catalytic performance and cost-effectiveness for electrical solutions [39].

For EC degradation, untreated MS has shown promising results by achieving degradation efficiencies between 58% and 100%, while the efficiency of chemically modified MS ranged 68%–100%. Removal of pharmaceuticals and other ECs using MS has been demonstrated across different research efforts [40–43]. However, the need to optimize operational conditions for improved efficiency and rate removal remains an interesting knowledge gap worth exploring. Degradation of industrial chemicals, including food additives, pesticides, and antimicrobials, using MS has proven effective in heterogeneous oxidation and photocatalytic processes. While promising results have been achieved, optimizing operational conditions to improve degradation kinetics remains another interesting knowledge gap urging attention.

MS modifications seem to be crucial for enhancing their properties for different applications, including contaminant removal and land reclamation. Modified steel slag has been found with potential in degrading pharmaceuticals offering insights into pollutant removal through adsorption and catalytic mechanisms [44–52]. MS also offer promising solution for degrading personal care products in wastewater. Different studies have explored removal of UV filters, parabens, and artificial sweeteners using unmodified MS incorporated in wetlands with interesting achievements, but the systematic evaluation of these systems remains an interesting avenue for future research urging for attention. Effective treatment and preparation of steel slag, including milling, sieving, and oxidative digestion, are essential for optimizing its catalytic performance in removing personal care product [40,53]. The removal of these ECs using MS has not been extensively investigated with different challenges remaining to be addressed to achieve high removal rates. Ongoing research and optimization efforts such as modification of MS materials to enhance efficacy in addressing the presence of these contaminants in wastewater treatment is another interesting avenue for future research.

MS have proven effective as Fenton-like catalysts showing remarkable success in degrading a wide range of industrial chemicals [54–56]. Moreover, MS also exhibits excellent adsorption capacities for heavy metals, arsenic, and other metalloids [57–61]. The modification of MS has been suggested to enhance their adsorption efficiency, particularly for the removal of heavy metals. Looking for different MS types displaying selectivity in removing specific contaminants is another interesting avenue for future research worth exploring. For example, iron oxide-rich slags effectively remove arsenic, while calcium-rich slags show promise in boron removal.

While MS application in wastewater treatment offers substantial benefits, it comes with challenges such as variations in slag composition, kinetics, or the need for efficient regeneration methods. These challenges emphasize the economic viability and sustainability of using MS for contaminant removal. Guided by the circular economy (CE) principles, modification techniques have been meticulously studied to enhance the catalytic and adsorption properties of MS. Moreover, CE strategies have played a pivotal role, emphasizing the imperative aspects of reusing, regenerating, and recycling slag-based materials, thereby transforming what was once considered waste into invaluable resources. Extensive research has explored various MS conditioning treatments, vital for enhancing efficiency in diverse environmental applications. Additionally, composite modification techniques involve blending MS with other materials have proved highly effective in enhancing performance. These treatments target modifications of MS properties for specific wastewater treatment processes [44–49,52,62], adsorption [63–66], photodegradation, selective removal, and regeneration techniques. These applications underscore the versatility of MS materials in addressing complex environmental challenges, showcasing their immense potential in sustainable wastewater treatment and environmental remediation.

5.5 CHALLENGES AND PERSPECTIVES FOR ENVIRONMENTAL APPLICATIONS

The number and type of reactive sites in a photocatalyst influence its performance for degrading pollutants through the photocatalytic process and are determined by the material's morphology, porosity, and particle size [16]. The photocatalyst surface area possesses a key role in improving efficiency because it is correlated with reactive sites availability. It is expected that materials with great surface area will show a greater production of OH radicals. Reactant adsorption on the photocatalyst depends on its electronegativity/surface charge with morphology, porosity, and particle size directly impacting the process. Particle size also impacts radiation absorption, when particle size reaches ≤ 10 nm, the bandgap energy increases. However, as material particle size approaches the nanoscale, its surface area increases, resulting in the increase of reactive sites. The smallest particle sizes and surface areas of nonferrous slags have been found for EAFS (particle size ~30–45 µm, surface area 0.7–4.8 m^2/g). For non-ferrous slags, copper slag shows the lowest values (particle size 6–25 mm surface area 0.67–1.37 m^2/g). EAFS and copper slags are, therefore, considered good candidates to be used as photocatalysts. By comparing with the most used photocatalysts (e.g., TiO_2, ZnO, CdS, Fe_2O_3, g-C_3N_4), MS exhibit very different values (particle size 2.8–200 nm, surface area 30–200 m^2/g) suggesting that photocatalysts from analytical grade precursors exhibit better values which directly improves performance. Even comparing EAFS and copper slags, iron oxide from analytical grade precursors exhibits up to 40 times larger smaller particle sizes. Improving physical properties of ferrous and non-ferrous slag using different techniques (e.g., heat treatment, acid, alkali modification) or physical modifications is a pending avenue of future research worth exploration. For example, a Ti-containing BFS was modified at 1,400°C for 1 h and subsequently water rapid cooling to produce a significant porosity increase with surface area in the 52.2–80.3 m^2/grange (53.7% surface area increase). By adding H_2O_2 the modified slag showed improvement in photocatalytic activity degrading 93.5% MB after 180 min [67]. The effect of mineralogical changes on the properties of alkali-activated binders in a BOFS was studied recently [18]. The study quantified particles' amorphous-crystalline proportions for three different size ranges. The relative variation in amorphous content between fine and coarse fractions was 45% directly impacted surface area and improving reaction kinetics. A high dose of photocatalyst, beyond a certain limit, was found to decrease the photodegradation rate by blocking radiation, increasing scattering and generating particle agglomeration, which reduces surface-active sites available for radiation exposure [68]. The optical and photoelectrochemical properties of slags utilized as photocatalysts are intricately shaped by the specific crystalline phases and chemical compositions of these materials. To enhance photocatalytic performance, incorporating additional crystalline phases and chemical motifs through innovative synthesis methods is a highly interesting knowledge gap identified in this study that is expected with led to key findings for overall effectiveness improvement of slag-based photocatalysis.

5.6 CONCLUSIONS

The mineralogical composition of MS allows their application as traditional photocatalyst or as Fenton-like heterogenous photocatalysts. MS are mainly constituted by iron oxides and, in smaller quantities, by mineralogical phases of transition metals that can be activated with UV or solar radiation to produce OH radicals to degrade organic pollutants. MS shows low values of textural properties (specific area and porosity) that limit reaction kinetics and physical, chemical, and thermochemical treatments have been explored to improve these properties. Likewise, its use as a raw material for composites production to increase ECs degradation kinetics has been studied. Thermochemical treatments and composite synthesis have been identified as the best options, effective to increase ECs removal rate.

Nevertheless, the search for technical and economic treatment alternatives that can be used to improve MS characteristics remain spending. Therefore, investigating this aspect more deeply, as well as optimizing operating conditions to improve their properties that accelerate degradation kinetics was found to remain as highly interesting pending research.

ACKNOWLEDGMENTS

The authors acknowledge the financial support provided through the IT103723 DGAPA-PAPIIT Project of UNAM and the DGAPA-UNAM postdoctoral scholarship of Karen Valencia-García.

REFERENCES

1. Shamsuddin, M. *Physical Chemistry of Metallurgical Processes, Second Edition.* Physical Chemistry of Metallurgical Processes (Springer Nature, Switzerland, 2021).
2. Criado, M., Ke, X., Provis, J. L. & Bernal, S. A. Alternative inorganic binders based on alkali-activated metallurgical slags. *Sustainable and Nonconventional Construction Materials using Inorganic Bonded Fiber Composites* (Elsevier Ltd, 2017). doi:10.1016/B978-0-08-102001-2.00008-5.
3. Netinger Grubeša, I., Barišić, I., Fucic, A. & Bansode, S. S. Ferrous slag. *Characteristics and Uses of Steel Slag in Building Construction* 15–30 (2016). doi:10.1016/b978-0-08-100368-8.00002-6.
4. Thomas, C., Rosales, J., Polanco, J. A. & Agrela, F. Steel slags. In *New Trends in Eco-Efficient and Recycled Concrete* 169–190 (Elsevier, 2018). doi:10.1016/B978-0-08-102480-5.00007-5.
5. Wang, G. C. Ferrous metal production and ferrous slags. *The Utilization of Slag in Civil Infrastructure Construction* 9–33 (2016). doi:10.1016/b978-0-08-100381-7.00002-1.
6. Van De Sande, J., Peys, A., Hertel, T., Rahier, H. & Pontikes, Y. Upcycling of non-ferrous metallurgy slags: Identifying the most reactive slag for inorganic polymer construction materials. *Resources, Conservation and Recycling* **154**, 104627 (2020).
7. Naidu, T. S., Sheridan, C. M. & van Dyk, L. D. Basic oxygen furnace slag: Review of current and potential uses. *Minerals Engineering* vol. 149. Preprint at https://doi.org/10.1016/j.mineng.2020.106234 (2020).
8. Yuksel, I. Blast-furnace slag. *Waste and Supplementary Cementitious Materials in Concrete* 361–415 (2018). doi:10.1016/B978-0-08-102156-9.00012-2.
9. Fernández-González, D. et al. The treatment of basic oxygen furnace (BOF) slag with concentrated solar energy. *Solar Energy* **180**, 372–382 (2019).
10. Guo, H. et al. Iron recovery and active residue production from basic oxygen furnace (BOF) slag for supplementary cementitious materials. *Resources, Conservation and Recycling* **129**, 209–218 (2018).
11. Ban, J. et al. Advances in the use of recycled non-ferrous slag as a resource for non-ferrous metal mine site remediation. *Environmental Research* **213** (2022).
12. Sohn, H. Y. Nonferrous Metals: Production and History. https://doi.org/10.1016/b978-0-12-803581-8.03608-0 (2017).
13. U.S. Department of the Interior & U.S. Geological Survey. *Mineral Commodity Summaries* 2023. (U.S. Geological Survey, Reston, VA, 2023).
14. Piatak, N. M. Environmental characteristics and utilization potential of metallurgical slag. In *Environmental Geochemistry: Site Characterization, Data Analysis and Case Histories: Second Edition* 487–519 (Elsevier, 2018). doi:10.1016/B978-0-444-63763-5.00020-3.
15. Mills, K. C. *The Influence of Structure on the Physico-Chemical Properties of Slags.* ISIJ International 33(1),148–155 (1993).
16. Buhunia, P., Dutta, K. & Vadivel, S. *Photocatalysts and Electrocatalysts in Water Remediation.* https://onlinelibrary.wiley.com/doi/book/10.1002/9781119855347 (2022).

17. Jiang, Y., Ling, T. C., Shi, C. & Pan, S. Y. Characteristics of steel slags and their use in cement and concrete– A review. *Resources, Conservation and Recycling* vol. 136 187–197. Preprint at https://doi.org/10.1016/j.resconrec.2018.04.023 (2018).

18. Lopez Gonzalez, P. L., Novais, R. M., Labrincha, J., Blanpain, B. & Pontikes, Y. The impact of granulation on the mineralogy of a modified-BOF slag and the effect on kinetics and compressive strength after alkali activation. *Cement and Concrete Composites* **140**, 105038, (2023).

19. Kieush, L. et al. Utilization of renewable carbon in electric arc furnace-based steel production: comparative evaluation of properties of conventional and non-conventional carbon-bearing sources. *Metals (Basel)* **13**, 722, (2023).

20. Skaf, M., Manso, J. M., Aragón, Á., Fuente-Alonso, J. A. & Ortega-López, V. EAF slag in asphalt mixes: A brief review of its possible re-use. *Resources, Conservation and Recycling* vol. 120 176–185. Preprint at https://doi.org/10.1016/j.resconrec.2016.12.009 (2017).

21. Piatak, N. M., Parsons, M. B. & Seal, R. R. Characteristics and environmental aspects of slag: A review. *Applied Geochemistry* vol. 57 236–266. Preprint at https://doi.org/10.1016/j.apgeochem.2014.04.009 (2015).

22. Araos Henríquez, P., Aponte, D., Ibáñez-Insa, J. & Barra Bizinotto, M. Ladle furnace slag as a partial replacement of Portland cement. *Construction and Building Materials* **289**, 123106, (2021).

23. Feng, Y. et al. Mechanical activation of granulated copper slag and its influence on hydration heat and compressive strength of blended cement. *Materials* **12(5)**, 772, (2019).

24. Oksri-Nelfia, L., Akbar, R. & Astutiningsih, S. A study of the properties and microstructure of high-magnesium nickel slag powder used as a cement supplement. In *IOP Conference Series: Materials Science and Engineering* vol. 829(1), 012007 (Institute of Physics Publishing, 2020).

25. Pan, D. et al. A review on lead slag generation, characteristics, and utilization. *Resources, Conservation and Recycling* vol. 146, 140–155. Preprint at https://doi.org/10.1016/j.resconrec.2019.03.036 (2019).

26. Wang, G. C. *The Utilization of Slag in Civil Infrastructure Construction,* Wang G.C (ed.), 35–61 (Elsevier, 2016). doi:10.1016/b978-0-08-100381-7.00003-3.

27. Kolmachikhina, O. B., Polygalov, S. E. & Lobanov, V. G. Research of the technology of joint processing of stale copper-smelting slags and pyrite cinders with the extraction of non-ferrous metals into a commercial product. In *IOP Conference Series: Earth and Environmental Science* 666(4), 042018(IOP Publishing Ltd., 2021). DOI 10.1088/1755-1315/666/4/042018

28. Nowinska, K. Mineralogical and chemical characteristics of slags from the pyrometallurgical extraction of zinc and lead. *Minerals* **10(4)**, 371, (2020).

29. Yang, Z. et al. Fluid flow characteristic of EAF molten steel with different bottom-blowing gas flow rate distributions. *ISIJ International* **60**, 1957–1967 (2020).

30. Shu, K. & Sasaki, K. Occurrence of steel converter slag and its high value-added conversion for environmental restoration in China: A review. *Journal of Cleaner Production* vol. 373. Preprint at https://doi.org/10.1016/j.jclepro.2022.133876 (2022).

31. Montoya-Bautista, C. V. et al. Photocatalytic H2 production and carbon dioxide capture using metallurgical slag and slag-derived materials. In *Handbook of Ecomaterials* vol. 3 1659–1677. (Springer International Publishing, 2019).

32. Makuła, P., Pacia, M. & Macyk, W. How to correctly determine the band gap energy of modified semiconductor photocatalysts based on UV-Vis spectra. *Journal of Physical Chemistry Letters* vol. 9, 6814–6817. Preprint at https://doi.org/10.1021/acs.jpclett.8b02892 (2018).

33. Zhang, Y. J., Zhang, L., Kang, L., Yang, M. Y. & Zhang, K. A new CaWO4/alkali-activated blast furnace slag-based cementitious composite for production of hydrogen. *International Journal of Hydrogen Energy* **42**, 3690–3697 (2017).

34. Bora, L. V. et al. Trash GGBFS-based geopolymer as a novel sunlight-responsive photocatalyst for dye discolouration. *Journal of the Indian Chemical Society* **99**, 100560, (2022).

35. Kaiqian Shu, Chitiphon Chuaicham, Yuto Noguchi, Longhua Xu & Keiko Sasaki. In-situ hydrothermal synthesis of Fe-doped hydroxyapatite photocatalyst derived from converter slag toward xanthate photodegradation and Cr(VI) reduction under visible-light irradiation. *Chemical Engineer Journal* **459**, 141474, (2023).

36. Montoya-Bautista, C. V., Acevedo-Peña, P., Zanella, R. & Ramírez-Zamora, R. M. Characterizati on and evaluation of copper slag as a bifunctional photocatalyst for alcohols degradation and hydrogen production. *Topics in Catalysis* **64**, 131–141 (2021).

37. Solís-López, M. et al. Assessment of copper slag as a sustainable Fenton-type photocatalyst for water disinfection. In *Water Reclamation and Sustainability* 199–227 (Elsevier Inc., 2014). doi:10.1016/B978-0-12-411645-0.00009-2.

38. Ji, R. et al. A review of metallurgical slag for efficient wastewater treatment: Pretreatment, performance and mechanism. *Journal of Cleaner Production* **380**, 135076 (2022).

39. Wang, F. P. et al. A review of modified steel slag application in catalytic pyrolysis, organic degradation, electrocatalysis, photocatalysis, transesterification and carbon capture and storage. *Applied Sciences (Switzerland)* **11(10)**, 4539, (2021).

40. Hussain, S. I. et al. Transport and attenuation of an artificial sweetener and six pharmaceutical compounds in a sequenced wetland-steel slag wastewater treatment system. *Water (Switzerland)* **15(15)**, 2835, (2023).

41. Nguyen, L. H. et al. Promoted degradation of ofloxacin by ozone integrated with Fenton-like process using iron-containing waste mineral enriched by magnetic composite as heterogeneous catalyst. *Journal of Water Process Engineering* **49**, 103000, (2022).

42. Khosravi Pour, L. et al. The remo val of tetracycline antibiotic by advanced oxidation method of sodium monopersulfate activated by steel industry slag from pharmaceutical effluent. *Journal of Water and Wastewater* **33**, 17–30 (2022).

43. Sun, R., Huang, R., Yang, J. & Wang, C. Magnetic copper smelter slag as heterogeneous catalyst for tetracycline degradation: Process variables, kinetics, and characterizations. *Chemosphere* **285**, 131560, (2021).

44. Chuaicham, C. et al. Enhancement of photocatalytic rhodamine B degradation over magnesium-manganese baring extracted iron oxalate from converter slag. *Separations* **10(6)**, 440, (2023).

45. Jiang, Y., Sun, L. Da, Li, N., Gao, L. & Chattopadhyay, K. Metal-doped ZnFe2O4 nanoparticles derived from Fe-bearing slag with enhanced visible-light photoactivity. *Ceramics International* **46**, 28828–28834 (2020).

46. Li, M. et al. Efficient activation of peroxymonosulfate by a novel catalyst prepared directly from electrolytic manganese slag for degradation of recalcitrant organic pollutes. *Chemical Engineering Journal* **401**, 126085, (2020).

47. Li, M. et al. Degradation of various thiol collectors in simulated and real mineral processing wastewater of sulfide ore in heterogeneous modified manganese slag/PMS system. *Chemical Engineering Journal* **413**, 127478, (2021).

48. Naser Mehrdadi, Afshin Takdastan, Laleh Khosravi Pour, Gholamreza Nabi Bidhendi & Masoume Taherian. The removal of azithromycin antibiotic by advanced oxidation method of sodium persulfate activated by steel industry slag from pharmaceutical effluent. *Journal of Water and Wastewater Science and Engineering* **8**, 29–40 (2023).

49. Safo, K., Noby, H., Matatoshi, M., Naragino, H. & El-Shazly, A. H. Statistical optimization modeling of organic dye photodegradation process using slag nanocomposite. *Research on Chemical Intermediates* **48**, 4183–4208 (2022).

50. Song, B. et al. Sulfur-zinc modified kaolin/steel slag: A particle electrode that efficiently degrades norfloxacin in a neutral/alkaline environment. *Chemosphere* **284**, 131328, (2021).

51. Song, Z., Gao, H., Liao, G., Zhang, W. & Wang, D. A novel slag-based Ce/TiO2@LDH catalyst for visible light driven degradation of tetracycline: performance and mechanism. *Journal of Alloys and Compounds* **901**, 163525, (2022).

52. Yañez-Aulestia, A. & Ramírez-Zamora, R. M. Effect of ascorbic acid to improve the catalytic performance of metallurgical copper slag in the photo-Fenton type process for hydroxyl radical production applied to the degradation of antibiotics. *Journal of Environmental Chemical Engineering* **11**, 109897, (2023).

53. Matthaiou, V. et al. Valorization of steel slag towards a Fenton-like catalyst for the degradation of paraben by activated persulfate. *Chemical Engineering Journal* **360**, 728–739 (2019).

54. García-Estrada, R., Arzate, S. & Ramírez-Zamora, R. M. Thiabendazole degradation by photo-NaOCl/Fe and photo-Fenton like processes, using copper slag as an iron catalyst, in spiked synthetic and real secondary wastewater treatment plant effluents. *Water Science and Technology* **87**, 620–634 (2023).

55. Herrera-Ibarra, L. M. et al. Treatment of textile industrial wastewater by the heterogeneous solar photo-Fenton process using copper slag. *Topics in Catalysis* **65**, 1163–1179 (2022).

56. Singh, S. & Garg, A. Characterisation and utilization of steel industry waste sludge as heterogeneous catalyst for the abatement of chlorinated organics by advanced oxidation processes. *Chemosphere* **242**, 125258, (2020).

57. Abdelbasir, S. M. & Khalek, M. A. A. From waste to waste: iron blast furnace slag for heavy metal ions removal from aqueous system. *Environmental Science and Pollution Research* **29**, 57964–57979 (2022).

58. J. García-Chirino, B.M. Mercado-Borrayo, R. Schouwenaars, J.L. González-Chávez & R.M. Ramírez-Zamora. Simultaneous Removal of Arsenic and Fluoride from Water using Iron and Steel Slags. In Zhu Y., Guo H., Bhattacharya P., Ahmad A., Bundschuch J., Naidu R. (eds) Environmental Arsenic in a changing word. CRC Press, London, UK (2018). https://doi.org/10.1201/9781351046633

59. Manchisi, J. et al. Ironmaking and steelmaking slags as sustainable adsorbents for industrial effluents and wastewater treatment: A critical review of properties, performance, challenges and opportunities. *Sustainability (Switzerland)* **12**, 2118, (2020).

60. Mercado-Borrayo, B. M., González-Chávez, J. L., Ramírez-Zamora, R. M. & Schouwenaars, R. Valorization of metallurgical slag for the treatment of water pollution: an emerging technology for resource conservation and re-utilization. *Journal of Sustainable Metallurgy* vol. 4 50–67. Preprint at https://doi.org/10.1007/s40831-018-0158-4 (2018).

61. Mercado-Borrayo, B. M., Schouwenaars, R., Litter, M. I. & Ramirez-Zamora, R. M. Adsorption of Boron by Metallurgical Slag and Iron Nanoparticles. Adsorption Science & Technology, 32(2-3), pp.117-123.

62. Tran Van, C. et al. Study on photocatalytic degradation of methylene blue by TiO2 synthesiszed from titanium slag using a new decomposition agent. *Vietnam Journal of Catalysis and Adsorption* **11**, 88–92 (2021).

63. Dhmees, A. S., Rashad, A. M., Eliwa, A. A. & Zawrah, M. F. Preparation and characterization of nano SiO2@CeO2extracted from blast furnace slag and uranium extraction waste for wastewater treatment. *Ceramics International* **45**, 7309–7317 (2019).

64. Shao, N. et al. An all-in-one strategy for the adsorption of heavy metal ions and photodegradation of organic pollutants using steel slag-derived calcium silicate hydrate. *Journal of Hazardous Materials* **382**, 121120, (2020).

65. Wang, H. et al. Efficient removal of mercury and chromium from wastewater via biochar fabricated with steel slag: Performance and mechanisms. *Frontiers in Bioengineering and Biotechnology* **10**, 961970, (2022).

66. Wang, S. et al. The mechanisms of conventional pollutants adsorption by modified granular steel slag. *Environmental Engineering Research* **26**, 1–10 (2021).

67. Cheng, Y. et al. A porous geopolymer containing Ti-bearing blast furnace slag: synthesis, characterization, and adsorption-photodegradation studies towards methylene blue removal under visible light condition. *Molecules* **28**, 3673, (2023).

68. Gusain, R., Kumar, N. & Ray, S. S. Factors influencing the photocatalytic activity of photocatalysts in wastewater treatment. in *Photocatalysts in Advanced Oxidation Processes for Wastewater Treatment* 229–270. (Wiley, 2020). doi:10.1002/9781119631422.ch8.

6 Adsorption of Pharmaceutical Derivatives from Aqueous Solutions Employing Naturally Available Diatomite as an Eco-Friendly Adsorbent

Jorge del Real-Olvera, Wendy N. Medina-Esparza, and Leonel Hernández-Mena

6.1 INTRODUCTION

One direct consequence of technological advances is a greater sensitivity of analytical tools which has allowed the detection of various pollutants present at low concentrations in water bodies. This group of newly discovered pollutants has recently been classified as "emerging contaminants" (ECs). ECs have been detected in wastewater, groundwater, and surface water including pharmaceuticals, endocrine disruptors, pesticides, and personal care products [1] generally at low concentrations, showing poor or no biodegradability, and tend to bioaccumulate [2]. The first reports regarding the quantification of pharmaceuticals in natural surface waters appeared in the early 1980s when salicylic acid and anticancer products were identified in water bodies [3]. Since then, analgesics and antipyretics have been among the pharmaceuticals most frequently detected in the environment posing a threat because of their adverse effects on human health and undesirable modifications to the biological balance of aquatic ecosystems [4]. Therefore, there is a significant need to find viable, eco-friendly, and cost-effective alternatives to remove these compounds from water [5].

Ciprofloxacin (CIP) is a quinolone antibacterial agent classified as second-generation fluoroquinolone with broad-spectrum action often used to treat human and animal bacterial infections [6]. The presence of CIP in wastewater and surface water is considered a significant environmental risk, even at very low concentrations, because it can increase antibiotic resistance of pathogenic bacteria and modify the biological balance of aquatic ecosystems [7]. Several studies have been carried out to

DOI: 10.1201/9781003441007-9

remove CIP from aqueous solutions including adsorption on different media [3,5,8] photocatalytic degradation [7], and electrocoagulation [9].

In the aftermath of the COVID-19 pandemic, acetaminophen (ACT, also known as paracetamol) became one of the most prescribed analgesics worldwide used to relieve pain and fever. ACT enters water bodies through discharges from hospitals or veterinary wastewater and is also present in domestic and industrial wastewater. The presence of ACT in wastewater is an outstanding concern because, in some locations, its detection frequency reaches 60%–90% at concentrations ranging from 0.1 to 1,500 µg/L posing a considerable threat to human health and environmental sustainability [10]. Several studies aiming at ACT removal from aqueous solutions including adsorption [11,12] and advanced oxidation processes [10,13] have been published recently.

Diatomaceous earth (DE) is a nonmetallic clay mineral formed by fossilized cell walls (frustules) of microscopic aquatic plants called diatoms. When diatoms die, the silicates absorbed from the environment make them sink into the water body bottom and form thick silicate sedimentary deposits. DE is composed of fine and porous aggregates with variable texture, and lacustrine or marine origin [3]. These large deposits of clay possess significant economic potential because they demonstrate thickness and material quality, a variety of geometric shapes, and mechanical properties (e.g., low density, high surface area, porosity, absorption capacity, and low thermal conductivity and chemical reactivity) as a filter medium used to purify diverse liquids generated during industrial processes [14].

Natural (raw) or modified (thermally or chemically treated) DE has been used to adsorb different pollutants from synthetic and real wastewater. However, very little is known about the use of DE for pharmaceutical adsorption with only a few reports available in the literature [10,15,16].

The aim of this study is to assess the feasibility of employing unmodified DE as a low-cost, eco-friendly adsorbent to remove CIP and ACT from aqueous solutions using two synthetic effluents: ultrapure water and treated domestic wastewater and identify the influence of pharmaceuticals' initial concentration, initial pH, adsorption time, and DE dose to better understand the adsorption processes.

6.2 MATERIALS AND METHODS

6.2.1 SOURCE AND PREPARATION OF THE ADSORBENT MATERIAL

Raw DE ($SiO_2 \cdot nH_2O$) was obtained from a deposit of clay mineral situated in Jalisco, Mexico (20° 18′ 54.50″ N; 103° 36′ 42.57″ W). A composite DE sample (1 kg) was created with individual samples collected from 20 different points within the deposit. The raw material was pulverized using a sphere mill and passed through metal sieves (150–700 µm). The 450–600 µm fraction was used for the adsorption experiments. The powdered DE was washed several times with deionized (DI) water to remove soluble impurities, sunlight dried for 8 h, stored in closed containers, and placed in a cold room (4°C) until use, as suggested in the literature [14].

DE chemical characterization was performed using Fourier-transform Infrared (FT-IR) spectroscopy. Spectra were obtained with a Spectrum GX spectrometer (Perkin Elmer, Waltham, MA, USA) carried out in the 4,000–550 cm^{-1} range using a resolution of 4 cm^{-1}. A powder X-ray diffractometer (XRD; D500 Model Kristallographie Siemens, Washington D.C., USA) was used with Cu Ka radiation (0.1542 nm wavelength). High-angle XRD data were accumulated from 5° to 40° at 0.02° increments and 1-min count times. DE images were obtained using transmission electron microscopy (TEM; JEOL Mag: 15Kx at 200 kV). For TEM observation, DE samples were suspended in ethanol and sonicated for 10 min, then the suspension was deposited dropwise on a copper grid coated with carbon conductive tape film and dried at room temperature until the ethanol evaporation. DE morphology was assessed by scanning electron microscopy (SEM; TESCAN, Brno-Kohoutovice, Czech Republic). The images were analyzed using the ImageJ 1.45 software to evaluate the material's porosity.

DE specific surface area (BET area) was determined using a Quantachrome Nova 2200e model sorptometer. DE particle zeta-potentials were determined using a Litesizer 500 instrument, and the results are reported as the arithmetic mean and standard deviation of three analyses. The solids were initially disseminated in a boric acid and potassium chloride (pH = 3) electrolyte buffer solution, a sodium and potassium phosphate buffer was used for pH = 6, and another potassium chloride and sodium hydroxide (NaOH) buffer solution was used for pH = 9 [3].

6.2.2 Synthetic Solutions of Pharmaceutical Derivatives in Ultrapure Water

CIP ($C_{17}H_{18}FN_3O_3$, mw 331.34 g/mol, water solubility at 20°C, 30 g/L), and ACT ($C_8H_9NO_2$, mw 151.163 g/mol, water solubility at 20°C, 14 g/L) were obtained from Sigma-Aldrich and its structural formulas are shown in Figure 6.1 along with their speciation as a function of pH [17].

To avoid cross-contamination, all the glassware materials used for this study were new and carefully washed with 5% Extran® 02 (Merck) followed by 5% nitric acid (Fermont), acetone (Jalmek), and deionized water before each experiment. In addition, an ultrasonic bath (Branson 5510) was used to ensure thorough cleaning of the glassware material and to prevent other pollutants interference.

Stock solutions of pharmaceuticals (1,000 mg/L) were prepared in ultrapure water each week to avoid degradation. The initial pH of the different working solutions was modified with either 0.1 M HCl or NaOH to assess the influence of the initial pH (pH range from 2 to 10) on the adsorption efficiency of DE. Working solutions with diverse CIP (5–50 mg/L) and ACT (30–80 mg/L) concentrations were prepared to investigate the influence of pharmaceuticals' initial concentration on adsorption efficiency. All the chemical reagents used were of analytical grade.

6.2.3 Synthetic Pharmaceuticals Solutions in Treated Domestic Wastewater

To verify and validate the adsorption process efficiency of pharmaceuticals on DE, treated domestic wastewater was spiked with pharmaceuticals. The treated wastewater was collected from a domestic wastewater treatment plant in Jalisco,

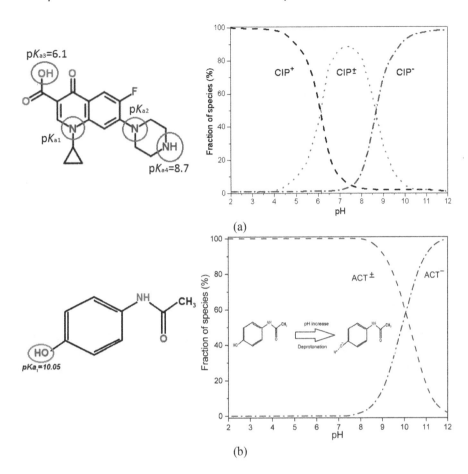

FIGURE 6.1 Chemical structure and speciation as a function of pH of (a) CIP and (b) ACT.

Mexico (20° 43′ 24.8″ N; 103° 23′ 52.2″ W). The wastewater treatment process consisted of fine screening, followed by settling, and anaerobic treatment. The effluent was characterized for chemical oxygen demand (COD), total nitrogen, total phosphorus, nitrate, and ammonia concentration and spiked to get the same pharmaceutical concentration described in Section 6.2.2.

CIP and ACT concentrations were analyzed using a UV-visible spectrophotometer (Hach DR 5000). Initially, a calibration curve for CIP was prepared ranging from 1 to 70 mg/L at 271 nm wavelength. For ACT the calibration curve ranged from 10 to 100 mg/L at 285 nm wavelength both calibration curves at neutral pH [3,13].

6.2.4 Adsorption Isotherm and Kinetic Studies

Adsorption trials were carried out in batch mode at a controlled temperature (25°C ± 1°C), using 2,000 mL beakers and magnetic stirring (Vante, MS7–H550). In all cases, suspensions were stirred to ensure complete homogenization.

The pharmaceutical concentrations were evaluated for 48 h continuously, and adsorption equilibrium was observed after 24 h in both cases. The adsorption process was followed through time: one 10 mL sample was taken every 10 min during the first 3 h. After that, samples were taken every hour until completing 12 h, and the final samples were taken after 24 and 48 h. All samples were filtrated using 0.45 μm membrane filters to avoid the presence of adsorbent particles.

The experimental conditions tested during the study included changes in initial pH (2–10), DE dosage (0.5–4.0 g/L), and working solution concentration. For CIP, initial concentrations ranged from 5 to 50 mg/L. For ACT, the initial concentration ranged 30–80 mg/L. The adsorption capacity at equilibrium (q_e) and at any time (q_t) were estimated using Equations (6.1) and (6.2), respectively:

$$q_e = \frac{(C_i - C_e)V}{W} \tag{6.1}$$

$$q_t\,(mg/g) = \frac{(C_i - C_t)V}{W} \tag{6.2}$$

where q_e and q_t are the adsorption capacity at equilibrium and at any time, respectively, C_i, C_t, and C_e are the pharmaceutical concentrations (mg/L) at initial, any time, and equilibrium, respectively, V is the volume (L), and W is the DE powder dose (g). Experimental results were fitted using the Freundlich and Langmuir isotherm models to depict the equilibrium relationships in all the tests performed at different conditions. Equation (6.3) was used to calculate CIP or ACT removal efficiency [18]:

$$Removal\ efficiency\ (\%) = \frac{(C_i - C_t)}{C_i} \times 100 \tag{6.3}$$

6.3   RESULTS

6.3.1   DE CHARACTERIZATION

DE exhibited low specific surface area (BET 29–38 m²/g) but was in the same order of magnitude as reported by other studies for similar materials (32 m²/g) [11]. SEM showed pore diameter within the 50–380 nm range, with an average of 171 ± 50 nm and a pore volume of 0.145 cm³/g. The zeta potential was found negative at all pH values tested, -38.3 ± 0.4 mV at pH 9, -32.7 ± 0.3 mV at pH 6, and -25.9 ± 0.2 mV at pH 3 [3].

Figure 6.2a shows a typical XRD spectrogram, data suggests DE mainly contains silica (SiO_2) and small amounts of Fe_2O_3, Na_2O, Al_2O_3, and CaO. The XRD patterns confirmed feldspar $Na(AlSi_3O_8)$ ($2\theta = 29°$, 23°, 22°), quartz (SiO_2) ($2\theta = 27°$), and cristobalite ($2\theta = 36°$, 39°), in agreement with literature [14]. XRD diffraction also suggested poorly crystallized DE. DE structure was confirmed by FT-IR (Figure 6.2b) where the band at 3,400 cm⁻¹ was identified as O–H stretching of physisorbed water and the band at 1,630 cm⁻¹ is characteristic of H–O–H bending

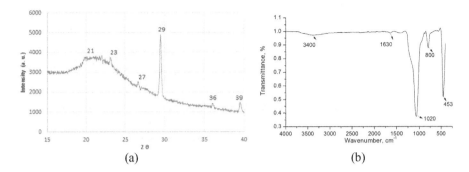

FIGURE 6.2   DE (a) XRD spectrogram, and (b) FT-IR spectra.

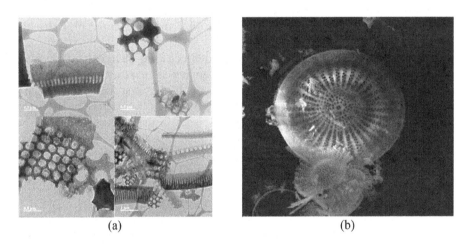

FIGURE 6.3   DE imagery using: (a) TEM analysis and (b) SEM analysis.

vibration. DE representative components are displayed in bands at 1,020 cm^{-1} (siloxane (Si–O–Si) group), 800 cm^{-1} (quartz and free silica), and 453 cm^{-1} (Si–O–Si and/ or Si–O–Al) [14].

Figure 6.3a shows TEM images for DE suggesting a heterogeneous array distribution on frustules. The average frustule size was estimated at 10.13 μm with an average frustule width of 2.84 μm. TEM information agrees with SEM images shown in Figure 6.3b. Using SEM imagery, it was found that most DE are circular consistent with the *Discostella* genera, in agreement with previous studies [3,14].

6.3.2 ADSORPTION KINETIC PROCESS

Different kinetics models frequently reported in the literature were used to evaluate the process adsorption mechanisms. Adsorption kinetics data were fit using both pseudo-first- and pseudo-second-order models. The integrated form of the Lagergren pseudo-first-order model is widely used, expressed using limiting conditions from $t=0$ to $t=t$ and from $q=0$ to $q=q_t$ which provides the following linear function:

$$\log(q_e - q_t) = \log q_e - \frac{k_1 t}{2.303} \tag{6.4}$$

where k_1 is the pseudo-first-order model rate constant (h^{-1}), q_t is the adsorption capacity at any time (mg/g), and q_e is the equilibrium adsorption capacity (mg/g).

For the pseudo-second-order kinetic mechanism, the linearized form considering the same boundary conditions (e.g., from $t=0$ to $t=t$ and from $q=0$ to $q=q_t$) is:

$$\frac{t}{q_t} = \frac{1}{k_2 q_e^2} + \frac{t}{q_e} \tag{6.5}$$

where k_2 is the pseudo-second-order kinetics rate constant (g/mg h).

The sorption of pharmaceuticals on DE at three different initial pH values is shown in Figure 6.4.

In both cases, the adsorption capacity was examined using 48 h contact time in all experiments, but results suggested only 24 h were needed to reach adsorption equilibrium. The first hours were found critical because the greatest adsorption capacity occurs during this time. Adsorption in both (CIP and ACT) was fast at lower pH values suggesting the involvement of cation species in the process, in agreement with other studies that suggest contact time between 24 and 72 h is needed to reach adsorption equilibrium for triclosan over DE [16]. In the case of the spiked treated wastewater, the presence of COD as high as 77 mg/L, total nitrogen 69 mg/L, total phosphorous 22.6 mg/L, nitrates 0.005 mg/L and ammonia nitrogen 67.2 mg/L did not show a significant effect on DE performance.

The kinetic parameter values for the two models proposed are summarized in Table 6.1 [3,18]. As shown, although regressions correlation coefficients (R^2) for

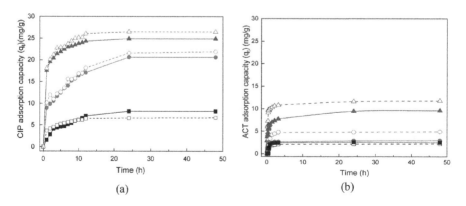

FIGURE 6.4 Adsorption of pharmaceuticals on DE as a function of pH. (a) CIP and (b) ACT using initial pH=3 ▲, pH=6 ●, and pH=9 ■ for synthetic water, and initial pH=3 △, pH=6 ○, and pH=9 □ for spiked treated domestic wastewater. Experimental conditions:1.5 g of DE with 30 mg/L of CIP, and 3 g of DE with 30 mg/L of ACT.

TABLE 6.1

Pseudo-Second- and Pseudo-First-Order Kinetics Parameters for pH Values

CIP		pH 3	pH 6	pH 9
Pseudo-first-order	q_e (mg/g)	18.311	11.416	13.026
	k_1 (1/h)	0.2207	0.1746	0.1788
	R^2	0.9191	0.8371	0.9161
Pseudo-second-order	q_e (mg/g)	25.498	21.251	12.225
	k_2 (g/mg h)	0.0343	0.0691	0.0336
	R^2	0.9966	0.9973	0.9723
ACT		**pH 3**	**pH 6**	**pH 9**
Pseudo-first-order	q_e (mg/g)	6.644	3.963	2.87
	k_1 (1/h)	0.0756	0.3658	0.2088
	R^2	0.9503	0.9672	0.9666
Pseudo-second-order	q_e (mg/g)	11.753	11.525	10.53
	k_2 (g/mg h)	0.0277	0.0325	0.0223
	R^2	0.9782	0.9776	0.9777

both models were greater than 0.80 for the entire range of pH values, the pseudo-second-order model was found to better represent the kinetic behavior of both pharmaceuticals. In this study, the pseudo-second-order kinetic constant (k_2) range obtained for CIP was $0.007 \leq k_2$ (g/mg h) ≤ 0.198, and for ACT was $0.0056 \leq k_2$ (g/mg h) ≤ 0.149 with R^2 over 0.97. These results suggest that the adsorption kinetics process was preferred over chemisorption.

6.3.3 EFFECT OF INITIAL PH

The effect of initial pH on the adsorption of pharmaceuticals was found depending on the experimental conditions and chemical structure of the pollutants involved. In this study, the solubility of both pharmaceuticals was found to change with pH resulting in the presence of different functional groups as shown in Figure 6.1. Initial pH value affects pharmaceutical adsorption because it affects adsorption capacity and mechanism. However, the specific effects will depend on adsorbent and adsorbate types. The removal efficiency of pharmaceuticals on DE at different initial pH values, using 20 mg/L CIP and 2 g DE and 30 mg/L ACT and 3 g DE is shown in Figure 6.5.

The effect of initial pH on CIP adsorption on DE suggests that, at low pH values, the presence of protonated CIP+ species occurs diminishing when pH exceeds 6.0, the CIP pK_{a3} constant (carboxylic acid group). For the pH range 6.0–8.0 (pK_{a4}, nitrogen on piperazinyl ring), three different species have been reported, where CIP± zwitterion is the least soluble. Finally, as the pH value increases to higher than 8.0, CIP becomes slightly more soluble because of the presence of CIP− species [3,7]. The best CIP adsorption was found for acidic pH, probably because of the

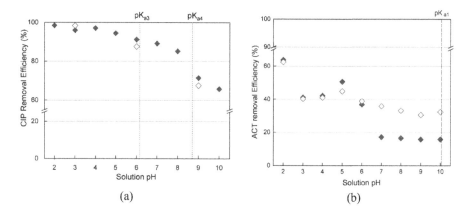

FIGURE 6.5   Adsorption efficiency of pharmaceuticals on DE at different initial pH values (a) CIP; (b) ACT in synthetic water (♦) and spiked treated domestic wastewater (◊).

relationship between total CIP and DE's surface charge. DE zeta potential was found -38.3 ± 0.4 mV at pH 9, -32.7 ± 0.3 mV at pH 6, and -25.9 ± 0.2 mV at pH 3. When the cationic CIP form (CIP+) is present, a negative DE surface will attract positive cationic CIP generating significant adsorption. At basic pH (pH = 9), anionic CIP form (CIP−) will likely produce repulsive interactions with negatively charged DE surfaces [7,14].

In agreement with Figure 6.1, different ACT chemical species are generated in the aqueous system when the initial pH value is modified with each species showing different water solubilities. At low pH (between 2 and 6) ACT± species is mainly generated which is poorly ionizable but more soluble in water than other species, significantly decreasing as pH value increases. When the pH value increases above 8, ACT− chemical species are expected at a significant proportion, which is less reactive in water because of the appearance of negative charges by deprotonation.

Results for the spiked treated effluent suggested ACT adsorption is favorable at low pH values supporting the hypothesis of cationic species participation in the process. In this case, cationic ACT adsorption on the negatively charged DE surface is proposed to occur through electrostatic attraction as the adsorption mechanism. As pH values increase, experimental conditions approach the ACT dissociation constant ($pK_a = 10.05$) until adsorption efficiency decreases significantly after the pH value reaches 8. The low performance at high pH values seems to be directly associated with the presence of anionic ACT species (ACT−) generating repulsive interactions with the negative DE surface [3].

6.3.4 EFFECT OF PHARMACEUTICALS CONCENTRATION

Figure 6.6 shows the removal efficiency of both investigated pharmaceuticals using different initial concentrations. CIP initial concentration was within the 5–50 mg/L range at initial pH = 3.0 and DE dose 2 g. ACT was tested within the 30–80 mg/L

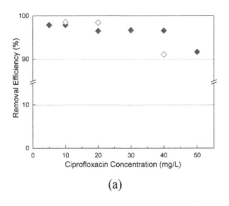
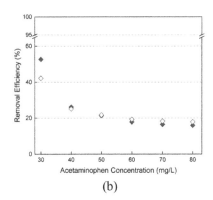

(a)   (b)

FIGURE 6.6   Effect of initial pharmaceuticals concentration on adsorption efficiency for (a) CIP and (b) ACT, for synthetic water (♦) and spiked treated domestic wastewater (◊).

range, using initial pH = 5 and DE dose 3 g, for synthetic water and spiked treated domestic wastewater.

The effect of initial pharmaceutical concentration on the adsorption using DE has been reported in several studies. Pharmaceutical concentration affects DE adsorption capacity depending on the specific chemical structure. As expected, the highest removal efficiency for both pharmaceuticals analyzed was found when low initial concentrations were tested. DE is capable of effectively adsorbing CIP from aqueous solutions, indicating its significant potential as an adsorbent by achieving adsorption equilibrium within 24 h. The best CIP removal efficiency was 97% using 2 g of DE, 20 mg/L of *CIP*, and pH = 2.0 [3]. Removal efficiency was found to be higher than 90% for the entire range of initial CIP concentrations tested, suggesting that unmodified DE is an excellent adsorbent for this pollutant.

The highest ACT removal efficiency was found when using low initial concentrations (i.e., 30 mg/L). By increasing the initial ACT concentration, DE adsorption capacity was affected. Other studies have shown that ACT adsorption is influenced by several factors including sorbent type and operating parameters. The key ACT uptake mechanism has been suggested to include π–π interactions, hydrogen bonds, and electrostatic interactions [19]. ACT adsorption efficiency was found greater than 50% for low ACT concentrations only in both types of water.

6.3.5   Effect of DE Dose

Figure 6.7 shows the effect of DE dose on the adsorption of both pharmaceuticals. The removal efficiency was assessed as a function of the DE dose using an initial CIP concentration of 30 mg/L at pH 3.0 and an initial ACT concentration of 30 mg/L at pH 5.0.

The dose of DE was found to have a crucial impact on adsorption capacity. In agreement with other studies higher DE doses lead to higher adsorption capacity values suggesting this material is a promising cost-effective adsorbent for pharmaceutical

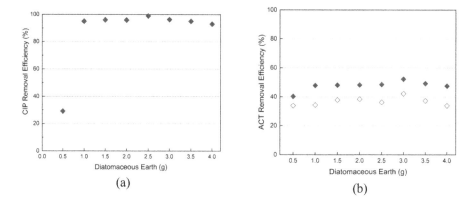

FIGURE 6.7 Effect of DE dose on adsorption efficiency of (a) 30 mg/L CIPat pH = 3; (b) 30 mg/L *ACT* at pH = 5, for synthetic water (♦) and spiked treated domestic wastewater (◊).

removal from aqueous phase [14]. A relatively low DE dose (1 g/L) generated significant CIP adsorption efficiency after 24 h (Figure 6.7a). Further increases in the DE dose (1–3 g/L) did not produce any significant difference in CIP adsorption. The high CIP removal is probably because of the surged strong affinity between the positively charged CIP molecules and the negatively charged DE surface, particularly when low initial pH is used. DE adsorption performance was different at the lowest dose (e.g., 0.5 g/L) probably because, under these conditions, all DE surface empty sites were occupied by CIP molecules and pollutants molecules remained in excess in aqueous solution.

As shown in Figure 6.7b, a relatively low DE dose (0.5 g/L) generated low ACT adsorption efficiency after 24 h, and no relevant changes in ACT adsorption efficiency were observed by increasing the DE dose from 1 to 4 g/L. The main ACT adsorption mechanism is related to the chemical affinity between ACT molecules and the DE surface, particularly when the process occurs at a low initial pH. The lack of effect of DE low doses (e.g., 0.5–1.0 g/L) is probably because, under these conditions, all active sites available on the DE surface area were occupied by ACT molecules, and still excess contaminant molecules existed in the aqueous solution.

6.3.6 Adsorption Isotherms

Langmuir and Freundlich isotherm models were used to assess the relationships between experimental data obtained under different operation conditions and theoretical models. Results are shown in Table 6.2.

Results in Table 6.2 suggest that the Langmuir model fairly fits the experimental data with an R^2 value of 0.95. Langmuir's model-specific constants (q_{max} and K_L) better explain sorption experimental results. From Table 6.2, q_{max} is DE maximum saturation, estimated to be 105.1 mg/g for CIP and 11.9 mg/g for ACT which supports results from Figure 6.7 in which a DE dose of 1.0 g was able to almost completely adsorb CIP and ACT and no effect was noticed for further increases in DE dose.

TABLE 6.2
Freundlich and Langmuir Isotherms for Pharmaceuticals Uptake on DE

CIP	Model			
	Langmuir		Freundlich	
	q_{max}(mg/g)	K_L (L/g)	n(dimensionless)	K_f(L/g)
	105.108	1.49	5.586	0.479
R^2	0.9526		0.7092	
ACT				
	Langmuir		Freundlich	
	q_{max}(mg/g)	K_L (L/g)	n(dimensionless)	K_f(L/g)
	11.904	0.058	12.195	0.0251
R^2	0.9657		0.8441	

The Langmuir model describes monolayer adsorption on a set of adsorption sites with the same sorption energies independent of surface coverage and without interaction between adsorbed and incoming molecules [3]. This is the most likely process occurring for pharmaceutical adsorption on DE because the Langmuir model best describes experimental data.

6.4 CONCLUSIONS

In this study, the potential of CIP and ACT adsorption shown by untreated DE in synthetic water and spiked treated domestic wastewater was investigated. DE characterization revealed that the adsorption process is complex and strongly depends on the contaminant's properties. When dispersive interactions (repulsion) between adsorbate and adsorbent were dominant, large DE surface area volume and micropores were proposed not occupied by pharmaceuticals.

Initial pH and pharmaceutical concentration were found key factors in the adsorption process. In both cases, the adsorption of pharmaceuticals by DE was considerable under acidic conditions, but minor under neutral and alkaline conditions. Furthermore, higher DE doses showed greater adsorption capacity because of greater surface area and porosity exposure to pharmaceuticals. An increase in initial pharmaceutical concentration generated a reduction in adsorption capacity. According to the literature, several factors can improve the adsorption capacity of pharmaceuticals at high concentrations but, to achieve this goal, adsorbents with enhanced physical and chemical characteristics are required.

The use of adsorbents without any modification is encouraged to reduce operating costs and contribute to the circular economy concept. The results obtained in this study are relevant, suggesting that more research is needed to optimize the process and develop increasingly efficient adsorbents for pharmaceutical removal from aqueous systems.

REFERENCES

1. Philip, J.M.; Aravind, U.K.; Aravindakumar, C.T. (2018). Emerging contaminants in Indian environmental matrices. A review. *Chemosphere*, Vol. 190, 307–326.
2. Rout, P.R.; Zhang, T.C.; Bhunia, P.; Surampalli, R.Y. (2021). Treatment technologies for emerging contaminants in wastewater treatment plants: A review. *Science of the Total Environment*, Vol. 753, 141990.
3. García-Alonso, J.A.; Sulbarán-Rangela, B.C.; Bandala, E.R.; del Real-Olvera, J. (2019). Adsorption and kinetic studies of the removal of ciprofloxacin from aqueous solutions by diatomaceous earth. *Desalination and Water Treatment*, Vol. 162, 331–340.
4. Nabgan, W.; Jalil, A.A.; Nabgan, B.; Ikram, M.; Ali, M.W; Kumar, A.; Lakshminarayana, P. (2022). A state of the art overview of carbon-based composites applications for detecting and eliminating pharmaceuticals containing wastewater. *Chemosphere*, Vol. 288, 132535.
5. Vinayagama, V.; Murugana, S.; Kumaresana, R.; Narayanana, M.; Sillanp, M.; Vof, D.V; Kushwahag, O.S.; Jenish, P.; Potdari, P.; Gadiya, S. (2022). Sustainable adsorbents for the removal of pharmaceuticals from wastewater: A review. *Chemosphere*, Vol. 300, 134597.
6. Peng, S.; Wei, Y.; Huang, Y.; Wei, L.; Chen, P. (2023). Highly efficient adsorption of antibiotic ciprofloxacin hydrochloride from aqueous solution by diatomite-basic zinc chloride composites. *Environmental Science and Pollution Research*, Vol. 43, 98490–98501.
7. Jia, Z.; Li, T.; Zheng, Z.; Zhang, J.; Liu, J.; Li, R.; WangY.; Zhang, X.; Wang, Y.; Fan, C. (2020). The BiOCl/diatomite composites for rapid photocatalytic degradation of ciprofloxacin: Efficiency, toxicity evaluation, mechanisms and pathways. *Chemical Engineering Journal*, Vol. 380, 122422.
8. Stefanelli, E.; Vitolo, S.; Di Fidio, N.; Puccino, M. (2023). Tailoring the porosity of chemically activated carbons derived from the HTC treatment of sewage sludge for the removal of pollutants from gaseous and aqueous phases. *Journal of Environmental Management*. Vol. 345, 118887.
9. Aljaberi, F.Y.; Ahmed, S.A.; Makki, H.F.; Naje, A.S.; Zwain, H.M.; Salman, A.D.; Juzsakova, T.; Viktor, S.; VanB.; Le, P.C.La, D.D.Chang, S.W. (2023). Recent advances and applicable flexibility potential of electrochemical processes for wastewater treatment. *Science of the Total Environment*, Vol 867, 161361.
10. Mousel, D.; Bastian, D.; Firk, J.; Palmowski, L.; Pinnekamp, J. (2021). Removal of pharmaceuticals from wastewater of health care facilities. *Science of the Total Environment*, Vol. 751, 141310.
11. Natarajan, R.; Banerjeea, K.; Kumarb, P.S.; Somannaa, T.; Tannania, D.; Arvinda, V.; Raja, R.; Voc, D.; Saikiaa, K.; Vaidyanathan, V.K. (2021). Performance study on adsorptive removal of acetaminophen from wastewater using silica microspheres: Kinetic and isotherm studies. *Chemosphere*, Vol. 272, 129896.
12. Nguyen, D.T.; Tran, H.N.; Juang, R.S.; Dat, N.D.; Tomul, F.; Ivanets, A.; Woo, S.H.; Hosseini-Bandegharaei, A.; Nguyen, V.P.; Chao, H.P. (2020). Adsorption process and mechanism of acetaminophen onto commercial activated carbon. *Journal of Environmental Chemical Engineering*, Vol. 8, 104408.
13. Brillas, E.; Peralta-Hernández, J.M. (2023). Removal of paracetamol (acetaminophen) by photocatalysis and photoelectrocatalysis. A critical review. *Separation and Purification Technology*, Vol. 309, 122982.
14. Reka, A. A.; Smirnov, P.V.; Belousov, P.; Durmishi, B.; Abbdesettar, L.; Aggrey, P.; Kabra-Malpani, S.; Idrizi, H. (2022). Diatomaceous earth: A literature review. *Journal of Natural Sciences and Mathematics of UT*, Vol. 7, 256–268.

15. Kiari, M., Berenguer, R., Montilla, F., Morallón, E. (2020). Preparation and characterization of montmorillonite/pedot-pss and diatomite/pedot-pss hybrid materials. Study of electrochemical properties in acid medium. *Journal of Composites Science*, Vol. 4, 51.
16. Luo, Z., He, Y., Zhi, D., Luo, L., Sun, Y., Khan, E., Wang, L., Peng, Y., Zhou, Y., Tsang, D.C.W. (2019). Current progress in treatment techniques of triclosan from wastewater: A review. *Science of the Total Environment*, Vol. 696, 133990.
17. Zhai, M.; Fu, B.; Zhai, Y.; Wang, W.; Maroney, A.; Keller, A.A.; Wang, H.; Chovelon, J.M. (2023). Simultaneous removal of pharmaceuticals and heavy metals from aqueous phase via adsorptive strategy: A critical review. *Water Research*, Vol. 236, 119924.
18. Murphy, O.P.; Vashishtha, M.; Palanisamy, Kumar, K.V.A Review on the adsorption isotherms and design calculations for the optimization of adsorbent mass and contact time. (2023). *ACS Omega*, Vol. 8, pp. 17407–17430.
19. Igwegbe, C.A.; Aniagor, C.O.; Oba, S.N.; Ighalo, J.O. (2021). Environmental protection by the adsorptive elimination of acetaminophen from water: A comprehensive review. *Journal of Industrial and Engineering Chemistry*, Vol. 104, 117–135.

7 Engineered Hydrochar Materials for Water Pollutants Removal

Kannan Nadarajah, Thusalini Asharp, and Loveciya Sunthar

7.1 INTRODUCTION

Water is a limited resource considered vital for all living organisms. Emerging sources of pollutants including heavy metals, dyes, emerging contaminants, nanoparticles, and microplastics threaten water quality significantly [1]. This indicated level of risk is answered with many facts: increasing pollution, intensive industrial operations, and introduction of new technologies. Therefore, suitable water use is considered significant for ecosystem dynamics. Among newly introduced water treatment technologies, pollutant removal from water using hydrochar is considered new [2]. Hydrochar is a partially carbonized material derived from biomass via hydrothermal carbonization (HTC) [2]. In the HTC process, biomass is enriched with surface functional groups due to the presence of hot compensated water. Hydrochar has been used for the removal of several contaminants from water [3].

However, there are some problems associated with the use of hydrochar for the removal of pollutants from water: low surface area, low porosity, and uneven pore distribution. Hence, there is a need to enhance the functional properties of hydrochar through various processes: activation with acids and bases, steam activation, microwave treatment, or radiation. The end products of these processes known as engineered hydrochar (EHC) are set with enhanced functional properties: surface functional groups (development and distribution), pore size and distribution, and surface area, which lead to the effective removal of pollutants. However, it is important to note that EHC has been systematically used for emerging contaminants (e.g., personal care products, pesticides, fertilizers, detergents, and pharmaceuticals), heavy metals (e.g., arsenic, lead, cadmium, and mercury) and organic dyes (e.g., methyl orange, methyl blue, Congo red and malachite green) [4,5]. However, the adsorption science of EHC remains to be understood deeply for various environmental applications. This chapter is, therefore written to facilitate readers for a better understanding of EHC and its usage for removing pollutants from water.

DOI: 10.1201/9781003441007-10

7.2   WATER POLLUTION

Water is ubiquitous and vital to life. Quality degradation of water resources is one of the main problems in the world. Several anthropogenic activities are reported in terms of rapid deterioration of water quality: agricultural activities, the addition of untreated or partially treated industrial wastewater, improper waste disposal practices, and waste sewage management. In addition to this atmospheric deposition of waste materials and heavy metals, stormwater runoff, and climate change also have major effects on water pollution. According to a report by Sri Lanka's Central Environmental Authority (CEA), water pollution in the country is a significant concern. Only 17% of sewage, generated from urban areas, is treated before reaching the waterbodies [6].

Contamination of drinking water sources is another significant concern because of its high level of impact on public health. Polluted drinking water may contain pollutants and synthetic chemicals. Around 12% of rural households in Sri Lanka had access to unsafe drinking water in the year 2020. It threatens the human health, economic development, and social prosperity of a community. Necessary actions should be taken to address this issue and ensure pollution control and water resources conservation. There are a number of water-related health issues recorded in Sri Lanka during the past, including widespread chronic kidney disease (CKD) in dry rural regions [7].

According to a CKD epidemiology review, over 800 million individuals (over 10% population worldwide) are affected by CKD [8]. In the study, a positive relationship between arsenic content and total dissolved solids (TDS) in drinking water with CKD occurrence was identified, through a reported study on the impact of water quality on CKD of unknown etiology. Another study on water pollution in the world found that 485,000 diarrhea-related deaths are caused by contaminated drinking water. Since safe water is required for drinking, immense attention must be given to groundwater (and other sources) quality. Figure 7.1 graphically explains the sources of water pollution.

Taiwan Environmental Protection Administration reported that 16% out of 21 rivers in Taiwan are severely polluted, while 22% are moderately polluted [9]. Approximately, 780 million individuals lack access to safe and clean water worldwide [10]. Polluted water causes of prevalence of diseases with sewage contamination generating most water pollution and, notable extended impacts not only to rivers but also to groundwater [11]. Water pollution impacts human health, social value, urban and domestic aspects, and productivity. Figure 7.2 displays a comprehensive summary of the impacts of water pollution.

7.3   OVERVIEW OF REMOVAL METHODS
FOR WATER POLLUTANTS

As environmental pollution has been a major concern in recent years, it is essential to discuss pollutant removal methods. Different methods and techniques are applicable including physical, chemical, and biological methods, depending on the nature and extent of waste, the effectiveness and efficiency of removal performance, and its impact on the environment [7]. Among all the removal techniques, adsorption using biosorbents (e.g., biochar, hydrochar, EHC, activated carbon) is a low-cost

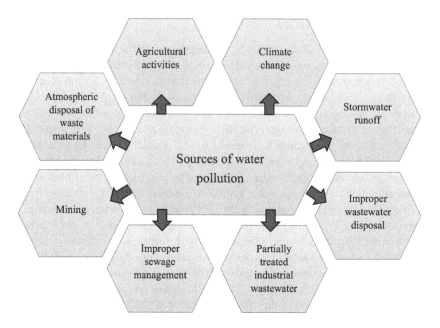

FIGURE 7.1   Water pollution sources.

and eco-friendly water treatment method [12]. Following a summary of methods commonly used for water pollutants removal is included.

7.3.1 PHYSICAL AND CHEMICAL METHODS

Physical methods are commonly used as initial water treatment processes. Commonly, it involves the removal of large particles [7] using gross processes such as netting and sedimentation, but other finer options also belong to this group such as membrane filtration (micro-filtration, ultra-filtration, nano-filtration, and reverse osmosis) and distillation, which are some notable and unique examples of physical methods for water pollutant removal [7].

Treatment procedures involving the addition of chemical substances and reactions are collectively known as chemical treatment processes. Coagulation, oxidation, photodegradation, solvent extraction, and ion exchange are some examples of chemical treatments. Oxidizers, reducers, acids, and bases are commonly utilized to transform hazardous waste into less harmful substances during chemical treatment. For example, chromate reduction (Cr^{6+} to Cr^{3+}); cyanide oxidation to cyanate at basic pH; or hydroxides and sulfates precipitation in an aqueous solution.

7.3.2 ADSORPTION

Adsorption is a low-cost treatment method that uses the functional properties of adsorbents for the removal of different pollutants. Furthermore, physical, chemical,

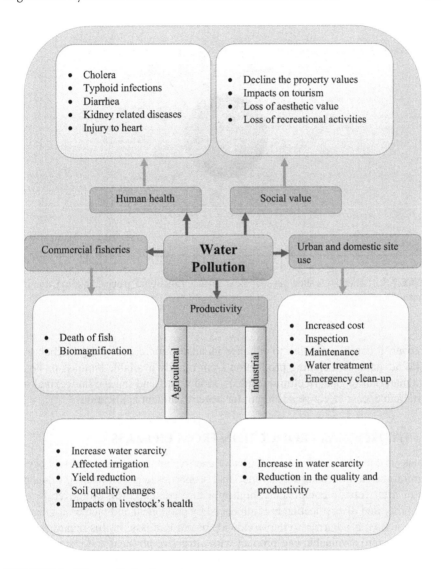

FIGURE 7.2   Water pollution impacts.

physiochemical, or other possible methods can enhance adsorption effectiveness in pollutant removal. Adsorbents are abundant and sustainable resources in nature [13]. Adsorption has been identified as superior to other techniques in terms of simple design, convenient operation, and insensitivity to toxic substances [14]. Structural components and surface functional groups play an important role in the adsorption process. The possible bonds and interactions involved in the adsorption process and a few surface functional groups are shown in Figure 7.3.

Hydrochar has been studied for its potential to remove organic pollutants, heavy metals, pharmaceuticals, and emerging contaminants from water and wastewater [15].

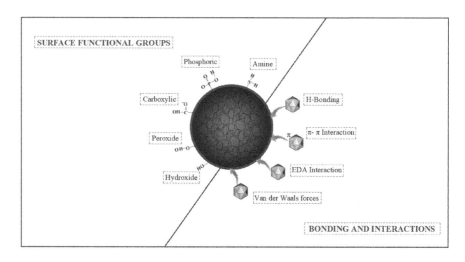

FIGURE 7.3   Schematics showing adsorbent surface functional groups, bonding, and interactions with pollutants.

Moreover, it can be modified to enhance its adsorption capacity. Hydrochar shows several advantages including production from readily available biomass or organic waste materials (e.g., agricultural residues, food waste, and sewage sludge) making it a sustainable, cost-effective adsorbent for water treatment applications.

7.4   HYDROCHAR PRODUCTION FROM BIOMASS

Biomass is considered a common source of energy, but it is an not ideal fuel because of limitations in its structural and functional properties (e.g., low heating value, presence of high volatile components, high moisture content) [16]. Moreover, biomass availability and diversification are influenced by geological and meteorological conditions, acting as a barrier to the worldwide use of biomass. In this regard, biomass conversion into a valuable end-product with attractive physical and chemical properties is attracting significant attention. Different physical, biological, and thermochemical pre-treatment processes have been tested to enhance biomass properties including pyrolysis, HTC, gasification, densification, palletization, anaerobic digestion, fermentation, and torrefaction [17].

In recent decades, an upsurge in interest has been given to thermochemical treatments over other methods. In this regard, HTC plays a major role in good-quality end-product production (hydrochar). In HTC, thermochemical conversion of feedstock with high moisture content takes place at low temperature (180°C–250°C) and high pressure [13] under closed conditions. The main parameters determining the yield and quality of hydrochar are temperature, pressure, holding time, amount of water used, and biomass nature. Among these, the effect of reaction temperature and holding time is highly significant [17]. One of the earlier works reported that the yield of hydrochar varies from 40% to 70% [13]. Figure 7.4 shows

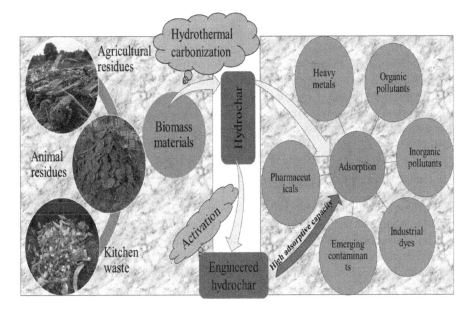

FIGURE 7.4   Hydrochar and EHC preparation and applications for water pollutant removal.

TABLE 7.1
Yield Performance of Various Biomass Materials under HTC

Biomass Source	HTC Temperature	Holding Time	Yield (%)	Reference
Hickory wood, bagasse, and bamboo	200°C	5 h	27.8–48.4	[18]
Corn stalk and *Tamarix ramosissima*	250°C	4 h	35.5–38.1	[19]
Corn stover	180°C–260°C	4 h	34.9–16.7	[20]
Eucalyptus bark	220°C–300°C	2 h	46.4–40.0	[21]
Sewage sludge and rape straw	220°C–260°C	1 h	56.1–47.2	[22]
Pineapple and watermelon peels	250°C 180°C	2 h 1.5 h	44 57	[23]
Oil palm empty fruit bunch	150°C–190°C	1–4 h	85.4–81.9	[24]
Corn silage	220°C	20 min	52.5	[25]
Coconut fiber	250°C	20 min	45.3	[26]
Oil palm frond fiber	210°C–250°C	20 min	67.8–52.7	[24]

a summary of hydrochar preparation and applications in water treatment. Ongoing research is focused on production process optimization, understanding adsorption mechanisms, and exploring potential environmental applications. Table 7.1 summarizes the yield performance of various biomass materials under HTC, including the effect of biomass type, water/biomass ratio, reaction temperature, and holding time used. As shown, an increase in temperature generated a biomass yield reduction, while holding time also showed a significant impact, from approximately 30% to 80%, approximately, helping to better understand hydrochar science. However, the research related to the combined effect of biomass type, temperature, and holding time on the yield profile of hydrochar is still at the developmental stage. Therefore, it is important to expand this research avenue to develop optimized processing conditions for different biomass materials to get reasonable hydrochar yield from the HTC process.

7.5   ENGINEERED HYDROCHAR

Hydrochar production is mainly influenced by several factors: temperature, residence time, feedwater pH, and heating rate. The adsorption properties of hydrochar mainly depend on surface functional groups, porous nature, and specific surface area. Unfortunately, hydrochar poses lower porosity and surface area due to hydrocarbon formation clogging pores and reducing the adsorption process [27]. Therefore, improving surface area and porosity to improve adsorption capacity is essential. Hydrochar can be modified via biological, physical, chemical, and chemical-physical combination approaches resulting in EHC with improved removal efficiency for contaminants. Table 7.2 summarizes different modification techniques, modification processes, and EHC improvements. A significant need for in-depth investigation focused on EHC remains to enhance environmental remediation including information on feedstock selection, preparation method, activator selection, activation temperature, and soaking time [28] which play a major role in changing EHC performance. This section elaborates on available modification techniques to produce EHC.

7.5.1   CHEMICAL MODIFICATION

Hydrochar chemical activation is typically performed using chemical reagents, followed by heat treatment (500°C–850°C) in an inert environment for 1–24 h, and finally rinsing with deionized or distilled water to remove chemical excess. It usually increases specific surface area and/or structural modifications, or composition changes due to the loss of oxygen-containing functional groups [29], significantly influenced by the nature of the chemical agent, activation temperature, and impregnation ratio. Several chemical reagents, including acids, bases, salts, and oxidizing agents, have been successfully used for the modification process [30]. In the case of acids, both strong (e.g., HCl, HNO_3) and weak acids (H_3PO_4) have been used to enhance surface functional properties, pore volume, and pore size. For instance, poplar sawdust hydrochar was modified with nitric acid to significantly increase specific surface area and pore volume (5.6 and 0.8 folds, respectively) compared to raw hydrochar [31]. Phosphoric acid poses several advantages over other chemical reagents because of its lower corrosivity, eco-friendliness, and less toxic by-products [32]. Furthermore, H_3PO_4 can break down lignocellulosic, aliphatic, and aromatic materials while producing phosphate and polyphosphate cross bridges to prevent shrinkage or contraction during pore formation [32]. It facilitates biomass dehydration and lowers the temperature necessary for biomass breakdown during HTC, improving

TABLE 7.2

Techniques for Hydrochar Modification

Feed Stock	Modification Method	Conditions	Improvement				References
			Pore Volume (cm³/g)		BET Surface Area (m²/g)		
			Before	After	Before	After	
Grape seed	Chemical	H_3PO_4 (1:2 (w/w), 500°C, 2 h)	–	0.17–0.20	8	590–654	[41]
		KOH (1:4 (w/w), 750°C, 1 h)	–	0.57–0.74		1,215–2,194	
		$FeCl_3$ (1:3 (w/w), 750°C, 1 h)	–	0.12–0.17		312–417	
Teak sawdust	Chemical	$ZnCl_2$ (1:1.75 (w/w), 800°C, 4 h)	0.015	1.02	7.33	1.757	[42]
Olive stone	Chemical	KOH (1:3 (w/w), 750°C, 1 h)	–	0.96	18	2,122	[29]
		$FeCl_3$ (3:1(w/w), 750°C, 1 h)	–	0.18		383	
		H_3PO_4 (1:3 (w/w), 500°C, 2 h)	–	0.5		1,155	
Rice straw	Chemical	$FeCl_3$ (1:3 (w/v), 200°C, 2 h)	–	–	39.9	44.3	[2]
Corn cob straw	Chemical	HCl-Polyethyleneimine (1:10 (w/v), 30°C, 1 h)	0.0029	0.0046	2.09	2.10	[43]
		KOH-Polyethyleneimine (1:10 (w/v), 30°C, 1 h)		0.0046		3.9	
Sugar cane bagasse	Chemical	NaOH (1:20 (w/v), 30°C, 2 h)	0.06	0.084	7.85	15.	[44]
Industry waste	Chemical	KOH (1:15 (w/v), 20°C, 24 h)	–	0.28	–	54.6	[45]
Paper waste sludge	Chemical modification	$FeCl_3$ (1:3.3 (w/v), 80°C, 4)	0.009	0.029	2.23	4.16	[46]
Wheat straw	Chemical	$FeCl_3$ (1:60 (w/v), 200°C, 6 h)	0.0021	0.3337	1.2	52.74	[47]

(*Continued*)

TABLE 7.2 (Continued)
Techniques for Hydrochar Modification

Feed Stock	Modification Method	Conditions	Improvement				References
			Pore Volume (cm^3/g)		BET Surface Area (m^2/g)		
			Before	After	Before	After	
Pinecone	Chemical	$KHCO_3$ and K_2FeO_4 (5:5:2 (w/w), 500°C, 2 h)	–	–	6.05	703.9	[48]
Sawdust	Chemical	H_2O_2 (1:60 (w/v), 25°C, 2 h)	–	–	9.76	10.8	[49]
Date pits	Physical	CO_2 (750°C, 3 h)	0.028	0.270	20.8	565.5	[50]
Olive stones			0.024	0.325	18.8	606.0	
Poplar sawdust	Physical modification	N_2- and air mixture (700°C, 3 h)	–	–	7.5	358.6	[51]
Almond shell	Physical modification	Synthetic air mixture (220°C, 3 h)	0.012	0.456	4	313	[52]
Olive stones	Chemical–physical combination	H_3PO_4 (1:2 (w/w), RT, 8 h) O_2 (500°C, 1 h)	0.7	0.35	1225	695	[53]

RT, room temperature.

carbon retention and lowering fibrocyte decomposition temperature. Many research activities focused on the use of weak acids as catalysts to change the functional properties of hydrochar for different environmental applications. However, the effect of impregnation of biomass with slightly strong acids for different durations under controlled conditions is still to be studied for the structured production of EHC. Therefore, comprehensive research activities into this concept are much needed to expand the science of EHC for productive pollution control applications.

Sodium (NaOH) and potassium hydroxide (KOH) are the major alkaline reagents used in the activation process. KOH activation produces carbon material with a larger surface area by promoting KOH breakdown to produce K_2O and subsequent reduction to metallic K [33]. This process increases the overall surface area by widening the pores between carbonaceous layers. According to past studies, KOH plays a major role in micropore development that ultimately enhances surface area and pore volume on EHC [34]. NaOH activation promotes mesopore development through NaOH breakdown to produce Na_2CO_3, which further degrades into Na, CO, and CO_2 at high temperatures. After production, Na is entrapped in the pores and facilitates further pore development [35]. The information on optimum conditions regarding the ratio, concentration of alkaline components, duration of impregnation, and impregnation temperature is still lacking, and it requires further research to explore more scientifically the production and characterization of hydrochar materials from different types of biomasses. Furthermore, the combination of different types of ethers, conjugate bases, and N-dopants with alkaline materials needs to be investigated further to improve the adsorption capacity of EHC for different water pollutants.

Metallic salts (e.g., ferric chloride, and zinc chloride) have been extensively used as chemical reagents to improve the surface characteristics of hydrochar. The use of salt facilitates micropores, mesopores, and surface area development, which untimely improves adsorption capacity for environmental remediation. The hydrochar properties can be further improved by pre-treatment using chemical reagents prior to HTC, acid washing after impregnation, and activation at elevated temperatures after impregnation. This helps eliminate unwanted inorganic fractions (e.g., salts and minerals, tannins, and volatile components) in carbonaceous compounds to improve pore volume and surface area as well as to increase active sites available for adsorption. Although special consideration needs to be taken with regard to temperature, activation time, and acidic material concentration since high activation temperatures and prolonged activation times will collapse developed pores and might convert micropores to mesopores [36]. The lack of information about the best experimental conditions for EHC production is an avenue of future research worth exploring to generate the best material possible without exceeding the energy requirements that may also produce low-performing products.

7.5.2   N-DOPED HYDROCHAR

The production of N-doped hydrochar has recently attracted attention. The performance of hydrochar may be improved by N-doping through a practical technique that adds N-heteroatoms and amino groups, increasing the binding capacity by creating stable complexes or hydrogen bonds with pollutants [37]. Most of the N-contained compounds can be used for N-doping. In addition, N-doping into the carbonaceous framework will increase hydrophilicity and surface-active sites of hydrochar. Urea $(CO(NH_2)_2)$, melamine $(C_3H_6N_6)$, polyethyleneimine (PEI) $(C_2H_5N)_n$, and NH_4Cl are some N-dopants used. Among those, PEI is commonly used because it consists of an amino spatial branch chain, which supplies abundant amino groups for binding pollutants successfully [37]. Many studies have shown that N-dopants are advantageous in improving the surface quality of carbonaceous materials by producing numerous C-N structures forming novel functional groups in hydrochar [38]. The custom design of EHC using N-dopants for specific pollutants is still not studied well. Hence, it is essential to make efforts to develop different structured protocols of EHC with N-dopants for different water pollutants. It will help to enhance the commercial application of the EHC with N-dopants for water treatment.

7.5.3 PHYSICAL MODIFICATION

Physical activation methods are a less used technique compared with chemical activation. When compared to chemical processes, physical activation generates well-developed micropores through inexpensive processes avoids the use of chemical reagents, washes after activation and does not develop secondary pollution [39]. However, activation efficiency is greatly influenced by activation temperature, time, and gas flow. CO_2, N_2 mix, air, and water steam are commonly used activation agents. The thermal air oxidization (AO) process produces micropore development and/or micropore enlargement to mesopores [40] by unclogging pores by partial pore wall decomposition and trapped tars volatilization at high temperatures. Physical activation is really an energy-consuming process. It requires energy during the production of steam and thermal activation at higher temperatures. Therefore, research into energy-efficient activation techniques and usage of renewable energy sources need to be studied properly to make it effective. Moreover, the functional group mapping of the modified hydrochar under different treatment conditions is also to be investigated further for the effective production of EHC for water treatment applications.

7.6   EHC SURFACE FUNCTIONAL GROUPS

Studying surface functional groups is essential to predicting EHC chemical characteristics [54]. Changes in surface functional groups during the activation process are typically analyzed and evaluated using Fourier Transform Infrared (FTIR). In FTIR, different functional groups are assigned for specific band ranges. For instance, the band range of common functional groups in unmodified hydrochar such as hydroxyl (^-OH), carbonyl (C=O), and carboxyl (–COOH) are assigned to 3,600–3,200 cm^{-1}, 1,740–1,650 cm^{-1}, and 1,590–1520 cm^{-1}, respectively [54]. Figure 7.5 describes typical functional group changes with activation. Unlike in unmodified hydrochar, EHC usually shows improved oxygen-containing functional groups and novel functional groups that were not present in unmodified hydrochar. The addition of new functional groups differs according to the nature of the chemical regent used. For instance, usage of N-dopants usually generates an imine group (C=N) on the hydrochar's carbonaceous skeleton, and a new Fe-O (600 cm^{-1}) band is formed related to

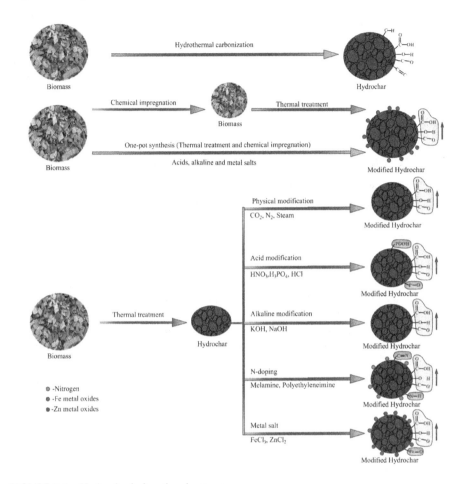

FIGURE 7.5 Hydrochar's functional groups.

$FeCl_3$ addition [47]. FTIR analysis can also be used to determine the aromatic degree of carbonaceous components using bands on 1610 cm^{-1} and 833 cm^{-1} corresponding to aromatic C=C and C-H stretching vibration [55].

FTIR analysis suggested that the presence of $ZnCl_2$ during HTC of the biomass produced EHC with increased O-containing functional groups (strengthened peaks assigned to C-O) and aromaticity [56]. A study on N-doped hydrochar production reported that amino groups in melamine reacted with intermediates (e.g., aldehydes) to produce lignocellulosic materials reducing the C-O functional group of EHC [57].

Chemical reagent concentration has been found to influence existing functional groups. For instance, C-C bonds abundance decreased when nitric acid concentration was increased while oxygen-containing functional groups abundance increased with C-O increasing (33.3%–33.5%) and C=O and –COOH abundance increased by 14.1% and 35.8%, respectively [58] which are known for their efficient removal of pollutants from water. Hence, surface functional groups play a vital role in pollutant removal from water by EHC. However, the scientific knowledge about the involvement of different surface functional groups of the EHC in the removal of different organic and inorganic pollutants from water is still at the embryonic stage. Therefore, research works are yet needed to fix the above-said problem. Moreover, appropriate quantification techniques need to be used for the quantification of surface functional groups present on the surface of the EHC used for the removal of pollutants from water.

7.7 RECENT EHC RESEARCH FOR POLLUTANT REMOVAL

EHC has a wide range of applications for water treatment because of its cost-effectiveness and eco-friendliness. Varying concentrations of heavy metals, emerging contaminants, and industrial dyes have been successfully removed using EHC [36]. In general, compared with EHC, unmodified hydrochar shows lower adsorption capacity because of its lower functional group density, reduced specific surface area, and lower pore volume. Figure 7.6 shows the major properties of EHC and the associated mechanism for pollutant removal. As shown, surface area and pore volume make adsorption sites and paths for pollutant removal, which can be improved by engineered modification.

Promising regeneration and reusability potential for EHC highlights its sustainability and cost-effectiveness as an effective adsorbent complying with the circular economy concept. Restoring the adsorption capacity of spent hydrochar has been explored unsuccessfully using different techniques (e.g., thermal treatment, solvent extraction, and electrochemical methods). Using EHC leads to eliminating limitations

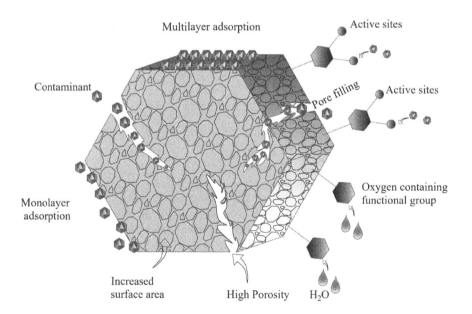

FIGURE 7.6 Engineered hydrochar and its characteristics.

of individual water treatment methods. Successful pilot-scale demonstrations of EHC indicate the potential for large-scale water treatment plants and address challenges posed by different contaminants. The advancement of sustainability and efficiency of water treatment technologies can be enhanced through continued research and development in this field.

Table 7.3 summarizes contaminant removal, adsorption capacity, adsorption mechanism, and modification type for different pollutants using EHC. Further research is needed to better understand more realistic situations to develop appropriate activation methods capable of effectively adsorbing specific pollutants.

7.7.1   HEAVY METALS

Chemical-modified EHC has been found to increase adsorptive performance for a variety of heavy metals including copper (Cu), cadmium (Cd), lead (Pb), chromium (Cr), arsenic (As), and mercury (Hg). Pb adsorption capacity using H_3PO_4-activated EHC increased by three-fold suggesting that Pb adsorption by H_3PO_4 activation was enhanced by increased BET surface area and oxygen-containing functional groups, as suggested by a reduction of C=O corresponding bands in FTIR by 8% after adsorption [55]. As shown in Table 7.3, EHC using either $KHCO_3$ or K_2FeO_4 was found effective in adsorbing Cd (adsorption capacity 128 mg/g) higher than commercial powdered activated carbon [59]. The high efficacy was associated with an increased number of micropores present in EHC and the formation of Fe–O groups after activation [60]. Major mechanisms involved in heavy metals adsorption are described in Table 7.3 as surface complexation, cationic–π interaction, mass diffusion, co-precipitation, and ion exchange enhanced by increased surface area and oxygen-containing functional groups. However, the involvement of specific functional groups: hydrophilic or hydrophobic, in the different adsorptive mechanisms of heavy metal removal by EHC should thoroughly be studied scientifically, since there is very limited information, related to this, available in the literature.

7.7.2   DYES

Bamboo powder hydrochar was activated using maleic anhydride and used to adsorb methylene blue showing improved adsorption performance at higher concentrations resulting in increased hydroxyl and carboxyl group density, which improved EHC's ability to adsorb the dye through π–π interaction, electrostatic interaction, and hydrogen bonds [61]. The activation process boosts the addition and improvement of functional groups (e.g., hydroxyl, carboxyl, and amine) on EHC serving as a hydrogen-bonding donor with dyes [36]. In addition, solution pH affects the degree of ionization, dye specification, and EHC surface charge impacting electrostatic interaction. As pH rises, zeta potential lowers and the surface functional groups in carbonaceous materials become negatively charged owing to deprotonation, resulting in electrostatic attraction, and increasing adsorption [62]. Therefore, to improve EHC adsorption performance, solution pH needs to be regulated based on the nature of the used dye.

TABLE 7.3
Contaminant Removal Using Engineered Hydrochar and Adsorption Mechanism

Feedstock	Activator	Contaminant Removed	Dosage (g/L)	Adsorption Capacity (Q_e mg/g)	Sorption Process	Mechanism	Reference
Olive stone	$H_3PO_4 + O_2$	Diclofenac sodium	0.2	275	Freundlich	$\pi-\pi$ interaction Hydrogen bonding	[53]
Teak sawdust	$ZnCl_2$	Tetracycline hydrochloride	2	258	Langmuir	Van der Waals force Pore filling $\pi-\pi$ interaction Hydrogen bonding Electrostatic attraction	[42]
Brewer's spent grain	KOH	Acetaminophen	1	318	Langmuir	$\pi-\pi$ interaction Hydrogen bonding	[64]
Sugar cane bagasse	NaOH	Sulfamethoxazole	0.1	400	Freundlich	$\pi-\pi$ interaction Hydrogen bonding	[65]
Paper waste sludge	$FeCl_3$	NH_4^+	2	24.10	Langmuir	Cation exchange Electrostatic attraction Surface complexation	[46]
	Polyethyleneimine	Pb^{2+}	0.25	214	Langmuir	Surface complexation Ion exchange Hydrogen bond	[55]
Palm leaves (*Phoenix dactylifera*)	H_2O_2	Pb^{2+}	1	107.49	Langmuir	Surface complexation Electrostatic interaction	[66]

(Continued)

TABLE 7.3 (Continued)
Contaminant Removal Using Engineered Hydrochar and Adsorption Mechanism

Feedstock	Activator	Contaminant Removed	Dosage (g/L)	Adsorption Capacity (Q_e mg/g)	Sorption Process	Mechanism	Reference
Pinecone	$KHCO_3$ and K_2FeO_4	Cr^{6+}	0.5	128.15	Langmuir and Sips	π-π stacking, Complexation	[60]
Banana peel	H_3PO_4	Pb^{2+}	1	275.11	Redlich-Peterson	Co-precipitation Surface complexation	[67]
		Cd^{2+}	1	94.52		Co-precipitation Surface complexation	
Rice straw	$FeCl_3$	Pb^{2+}	8	6.75	Langmuir	Surface complexation, Cationic-π interaction Mass diffusion	[2]
		Cu^{2+}	8	4			
Pharmaceutical industry waste	KOH	Methylene blue	2.5	112	Langmuir	Van der Waals force Pore filling π-π interaction Hydrogen bonding Electrostatic attraction	[45]
Wheat straw	$FeCl_3$	Rhodamin B	1	80	Langmuir	–	[47]
Bamboo sawdust	HCl and NaOH	Methylene blue	0.8	655.76	Langmuir	Electrostatic attraction Ion exchange	[63]
Sugarcane bagasse	NaOH	Methylene blue	0.8	357.14	Langmuir	Electrostatic attraction, Hydrogen bond π-π interaction	[44]

As shown in Table 7.3, significant surface area (e.g., from 7.9 to 31.6 m²/g) and pore volume (e.g., from 0.031 to 0.110 cm³/g) enlargement was observed in the presence of an acidic medium for EHC. HTC in an acidic medium followed by alkaline treatment produced EHC effective for methylene blue adsorption in wastewater [63]. Therefore, novel research is needed to improve adsorption capacity testing other regents (acid, alkaline, salts, and N-dopants) and different treatments (physical and chemical treatment) to increase the adsorption capacity.

7.7.3 EMERGING CONTAMINANTS (ECs)

ECs are components generated from anthropogenic activity that cause environmental harm even at low concentrations. Personal care products, food additives, pesticides, pharmaceuticals, surfactants, veterinary products, and steroids are some examples of those contaminants. EHC has been used to reduce the impact of ECs and current research is now focused on developing strategies to understand the adsorptive mechanisms. Physical and chemical interactions between EHC and ECs have been found to contribute to the removal process as well as the presence of oxygen-containing functional groups [4]. For example, studies have reported significant removal percentages for pharmaceuticals (e.g., ibuprofen, diclofenac, and triclosan) with high adsorption capacity values (64, 43, and 98 mg/g, respectively) using EHC derived from olive oil waste [4]. Adsorption capacity as high as 258 mg/g was found for tetracycline using $ZnCl_2$–activated hydrochar. Authors suggested that π–π interaction, hydrogen bonding, and electrostatic attraction with EHC were the main mechanisms involved [42]. More detailed information on EHC applications for EC adsorption in water is shown in Table 7.3. Moreover, based on the information presented in Table 7.3, the interaction between the ECs and the EHC is still to be studied further to locate the most dominant mechanism of EHC for the removal of specific groups of ECs from water. As the active structure of ECs is highly sensitive to the nature of the medium in which adsorption, by EHC, takes place, a wide range of all parameters (pH, temperature, free H^+ concentration, turbidity, etc.) influencing the adsorptive process should be checked to optimize the adsorptive removal.

7.8 CONCLUSION

Water pollution by various sources is a pressing issue that needs to be fixed for the sustainable use of safe water for all living things on this planet. The new sources of contaminants are threatening the water sources significantly. However, the removal methods for emerging sources are highly expensive for practical usage. Therefore, low-cost adsorbents, at this point, draw significant attention. EHC is one of the active, low-cost adsorbents for the removal of various pollutants from wastewater: ECs, organic dyes, and heavy metals. Anyhow, the functional properties of hydrochar must systematically be customized for the effective removal of pollutants by various methods. This chapter, therefore, tends to provide the cream of EHC usage for water treatment by addressing key points with the help of published research literature. This will help researchers understand the mechanism of EHC for the removal of various pollutants from water in an eco-friendly manner.

REFERENCES

1. Zamora-Ledezma, C. et al. Heavy metal water pollution: A fresh look about hazards, novel and conventional remediation methods. *Environ. Technol. Innov.* **22**, 101504 (2021).
2. Nadarajah, K., Bandala, E. R., Zhang, Z., Mundree, S. & Goonetilleke, A. Removal of heavy metals from water using engineered hydrochar: Kinetics and mechanistic approach. *J. Water Process Eng.* **40**, 101929 (2021).
3. Wang, T., Zhai, Y., Zhu, Y., Li, C. & Zeng, G. A review of the hydrothermal carbonization of biomass waste for hydrochar formation: Process conditions, fundamentals, and physicochemical properties. *Renew. Sustain. Energy Rev.* **90**, 223–247 (2018).
4. Delgado-Moreno, L. et al. New insights into the efficient removal of emerging contaminants by biochars and hydrochars derived from olive oil wastes. *Sci. Total Environ.* **752**, 141838 (2021).
5. Qian, W. C., Luo, X. P., Wang, X., Guo, M. & Li, B. Removal of methylene blue from aqueous solution by modified bamboo hydrochar. *Ecotoxicol. Environ. Saf.* **157**, 300–306 (2018).
6. Kumar, S., Meena, H. M. & Verma, K. Water pollution in India: Its impact on the human health: Causes and remedies. *Int. J. Appl. Environ. Sci.* **12**, 275–279 (2017).
7. Kordbacheh, F. & Heidari, G. Water pollutants and approaches for their removal. *Mater. Chem. Horizons.* Damghan University. **2**, 139–153. (2023). doi: 10.22128/MCH.2023.684.1039.
8. Gobalarajah, K. et al. Impact of water quality on chronic kidney disease of unknown etiology (CKDu) in Thunukkai Division in Mullaitivu District, Sri Lanka. *BMC Nephrol.* **21**, 1–11 (2020).
9. Savira, M. et al. Long-term river water quality trends and pollution source apportionment in Taiwan. *Water.* 1–17 (2018). doi:10.3390/w10101394.
10. World Wildlife Fund (WWF). Water for our future: America's regional process event. *World Wildl. Fund* (2014). https://files.worldwildlife.org/wwfcmsprod/files/Publication/file/4nrdaknq9n_10_20_World_Water_Forum_WP_FIN_15_122.pdf?_ga=2.239488133.700213726.1713824653-820966953.1713824653
11. Lin, L., Yang, H. & Xu, X. Effects of water pollution on human health and disease heterogeneity: A review. *Front. Environ. Sci.* **10**, 880246. (2022).
12. Kumar, N. S., Shaikh, H. M., Asif, M. & Al-Ghurabi, E. H. Engineered biochar from wood apple shell waste for high-efficient removal of toxic phenolic compounds in wastewater. *Sci. Rep.* **11**, 1–17 (2021).
13. Zhang, Z., Zhu, Z., Shen, B. & Liu, L. Insights into biochar and hydrochar production and applications: A review. *Energy* **171**, 581–598 (2019). doi:10.1016/j.energy.2019.01.035.
14. Krishna, B. G., Tiwari, S., Ghosh, D. S. & Rao, M. J. Environmental applications of activated carbon. In Verma C., Quraishi M.A. (eds), *Activated Carbon: Progress and Applications* 92–133 (The Royal Society of Chemistry, 2023).
15. Fang, J., Zhan, L., Ok, Y. S. & Gao, B. Mini review of potential applications of hydrochar derived from hydrothermal carbonization of biomass. *J. Ind. Eng. Chem.* **57**, 15–21 (2018).
16. Kambo, H. S. & Dutta, A. A comparative review of biochar and hydrochar in terms of production, physico-chemical properties and applications. *Renew. Sustain. Energy Rev.* **45**, 359–378 (2015).
17. Sharma, H. B., Sarmah, A. K. & Dubey, B. Hydrothermal carbonization of renewable waste biomass for solid biofuel production: A discussion on process mechanism, the influence of process parameters, environmental performance and fuel properties of hydrochar. *Renew. Sustain. Energy Rev.* **123**, 109761 (2020).
18. Sun, Y. et al. Effects of feedstock type, production method, and pyrolysis temperature on biochar and hydrochar properties. *Chem. Eng. J.* **240**, 574–578 (2014).

19. Xiao, L., Shi, Z., Xu, F. & Sun, R. Bioresource technology hydrothermal carbonization of lignocellulosic biomass. *Bioresour. Technol.* **118**, 619–623 (2012).
20. Zhang, Y., Jiang, Q., Xie, W., Wang, Y. & Kang, J. Effects of temperature, time and acidity of hydrothermal carbonization on the hydrochar properties and nitrogen recovery from corn stover. *Biomass Bioenergy* **122**, 175–182 (2019).
21. Gao, P. et al. Preparation and characterization of hydrochar from waste eucalyptus bark by hydrothermal carbonization. *Energy* **97**, 238–245 (2016).
22. Zhai, Y., Peng, C., Xu, B., Wang, T. & Li, C. Hydrothermal carbonisation of sewage sludge for char production with different waste biomass: Effects of reaction temperature and energy recycling. *Energy* **127**, 167–174 (2017).
23. Azaare, L., Commeh, M. K., Smith, A. M. & Kemausuor, F. Co-hydrothermal carbonization of pineapple and watermelon peels: Effects of process parameters on hydrochar yield and energy content. *Bioresour. Technol. Rep.* **15**, 100720 (2021).
24. Zakaria, M. R., Norrrahim, M. N. F., Hirata, S. &Hassan, M. A. Hydrothermal and wet disk milling pretreatment for high conversion of biosugars from oil palm mesocarp fiber. *Bioresour. Technol.* **181**, 263–269 (2015).
25. Oliveira, I., Blöhse, D. & Ramke, H. G. Hydrothermal carbonization of agricultural residues. *Bioresour. Technol.* **142**, 138–146 (2013).
26. Liu, Z., Quek, A., Kent Hoekman, S., Srinivasan, M. P. & Balasubramanian, R. Thermogravimetric investigation of hydrochar-lignite co-combustion. *Bioresour. Technol.* **123**, 646–652 (2012).
27. Masoumi, S., Borugadda, V. B., Nanda, S. & Dalai, A. K. Hydrochar: A review on its production technologies and applications. *Catalysts* **11**, 939 (2021).
28. Akhil, D. et al. Production, characterization, activation and environmental applications of engineered biochar: A review. *Environ. Chem. Lett.* **19** (2021).
29. Diaz, E., Sanchis, I., Coronella, C. J. & Mohedano, A. F. Activated Carbons from Hydrothermal Carbonization and Chemical Activation of Olive Stones: Application in Sulfamethoxazole Adsorption. *Resources* **11**, 43 (2022).
30. Hokkanen, S., Bhatnagar, A. & Sillanpää, M. A review on modification methods to cellulose-based adsorbents to improve adsorption capacity. *Water Res.* **91**, 156–173 (2016).
31. He, H., Hua, Y., Chu, Q., Feng, Y. & Yang, L. Chemical aging of hydrochar improves the Cd^{2+} adsorption capacity from aqueous solution. *Environ. Pollut.* **287**, 117562 (2021).
32. Tsang, D. C. W. & Sun, Y. *Biochar Applications for Wastewater Treatment.* (John Wiley & Sons, 2023).
33. Xi, Y. et al. Renewable lignin-based carbon with a remarkable electrochemical performance from potassium compound activation. *Ind. Crops Prod.* **124**, 747–754 (2018).
34. Tu, W. et al. A novel activation-hydrochar via hydrothermal carbonization and KOH activation of sewage sludge and coconut shell for biomass wastes: Preparation, characterization and adsorption properties. *J. Colloid Interface Sci.* **593**, 390–407 (2021).
35. Islam, A., Ahmed, M. J., Khanday, W. A., Asif, M. & Hameed, B. H. Mesoporous activated carbon prepared from NaOH activation of rattan (*Lacosperma secundiflorum*) hydrochar for methylene blue removal. *Ecotoxicol. Environ. Saf.* **138**, 279–285 (2017).
36. Cavali, M. et al. A review on hydrothermal carbonization of potential biomass wastes, characterization and environmental applications of hydrochar, and biorefinery perspectives of the process. *Sci. Total Environ.* **857**, 159627 (2023).
37. Qu, J. et al. A novel PEI-grafted N-doping magnetic hydrochar for enhanced scavenging of BPA and Cr(VI) from aqueous phase. *Environ. Pollut.* **321**, 121142 (2023).
38. Yan, S. et al. One-pot synthesis of porous N-doped hydrochar for atrazine removal from aqueous phase: Co-activation and adsorption mechanisms. *Bioresour. Technol.* **364**, 128056 (2022).

39. Shen, R., Lu, J., Yao, Z., Zhao, L. & Wu, Y. The hydrochar activation and biocrude upgrading from hydrothermal treatment of lignocellulosic biomass. *Bioresour. Technol.* **342**, 125914 (2021).
40. Xiao, F. A review of biochar functionalized by thermal air oxidation: A review of biochar functionalized by thermal air oxidation. *Environ. Funct. Mater.* (2022). doi:10.1016/j.efmat.2022.03.001.
41. Diaz, E., Manzano, F. J., Villamil, J., Rodriguez, J. J. & Mohedano, A. F. Low-cost activated grape seed-derived hydrochar through hydrothermal carbonization and chemical activation for sulfamethoxazole adsorption. *Appl. Sci.* **9**, 5127 (2019).
42. Ngoc, D. M. et al. Tetracycline removal from water by adsorption on hydrochar and hydrochar-derived activated carbon: Performance mechanism, and cost calculation. *Sustainability* **15**, 4412 (2023).
43. He, X. et al. Science of the total environment enhanced adsorption of Cu(II) and Zn(II) from aqueous solution by polyethyleneimine modi fi ed straw hydrochar. *Sci. Total Environ.* **778**, 146116 (2021).
44. Zhou, F., Li, K. & Hang, F. Efficient removal of methylene blue by activated hydrochar prepared by hydrothermal carbonization and NaOH activation of sugarcane bagasse and phosphoric acid. *RSC Adv.* 1885–1896. (2022). doi:10.1039/d1ra08325b.
45. Liu, S. et al. Alkaline etched hydrochar-based magnetic adsorbents produced from pharmaceutical industry waste for organic dye removal. *Environ. Sci. Pollut. Res.* **30**, 65631–65645 (2023).
46. Nguyen, L. H. et al. Paper waste sludge-derived hydrochar modified by iron(III) chloride for enhancement of ammonium adsorption: An adsorption mechanism study. *Environ. Technol. Innov.* **21**, 101223 (2021).
47. Kohzadi, S., Marzban, N., Libra, J. A., Bundschuh, M. & Maleki, A. Removal of RhB from water by Fe-modified hydrochar and biochar – An experimental evaluation supported by genetic programming. *J. Mol. Liq.* **369**, 120971 (2023).
48. Qu, J. et al. Pinecone-derived magnetic porous hydrochar co-activated by $KHCO_3$ and K_2FeO_4 for Cr(VI) and anthracene removal from water. *Environ. Pollut.* **306**, 119457 (2022).
49. Wei, C. et al. Enhanced adsorption of methylene blue using H2O2-modified hydrochar. *Water Air Soil Pollut.* **233**, 422 (2022).
50. Bourafa, A., Berrich, E., Belhachemi, M., Jellali, S. & Jeguirim, M. Preparation and characterization of hydrochars and – CO_2 – activated hydrochars from date and olive stones. *Biomass Convers. Biorefinery* (2023). doi:10.1007/s13399-023-04225-6.
51. Huang, H. et al. Thermal oxidation activation of hydrochar for tetracycline adsorption: The role of oxygen concentration and temperature. *Bioresour. Technol.* **306**, 123096 (2020). doi:10.1016/j.biortech.2020.123096.
52. Ledesma, B., Olivares-marín, M., Álvarez-murillo, A., Roman, S. & Nabais, J. M. V. Method for promoting in-situ hydrochar porosity in hydrothermal carbonization of almond shells with air activation. *J. Supercrit. Fluids* **138**, 187–192 (2018).
53. Mehdi, E., El, A., Ojala, S. & Brahmi, R. Thermal treatment of H_3PO_4-impregnated hydrochar under controlled oxygen flows for producing materials with tunable properties and enhanced diclofenac adsorption. *Sustain. Chem. Pharm.* **34**, 101164 (2023).
54. Singh, B., Wang, T. & Camps-Arbestain, M. *Biochar Characterization for Water and Wastewater Treatments. Sustainable Biochar for Water and Wastewater Treatment* (Elsevier Inc., 2022). doi:10.1016/B978-0-12-822225-6.00003-8.
55. Jiang, Q., Xie, W., Han, S., Wang, Y. & Zhang, Y. Enhanced adsorption of Pb(II) onto modified hydrochar by polyethyleneimine or H_3PO_4: An analysis of surface property and interface mechanism. *Colloids Surf. A: Physicochem. Eng. Asp.* **583**, 123962 (2019).
56. Li, F. et al. One-pot synthesis and characterization of engineered hydrochar by hydrothermal carbonization of biomass with ZnCl2. *Chemosphere* **254**, 126866 (2020).

57. Lin, Z. et al. Nitrogen-doped hydrochar prepared by biomass and nitrogen-containing wastewater for dye adsorption: Effect of nitrogen source in wastewater on the adsorption performance of hydrochar. *J. Environ. Manage.* **334**, 117503 (2023).

58. Zheng, X. et al. Nitric acid-modified hydrochar enhance Cd^{2+} sorption capacity and reduce the Cd^{2+} accumulation in rice. *Chemosphere* **284**, 131261 (2021).

59. Zhang, Y. & Zhang, T. Biowaste valorization to produce advance carbon material-hydrochar for potential application of Cr(VI) and Cd(II) adsorption in wastewater: A review. *Water* **14**, 3675 (2022).

60. Qu, J. et al. Pinecone-derived magnetic porous hydrochar co-activated by $KHCO_3$ and K_2FeO_4 for Cr(VI) and anthracene removal from water. *Environ. Pollut.* **306**, 119457 (2022).

61. Li, B., Guo, J., Lv, K. & Fan, J.Adsorption of methylene blue and Cd(II) onto maleylated modified hydrochar from water. *Environ. Pollut.* **254**, 113014 (2019).

62. Li, H. Z. et al. Preparation of hydrochar with high adsorption performance for methylene blue by co-hydrothermal carbonization of polyvinyl chloride and bamboo. *Bioresour. Technol.* **337**, 125442 (2021).

63. Qian, W. C., Luo, X. P., Wang, X., Guo, M. & Li, B. Removal of methylene blue from aqueous solution by modified bamboo hydrochar. *Ecotoxicol. Environ. Saf.* **157**, 300–306 (2018).

64. de Araújo, T. P., Quesada, H. B., Bergamasco, R., Vareschini, D. T. & de Barros, M. A. S. D. Activated hydrochar produced from Brewer's spent grain and its application in the removal of acetaminophen. *Bioresour. Technol.* **310**, 123399 (2020).

65. Prasannamedha, G., Kumar, P. S., Mehala, R., Sharumitha, T. J. &Surendhar, D. Enhanced adsorptive removal of sulfamethoxazole from water using biochar derived from hydrothermal carbonization of sugarcane bagasse. *J. Hazard. Mater.* **407**, 124825 (2021).

66. Hammud, H. H., Karnati, R. K., Al Shafee, M., Fawaz, Y. & Holail, H. Activated hydrochar from palm leaves as efficient lead adsorbent. *Chem. Eng. Commun.* **208**, 197–209 (2021).

67. Ge, Q., Tian, Q., Wang, S., Zhang, J. & Hou, R. Highly efficient removal of lead/cadmium by phosphoric acid-modified hydrochar prepared from fresh banana peels: Adsorption mechanisms and environmental application. *Langmuir* **38**, 15394–15403 (2022).

8 Immobilized Microorganisms for Organic Matter and Nutrient Removal

Yaneth A. Bustos-Terrones

8.1 INTRODUCTION

Accelerated population growth and industrial and agricultural activities intensification have led to water pollution, primarily related to organic matter and nutrients [1,2]. The excessive accumulation of these pollutants poses a serious threat to aquatic and terrestrial ecosystems, negatively impacting water quality, biodiversity, and public health [3,4].

Conventional water treatment methods have been identified with several drawbacks such as the generation of large amounts of costly-to-handle wastes, high energy consumption, and environmental risks associated with the use of chemicals [5,6]. Furthermore, these methods are sensitive to changes in water conditions and entail high operational and maintenance costs [7] which has led to the development and adoption of advanced and sustainable approaches. Immobilized microorganisms, encapsulated in solid matrices, have emerged as an effective and versatile solution for enhancing water quality through controlled biological processes [8–10].

Immobilized microorganisms (IMs) refer to living microorganisms (e.g., bacteria, fungi, and/or other microbes) attached or confined within a solid matrix support [11]. This support allows microorganisms to remain in a specific location and perform their contaminant removal functions without freely moving along the medium [12]. Compared to free-living microorganisms, IMs play a crucial role in wastewater treatment with significant advantages, applications, and contributions to achieving a cleaner and healthier aquatic environment and are employed for various biotechnological processes [13–15]. As a result, IMs are valuable tools for removing organic matter and nutrients in polluted environments, generating processes with longer retention times, improved contaminant degradation efficiency, reduced residual sludge generation, easy product recovery, and high cost-effectiveness [16]. Furthermore, IMs represent an exciting promise in the organic matter and nutrient removal field, and their understanding and application are considered to play a crucial role in environmental conservation [9,17].

DOI: 10.1201/9781003441007-11

Katam and Bhattacharyya [18] mentioned that IMs offer more benefits compared to suspended co-culture in terms of treatment and biofuel production.

Wastewater often includes elevated nutrient (e.g., nitrogen and phosphorus) levels that may lead to eutrophication in aquatic ecosystems. Efficiently removing these contaminants is crucial for environmental preservation, water resource security, climate change mitigation, agricultural sustainability, biodiversity, and regulatory compliance, thus contributing to responsible environmental management and building of a sustainable future [11,19]. This chapter focuses on the role of IMs in organic matter and nutrient removal for various environmental applications. Different immobilization technologies employed are explored, along with advancements in optimizing removal systems. Furthermore, current, and potential applications for organic matter and nutrient removal in wastewater treatment are discussed.

8.2   USES OF IMS

IMs are used in a wide variety of applications because of their stability, reusability, and process control advantages. Some areas where immobilized microorganisms are employed include biotechnology and bioprocesses [6], bioremediation [7,20], biosensors [8], biotransformation [12], with scientific interest focused on reaction kinetics, microbial metabolism, and different aspects of cellular biology [13], and wastewater treatment [1] with key benefits reported in all cases [21].

Despite information on Ims' application for biofuel production [4], ethanol [3], and hydrogen [17] being available, wastewater treatment using IMs is a field increasingly gaining interest in recent years with several studies available where various contaminants were degraded in wastewater. For example, the removal of ammonium [16], microplastics [9], p-nitrophenol [11], pharmaceuticals [22], textile dyes [23], and organic contaminants and nutrients [19] has been reported using IMs with interesting results suggesting a high efficiency in water treatment. Among other advantages, IMs notably save space and costs by requiring more compact systems [1,24], offering superior stability, efficiency, and flexibility, although its characteristics depend on the process and treatment objectives [25,26].

8.3   MICROORGANISM TYPES

The use of IMs involves bacteria [27], fungi [28], microalgae [29], or other microorganisms [30] in attachment to solid supports or matrices to enhance their contaminant removal efficiency. The process may include nitrifying and denitrifying bacteria for nitrogen removal, anaerobic bacteria for organic matter degradation, autotrophic microorganisms for phosphorus and carbon removal, fungi for recalcitrant compounds, and sometimes mixed microbial communities, depending on the specific characteristics of the wastewater and treatment objectives [26]. Different strategies have been used to degrade organic contaminants [19] with some studies [1,6,11,30] suggesting using immobilized activated sludge because of its higher microbial diversity, resistance to changing conditions, retention efficiency, design flexibility, and high treatment efficiency.

8.3.1 MATERIALS FOR MICROORGANISM IMMOBILIZATION

Selecting the support material is crucial for immobilization process as it impacts viability of immobilized microorganisms and, consequently, affects wastewater treatment efficiency [31–33]. There are three types of available supports: organic, inorganic, and composite. Organic supports are classified as natural (e.g., chitin, agar, alginate, carrageenan) and synthetic (e.g., polyvinyl alcohol (PVA), polyvinyl-pyrrolidone (PVP), polypropylene amine, polyurethane, acrylamide). Inorganic supports include activated carbon, clay, zeolite, anthracite, ceramics, and porous glass. Composite supports usually represent a combination of organic and inorganic materials [31,34].

Among different supports, alginate gel can form microspheres when microorganisms are mixed with alginate and calcium [1]. Polyurethane foam has been reported to create biofilms for contaminant degradation [35]. Activated carbon has been suggested to adsorb microorganisms to remove contaminants and odors [36,37]. Cellulosic materials such as filter paper, sponges, and cotton have been reported to generate degrading biofilms; zeolites, microporous and adsorbent materials [38], while silica gels are known to allow microorganisms to grow within their pores [39]. Various studies [1,8,40,41] have used polymers as support materials to immobilize microorganisms for several reasons. Polymers provide a solid matrix to trap microorganisms, act as a protective shield against adverse environmental conditions, regulate nutrients and product diffusion, facilitate microorganisms' recovery and reuse, reduce waste generation, and possess design flexibility to adapt for different applications and microorganism types. Some researchers [31] suggested that polymers provide high diffusion rates, are environmentally friendly, stable in wastewater samples, possess good durability and mechanical strength, and, above all, are non-toxic to microorganisms. Table 8.1 shows examples of materials used for immobilizing microorganisms. Each material has its own characteristics and advantages, and choosing the most appropriate material will depend on the specific needs of the immobilization process and the desired application.

8.3.2 IMMOBILIZATION METHODS

The immobilization method employed significantly affects microorganisms' physiological responses [58,59] and depends on the process-specific needs, microorganisms characteristics, and environmental conditions [1]. For wastewater treatment, microorganism immobilization methods include encapsulation in polymeric matrices, adsorption on solid surfaces, biofilms on porous supports, and physical retention in gels. These approaches allow microorganisms to be kept in specific locations, enhancing their stability, pollutant degradation capacity, and resistance to adverse conditions, resulting in more efficient and sustainable water treatment [31].

Immobilization methods are classified as physical and chemical with both approaches offering specific advantages depending on the application requirements [35]. Physical immobilization methods involve retaining microorganisms in solid

TABLE 8.1
Examples of Supports Used for Immobilizing Microorganisms and Degraded Contaminants

Immobilizing Matrix	Microorganism	Degraded Pollutant	References
Sodium Alginate	Activated sludge	COD and TP	[1]
Sodium Alginate	Denitrifying bacteria, anaerobic bacteria	Nitrogen	[5]
Sodium alginate-kaolin	Activated sludge	Nitrate	[6]
Sodium Alginate	*E. coli*	Cd, As, Hg, Cr	[8]
Polyvinyl alcohol/ alginate	Activated sludge	P-nitrophenol	[11]
Sodium Alginate	Nitrifying bacteria	Ammoniacal nitrogen	[16]
Luffa	Microbial consortium	COD, phenol, suspended solids	[22]
Chitosan-sodium alginate	*Bacillus subtilis*	Ammonia	[25]
Polyurethane	Bacterial consortia	Total nitrogen	[27]
Polyvinyl alcohol	Activated sludge	Ammonia	[30]
Sodium alginate	*Pseudomonas* and *Stenotrophomonas*	2,3',4,4',5-pentachlorodiphenyl	[40]
Calcium alginate	*S. cerevisiae*	Ethanol	[41]
Biochar	Bacterial consortium	PAHs	[42]
Carbon silica	Bacterial cell	COD and BOD	[43]
Activated carbon	*Olivibacter jilunii*	Nutrients	[44]
Polyvinyl alcohol/ alginate	Ammonia-oxidizing bacteria	Nitrate, NH_4-N	[45]
Sodium Alginate	Sulfate-reducing bacteria	Heavy metal	[46]
Biochar	Bacterial consortium	Nonylphenol	[47]
Chitosan	*Lobosphaera*	Nutrients	[48]
Biochar	*Pseudomonas mendocina*	Nitrate	[49]
Biochar	*Vibrio sp.*	Diesel	[50]
Sodium Alginate	*Chlorella vulgaris, Chlamydomonas reinhardtii*	Nutrients	[51]
Sodium Alginate	*Chlorella vulgaris*	Nutrients	[52]
Polyacrylamide/sodium Alginate	*Pseudomonas, Bacillus, Acinetobacter*	Oil	[53]
Sodium alginate/biochar	*Chlorella pyrenoidosa*	Nutrients, TOC	[54]
Polyvinyl alcohol, polyethylene glycol	Activated sludge	Nutrients, organic matter	[55]
Chitosan	Activated sludge	p-Nitrophenol	[56]
Alginate/biochar	*Bacillus cereus*	Phenol	[57]

matrices, allowing them to grow and function in a confined environment, which facilitates their handling and subsequent recovery. Physical methods include gel encapsulation, porous membrane immobilization, synthetic fibers attachment, and magnetic particles adherence [60]. Chemical methods, on the other hand, are based on the generation of chemical bonds between microorganisms and chemical support, binding them together permanently. Both methods enable high organic contaminants and nutrient degradation efficiency in wastewater, improving the quality of treated water and reducing environmental and health risks [31,61].

8.3.3 FACTORS AFFECTING MICROORGANISM IMMOBILIZATION

The effective immobilization of microorganisms depends on critical factors such as supports or matrices type, physicochemical conditions, microorganisms' concentration and type, inhibitors presence, density charges, immobilization technique, and microorganism selection [29]. These elements must be optimized to ensure biotechnological and water treatment applications while maintaining process efficiency and stability [60]. Furthermore, process condition control is essential to prevent risks related to toxic substrates, pH and/or temperature changes, inadequate oxygen concentrations, and microorganisms' competition, as these can inhibit immobilized microorganisms' growth and activity [31]. Some studies, however, have suggested that hydrogels immobilized microorganisms exhibit greater tolerance to toxic contaminants due to protective encapsulation [61]. The potential of hydrogel immobilized microorganisms to tolerate toxic contaminants is an interesting avenue for future research worth exploring because it would provide with alternative treatment for a variety of toxic pollutants usually present in water or wastewater effluents.

8.4 ADSORPTION MODELS INFLUENCING IMS

There are various factors affecting the degradation kinetics of biological treatment using immobilized microorganisms including the microorganisms type used, substrate type, operational conditions, and reactor design [1]. Different mathematical models and experimental conditions have been employed to determine the degradation rate, microbial growth rate, and other fundamental variables for immobilized microorganism-based biological treatment systems [31]. For instance, a study utilized the bed-depth service time model, which describes the linear relationship between column service time and packed bed depth [62]. These authors also employed the Yoon-Nelson kinetic constants, where column behavior results can be predicted without the need for further experiments and the design of larger-scale adsorption columns. Langmuir models were used, and the authors also suggested that microorganisms immobilization holds great potential for pollutant removal from wastewater.

Langmuir isotherm model assumes monolayer adsorption, characterized by homogeneous adsorption, meaning that all sites have the same affinity for the adsorbate [52]. Redlich-Peterson and Freundlich's isotherms have also been used to model microorganisms' immobilization on solid surfaces [63]. Langmuir and Freundlich's isotherms are models usually adopted to study pollutant distribution (e.g., nutrients and organic matter) in an aqueous solution through the adsorbent phase. Equation (8.1) describes the Langmuir isotherm model, and the Freundlich isotherm model corresponds to equation (8.2) [64–66].

$$q_e = \frac{q_{max}K_LC_e}{1 + K_LC_e} \tag{8.1}$$

$$q_e = K_FC_e^{1/n} \tag{8.2}$$

where q_e is the adsorption amount per adsorbent unit mass in equilibrium, (mg/g); C_e is the pollutant equilibrium concentration in solution, (mg/L); q_{max} is the maximum adsorption capacity per unit mass of adsorbent, (mg/g); and K_L is the Langmuir's constant based on the affinity of the adsorbate binding site per adsorbent (L/g). K_F and n are the Freundlich constants, related to adsorption capacity and adsorption intensity. By plotting the experimental data of adsorption capacity (q_e) against equilibrium concentration (C_e) and fitting the data to the Langmuir isotherm equation, q_{max} and K_L values are obtained.

8.4.1 ORGANIC MATTER AND NUTRIENT DEGRADATION

Microorganisms, whether free or immobilized, require carbon, nitrogen, and phosphorus for their growth and function. During bioremediation using IMs, microorganisms primarily obtain food from wastewater, typically enriched with nutrients and organic matter [61,67]. Wastewater treatment using IMs for organic matter and nutrient degradation is an effective and environmentally friendly technique [31,68]. Several studies have shown high nutrient and organic matter removal efficiencies through IMs [48,51,52,55]. For example, Guo et al. [25] reported nutrients and organic matter removal using *Chlorella pyrenoidosa* supported on sodium alginate/biochar as a medium. The study found interesting removal rates of ammonia nitrogen (69.2%), total nitrogen (43.0%), total phosphorus (73.8%), and total organic carbon (81.0%) using that process. Mannacharaju et al. [43] assessed dissolved organic compounds (chemical oxygen demand [COD] and biochemical oxygen demand [BOD]) removal from wastewater using an immobilized bacterial cell-packed bed reactor. These authors found COD removal efficiency as high as 61% ± 4%, while 87% ± 3% removal was found for BOD. When continuous operation was tested, the treated wastewater was found to still contain 236 ± 21 mg/L COD and 12 ± 3 mg/L BOD. Lee et al. [51] achieved 95% nutrient removal using immobilized *Chlorella vulgaris* and *Chlamydomonas reinhardtii* in sodium alginate (4% w/v). Katam and Bhattacharyya [18] reported immobilization with activated sludge resulted in higher removal of carbon, nitrogen, and phosphorus (88%, 91%, and 93%, respectively) than using suspended microorganisms.

Shin et al. [55] obtained high organic matter (85%) and nutrients (99%) removals through activated sludge immobilization in polyvinyl alcohol (20% w/v) and polyethylene glycol (15% w/v). Mannacharaju et al. [69] assessed organic compound removal by immobilizing a microbial consortium in a carbon-silica matrix using a fluidized reactor with post-tanning wastewater. The system exhibited 43% ± 2.8% Total Chemical Oxygen Demand (COD_{tot}) removal efficiency and 42% ± 3.3% Dissolved Chemical Oxygen Demand (COD_{dis}) removal efficiency within 24 h in batch mode operation. When continuous mode operation was tested, it showed $43\% \pm 8.4\%$ COD_{tot} removal efficiency and $50\% \pm 8.4\%$ COD_{dis} removal efficiency, suggesting efficient degradation of organic compounds in wastewater.

In another study, Mujtaba et al. [70] reported the simultaneous removal of nutrients (nitrogen and phosphorus) and organic matter from municipal wastewater through immobilized *Chlorella vulgaris* and activated sludge. These authors reported efficient nutrients and organic matter (COD – Chemical Oxygen Demand) removal after two days of detention time. The highest nitrogen (99.8%), phosphorus (100%), and COD (90%–95%) removal occurred after two days. Similar results were obtained when repeating batch cycles three times. Bustos-Terrones et al. [1] evaluated organic matter and nutrient degradation using immobilized microorganisms in sodium alginate beads (2.5% w/v). These authors reported high total phosphorus (94.26%) and chemical oxygen demand (78.25%) removals.

Figure 8.1 illustrates the nutrient adsorption and assimilation mechanism by IMs in alginate (adapted from Banerjee et al. [52]). According to Figure 8.1, Banerjee et al. [52] suggested that NH_4^+ ions adsorption is based on the electrostatic attraction between protonated amino groups and negatively charged $COO-$ functional groups on alginate. Likewise, PO_4^{3-} ions and OH physical interaction is proposed to occur through hydrogen bonds. In the presence of an electrostatic environment within the polymeric matrix, the NH_4^+ ion forms a weak hydrogen bond with carbonyl's free oxygen. NH_4^+ and PO_4^{3-} weakly bond to the alginate's structure being assimilated by microorganisms over time. Simulation results in vacant positions on alginate (e.g., $COO-$ and OH groups) and leads to greater nutrient uptake by alginate beads. This process continues until microorganisms limit nutrient uptake, saturating the beads with adsorbed nutrients.

8.4.2 OXYGEN CONSUMPTION BY IMS

Immobilized microorganisms that are attached to solid support break down organic matter and/or nutrients as they consume oxygen in the biodegradation process. The rate of oxygen consumption (OUR) depends on the microorganisms used. When OUR decreases, it is a sign that the immobilized microorganisms are dying. In an aerobic process, oxygen must be continuously supplied to achieve acceptable efficiencies. Given the crucial role of oxygen in the growth and metabolism of microorganisms, it is necessary to understand both OUR by the microorganisms and the oxygen transfer rate (OTR) to the system. The concentration of dissolved oxygen (DO) in the system decreases during the rapid growth of microorganisms but then increases when they reach the stationary phase due to reduced oxygen demand. Dissolved oxygen in the bioreactor results in a balance between the consumption rate by microorganisms and the transfer rate from the gas phase to the liquid phase. Monitoring dissolved oxygen in the reactor is vital, as oxygen often becomes the factor regulating metabolic pathways in microbial cells.

Measurement of OUR is used to assess the metabolic activity of cells, estimate the concentration of viable cells in cell cultures, characterize organic composition in the

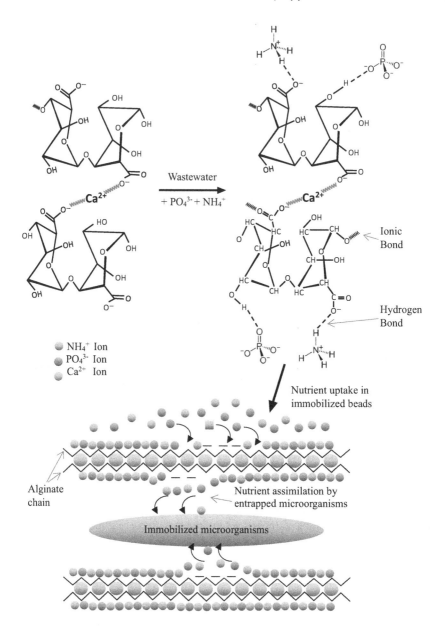

FIGURE 8.1 Mechanism of nutrients (NH_4^+ and PO_4^{3-}) adsorption and assimilation by alginate-trapped microorganisms.

substrate, and indicate potential hydrodynamic stresses, cell damage, and microbial death. Additionally, Bustos Terrones et al. [1] mention that when degrading organic pollutants, the oxygen consumption rate can be measured by monitoring dissolved oxygen levels in the water before and after treatment, indicating how much oxygen has been consumed by microorganisms to degrade organic matter. The rate of oxygen consumption (OUR, along with OTR) is one of the key parameters affecting the design, operation, and scaling of bioreactors, as the concentration of dissolved oxygen (DO) depends on transport processes. Bustos-Terrones et al. [1] conducted a study on OUR in the treatment using microorganisms immobilized in sodium alginate beads in domestic wastewater. They reported that OUR (mg O_2/L min) decreased completely after 24 h of treatment. This coincides with findings reported by Garcia-Ochoa et al. [71] who studied the kinetic modeling of OUR behavior in the bioreactor. They also described methods for measuring OUR, including the gas balancing method, dynamic method, and yield coefficient method. Furthermore, they mentioned that in biological processes, oxygen is a critical parameter that influences the process, as it is the only nutrient that must be continuously supplied, even in batch cultures.

8.5 CASE STUDY

A study was conducted on pollutant biodegradation in domestic wastewater using IMs in sodium alginate (SA). Microorganisms were obtained from a large-scale activated sludge plant. The activated sludge sample was allowed to settle for 2 h and the supernatant was then removed. The initial biomass concentration was 5,800 mg/L of volatile suspended solids (VSS). One liter of settled activated sludge was used, immobilized in 2.5% w/v SA (see Figure 8.2).

For immobilization, a mixture was prepared using a peristaltic pump and dripped into a $CaCl_2$ (3% w/v) solution. A fixed-bed tubular reactor was filled with Ims and employed to remove organic matter and nutrients while continuously aerated to support aerobic conditions. Domestic wastewater from the same treatment plant where the activated sludge was obtained was used as feeding effluent.

A reduction in organic matter, expressed as COD, from 198.4 to 42.5 mg/L was observed. Furthermore, total phosphorus concentration decreased from 2.48 to 0.14 mg/L with a residence time of 12 h suggesting that microorganisms immobilization on SA is an effective process for domestic wastewater treatment, achieving high rates of organic contaminants and nutrient removal with a notable reduction in sludge generation. Other studies also highlight the efficiency of IM treatments. For example, Shin et al. [55] investigated the physical and chemical characteristics of

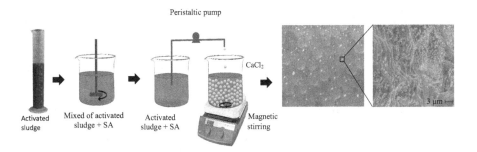

Peristaltic pump

$CaCl_2$

Activated sludge

Mixed of activated sludge + SA

Activated sludge + SA

Magnetic stirring

3 µm

FIGURE 8.2   Schematic view of the preparation of SA beads with IMs.

immobilized microorganisms (activated sludge) on polyvinyl alcohol. They reported 99% nitrification and denitrification and 85% organic matter removal. Banerjee et al. [52] suggested that *Chlorella vulgaris* immobilization on SA (3% w/v), efficiently removed nutrients (NH_4^+ and PO_4^{3-}) from wastewater.

8.6 CHALLENGES AND FUTURE INVESTIGATIONS

Microorganism immobilization is an essential technique for removing organic matter and nutrients from water to enhance water quality, and environmental sustainability, and support SDG 6 goals. Despite its numerous advantages and promising performance in contaminant removal efficiency, most work has been limited to laboratory-scale studies. Additional research is needed on the scale-up of these processes to assess their technology readiness level (TRL) and enhance real applications. We believe that the lack of information on scaling-up parameters for the use of IMs in the removal of nutrients is a significant knowledge gap worth exploring to identify scaling-up parameters or operation conditions that may lead to increased technology's TRL.

Other significant challenges faced by the technology are wastewater complexity (presence of different contaminants) and the need for efficient/standardized immobilization supports. Several opportunities for further research and development include performance enhancing and expanding IM technology to other applications. The development of new techniques also opens new opportunities using natural, underused, or waste materials, to follow up with the circular economy concept.

More research is needed to gain a deeper understanding of microorganism interaction mechanisms with materials. This includes investigating adsorption, dispersion, and their impact on contaminant degradation kinetics. Additionally, exploring the effects of microorganisms and immobilization materials for different environmental conditions, such as high temperatures, freeze/thaw cycles, and/or aggressive chemical environments, is crucial. Furthermore, the modification of IMs with nanoparticles and other additives to improve their performance and expand their applications is a highly interesting area of future research. Conducting life cycle assessments (LCA) of immobilization materials and their environmental impacts is also essential to identify improvement opportunities to enhance materials' sustainability. Likewise, another opportunity is to assess OUR and OTR behavior according to microorganism type, generate models that describe this behavior, and conduct variable analysis limiting treatment rate. Research in this field continues to advance, and continuous growth is expected in the coming years, along with increased commercial and field adaptation. Therefore, exploring new materials, selecting suitable microorganisms, and improving this technology are crucial for widespread use in the future to address pollution issues, improve water quality, combat climate change, and support sustainable development goals.

8.7 CONCLUSIONS

This chapter has provided a comprehensive overview of using IMs in the removal of nutrients and organic contaminants from water. The significance of environmental preservation through the application of IMs for removing organic matter and nutrients from water has been emphasized. Additionally, it reviews advancements in microorganisms immobilization using various materials and immobilization techniques. The interaction mechanism between microorganisms and immobilization supports has been highlighted. Notably, immobilized microorganisms offer advantages such as microbial cell stability, continuous operation, and resilience to adverse environmental conditions.

There are, however, factors affecting IMs posing a challenge for their full-scale application in water treatment. These factors can inhibit their growth and activity, including the presence of toxic substrates, pH and temperature changes, inadequate aeration conditions, and competition with other microorganisms. Different models have been evaluated for assessing pollutant degradation kinetics including adsorption isotherms to evaluate the microorganism's attachment to solid matrices, primarily Langmuir and Freundlich adsorption models have been used to study contaminant distribution. Finally, IM technology is identified as promising and deserves further research to prolong hydrological quality, maintain ecological balance, address climate change, and procure sustainable development.

REFERENCES

1. Bustos-Terrones, Y. A., Estrada-Vázquez, R., Ramírez-Pereda, B., Bustos-Terrones, V. & Rangel-Peraza, J. G. Kinetics of a fixed bed reactor with immobilized microorganisms for the removal of organic matter and phosphorous. *Water Environ. Res.* **92**, 1956–1965 (2020).
2. Liang, E. et al. Treatment of micro-polluted water with low C/N ratio by immobilized bioreactor using PVA/sintered ores@ sponge cube: Performance effects and potential removal pathways. *Sci. Total Environ.* **870**, 162003 (2023).
3. Ünsal, S. et al. Ethanol production by immobilized *Saccharomyces cerevisiae* cells on 3D spheres designed by different lattice structure types. *Process Biochem.* **125**, 104–112 (2023).
4. Costa, I. et al. Enzyme immobilization technology as a tool to innovate in the production of biofuels: a special review of the Cross-Linked Enzyme Aggregates (CLEAs) strategy. *Enzyme Microb. Technol.* **170**, 110300 (2023).
5. Li, L., Zhang, M., Jiang, W. & Yang, P. Study on the efficacy of sodium alginate gel particles immobilized microorganism SBBR for wastewater treatment. *J. Environ. Chem. Eng.* **10**, 107134 (2022).
6. Yu, X. et al. Immobilized-microbial bioaugmentation protects aerobic denitrification from heavy metal shock in an activated-sludge reactor. *Bioresour. Technol.* **307**, 123185 (2020).
7. Chwastowski, J. & Staroń, P. Immobilization of Phaffia rhodozyma cells in biopolymer for enhanced Cr(VI) bioremediation. *Colloids Surf. A: Physicochem. Eng.* **672**, 131698 (2023).
8. Wasito, H., Fatoni, A., Hermawan, D. & Susilowati, S. S. Immobilized bacterial biosensor for rapid and effective monitoring of acute toxicity in water. *Ecotoxicol. Environ. Saf.* **170**, 205–209 (2019).
9. Tang, K. H. et al. Immobilized enzyme/microorganism complexes for degradation of microplastics: A review of recent advances, feasibility and future prospects. *Sci. Total Environ.* **832**, 154868 (2022).
10. Shaidorova, G. et al. Evaluation of the safety of immobilized microorganisms *Lysobacter* sp. on inorganic media. *Eng. Proc.* **37**, 71 (2023).

11. Sam, S. P., Adnan, R. & Ng, S. L.Statistical optimization of immobilization of activated sludge in PVA/alginate cryogel beads using response surface methodology for p-nitrophenol biodegradation. *J. Water Process. Eng.* **39**, 101725 (2021).

12. Lapponi, M. J., Méndez, M. B., Trelles, J. A. & Rivero, C. W. Cell immobilization strategies for biotransformations. *Curr. Opin. Green Sustain. Chem.* **33**, 100565 (2022).

13. Sivakumar, M., Senthamarai, R., Rajendran, L. & Lyons, M. E. G. Reaction and kinetic studies of immobilized enzyme systems: Part-I Without external mass transfer resistance. *Int. J. Electrochem. Sci.* **17**, 221159 (2022).

14. Saravanan, A. et al. A comprehensive review on immobilized microbes-biochar and their environmental remediation: Mechanism, challenges and future perspectives. *Environ. Res.* **236**, 116723 (2023).

15. Wu, C. et al. Immobilization of microbes on biochar for water and soil remediation: A review. *Environ. Res.* **212**, 113226 (2022).

16. Vishakar, V. V., Haran, N. H., Vidya, C. & Mohamed, M. A. Removal of ammonia in water systems using cell immobilization technique in surrounding environment. *Mater. Today: Proc.* **43**, 1513–1518 (2021).

17. Woo, W. X. et al. An overview on cell and enzyme immobilization for enhanced biohydrogen production from lignocellulosic biomass. *Int. J. Hydrog. Energy* **47**, 40714–40730 (2022).

18. Katam, K. & Bhattacharyya, D. Simultaneous treatment of domestic wastewater and bio-lipid synthesis using immobilized and suspended cultures of microalgae and activated sludge. *J. Ind. Eng. Chem.* **69**, 295–303 (2019).

19. Bustos-Terrones, Y. A., Bandala, E. R., Moeller-Chávez, G. E. & Bustos-Terrones, V. Enhanced biological wastewater treatment using sodium alginate-immobilized microorganisms in a fluidized bed reactor. *Water Sci. Eng.* **15**, 125–133 (2022).

20. Wu, P. H., Hsieh, T. M., Wu, H. Y. & Yu, C. P. Characterization of the immobilized algae-based bioreactor with external ceramic ultrafiltration membrane to remove nutrients from the synthetic secondary wastewater effluent. *Int. Biodeterior. Biodegrad.* **164**, 105309 (2021).

21. Kiran, M. G., Pakshirajan, K. & Das, G. Heavy metal removal from aqueous solution using sodium alginate immobilized sulfate reducing bacteria: mechanism and process optimization. *J. Environ. Manage.* **218**, 486–496 (2018).

22. Murshid, S. & Dhakshinamoorthy, G. P. Application of an immobilized microbial consortium for the treatment of pharmaceutical wastewater: Batch-wise and continuous studies. *Chin. J. Chem. Eng.* **29**, 391–400 (2021).

23. Subramaniam, S., Velu, G., Palaniappan, M. & Jackson, K. M. Decolourization of azo dyes using immobilized bacterial isolates from termite mound ecosystem. *Total Environ. Res. Themes* **6**, 100041 (2023).

24. Jenjaiwit, S. et al. Removal of triclocarban from treated wastewater using cell-immobilized biochar as a sustainable water treatment technology. *J. Clean. Prod.* **320**, 128919 (2021).

25. Guo, J. et al. Effective immobilization of *Bacillus subtilis* in chitosan-sodium alginate composite carrier for ammonia removal from anaerobically digested swine wastewater. *Chemosphere* **284**, 131266 (2021).

26. Han, M., Zhang, C. & Ho, S. H. Immobilized microalgal system: An achievable idea for upgrading current microalgal wastewater treatment. *Environ. Sci. Ecotechnol.* **14**, 100227 (2022).

27. Gan, Y., Ye, Z., Zhao, Q., Li, L. & Lu, X. Spatial denitrification performance and microbial community composition in an up-flow immobilized biofilter for nitrate micro-polluted water treatment. *J. Clean. Prod.* **258**, 120913 (2020).

28. Gao, Y. et al. Immobilized fungal enzymes: Innovations and potential applications in biodegradation and biosynthesis. *Biotechnol. Adv.* **57**, 107936 (2022).

29. Zhang, J., Wei, J., Massey, I. Y., Peng, T. & Yang, F. Immobilization of microbes for biodegradation of microcystins: A mini review. *Toxins* **14**, 573 (2022).

30. Wang, J., Yang, H., Zhang, F., Su, Y. & Wang, S. Activated sludge under free ammonia treatment using gel immobilization technology for long-term partial nitrification with different initial biomass. *Process Biochem.* **99**, 282–289. (2020).

31. Bouabidi, Z. B., El-Naas, M. H. & Zhang, Z. Immobilization of microbial cells for the biotreatment of wastewater: A review. *Environ. Chem. Lett.* **17**, 241–257 (2019).

32. Gong, Y. Z., Niu, Q. Y., Liu, Y. G., Dong, J., & Xia, M. M. Development of multifarious carrier materials and impact conditions of immobilised microbial technology for environmental remediation: A review. *Environ. Pollut.* **314**, 120232 (2022).

33. Barbhuiya, S. & Das, B. B. Water-soluble polymers in cementitious materials: A comprehensive review of roles, mechanisms and applications. *Case Stud. Constr. Mater.* **19**, e02312 (2023).

34. Girijan, S. & Kumar, M. Immobilized biomass systems: an approach for trace organics removal from wastewater and environmental remediation. *Curr. Opin. Environ. Sci.* **12**, 18–29 (2019).

35. Zheng, Y. et al. A novel method for immobilizing anammox bacteria in polyurethane foam carriers through dewatering. *J. Water Process. Eng.* **53**, 103738 (2023).

36. Li, R. et al. Application of biochar immobilized microorganisms for pollutants removal from wastewater: A review. *Sci. Total Environ.* **837**, 155563 (2022).

37. Yang, J. et al. A dual bacterial alliance removed erythromycin residues by immobilizing on activated carbon. *Bioresour. Technol.* **384**, 129288 (2023).

38. Zamel, D. & Khan, A. U. Bacterial immobilization on cellulose acetate based nanofibers for methylene blue removal from wastewater: Mini-review. *Inorg. Chem. Commun.* **131**, 108766 (2021).

39. Brányik, T., Kuncová, G., Páca, J. & Demnerová, K. Encapsulation of microbial cells into silica gel. *J. Sol. Gel. Sci. Technol.* **13**, 283–287 (1998).

40. Ouyang, X. et al. Enhanced bioremediation of 2,3′,4,4′,5-pentachlorodiphenyl by consortium GYB1 immobilized on sodium alginate-biochar. *Sci. Total Environ.* **788**, 147774 (2021).

41. Perez, C. L. et al. Towards a practical industrial 2G ethanol production process based on immobilized recombinant *S. cerevisiae*: Medium and strain selection for robust integrated fixed-bed reactor operation. *Renew. Energ.* **185**, 363–375 (2022).

42. Qiao, K. et al. Removal of high-molecular-weight polycyclic aromatic hydrocarbons by a microbial consortium immobilized in magnetic floating biochar gel beads. *Mar. Pollut. Bull.* **159**, 111489 (2020).

43. Mannacharaju, M. et al. Bacterial cell immobilized packed bed reactor for the elimination of dissolved organics from biologically treated post-tanning wastewater and its microbial community profile. *Chemosphere* **320**, 138022 (2023).

44. Huang, J., Xiao, Y. & Chen, B. Nutrients removal by Olivibacterjil unii immobilized on activated carbon for aquaculture wastewater treatment: ppk1 gene and bacterial community structure. *Bioresour. Technol.* **370**, 128494 (2023).

45. Dong, Y., Zhang, Y. & Tu, B. Immobilization of ammonia-oxidizing bacteria by polyvinyl alcohol and sodium alginate. *Braz. J. Microbiol.* **48**, 515–521 (2017).

46. Bai, F., Liu, S., Ma, J. & Zhang, Y. Biodegradation of sulfate and elimination of heavy metals by immobilized-microbial bioaugmentation coupled with anaerobic membrane bioreactor. *Chem. Eng. J.* **473**, 145196 (2023).

47. Lou, L. et al. Adsorption and degradation in the removal of nonylphenolfrom water by cells immobilized on biochar. *Chemosphere* **228**, 676–684 (2019).

48. Vasilieva, S. et al. Bio-inspired materials for nutrient biocapturefrom wastewater: Microalgal cells immobilized on chitosan-based carriers. *J. Water Process. Eng.* **40**, 101774 (2021).

49. Zhang, W. et al. Efficient nitrate removal by *Pseudomonas mendocina* GL6 immobilized on biochar. *Bioresour. Technol.* **320**, 124324 (2021).

50. Zhou, H. et al. Enhanced bioremediation of diesel oil-contaminated seawater by a biochar-immobilized biosurfactant-producing bacteria Vibrio sp. LQ2 isolated from cold seep sediment. *Sci. Total Environ.* **793**, 148529 (2021).

51. Lee, H., Jeong, D., Im, S. & Jang, A. Optimization of alginate bead size immobilized with *Chlorella vulgaris* and *Chlamydomonas reinhardtii* for nutrient removal. *Bioresour. Technol.* **302**, 122891 (2020).

52. Banerjee, S., Tiwade, P. B., Sambhav, K., Banerjee, C. & Bhaumik, S. K. Effect of alginate concentration in wastewater nutrient removal using alginate-immobilized microalgae beads: Uptake kinetics and adsorption studies. *Biochem. Eng. J.* **149**, 107241 (2019).

53. Chen, L., Zhao, S., Yang, Y., Li, L. & Wang, D. Study on degradation of oily wastewater by immobilized microorganisms with biodegradable polyacrylamide and sodium alginate mixture. *ACS Omega* **4**, 15149–15157 (2019).

54. Guo, Q. et al. Application of *Chlorella pyrenoidosa* embedded biochar beads for water treatment. *J. Water Process. Eng.* **40**, 101892 (2021).

55. Shin, D. C., Kim, J. S. & Park, C. H. Study on physical and chemical characteristics of microorganism immobilized media for advanced wastewater treatment. *J. Water Process. Eng.* **29**, 100784 (2019).

56. Ng, S. L., Yong, K. J. L., Pushpamalar, J. & Sam, S. P. Carboxymethyl sago pulp/chitosan hydrogel as an immobilization medium for activated sludge for p-nitrophenol biodegradation. *J. Appl. Polym. Sci.* **136**, 47531 (2019).

57. Li, J., Jia, Y., Zhong, J., Liu, Q. & Li, H. Agranovski, I. Use of calcium alginate/biocharmicrosphere immobilized bacteria *Bacillus* sp. for removal of phenol in water. *Environ. Chall.* **9**, 100599 (2022).

58. Ashikin, N. A. L. N. et al. Optimization and characterization of immobilized *E. coli* for engineered thermostablexylanase excretion and cell viability. *Arab. J. Chem.* **15**, 103803 (2022).

59. Elmerhi, N. et al. Enzyme-immobilized hierarchically porous covalent organic framework biocomposite for catalytic degradation of broad-range emerging pollutants in water. *J. Hazard. Mater.* **459**, 132261 (2023).

60. Liu, D. et al. Immobilization of biomass materials for removal of refractory organic pollutants from wastewater. *Int. J. Environ. Res. Public Health.* **19**, 13830 (2022).

61. Mehrotra, T. et al. Use of immobilized bacteria for environmental bioremediation: A review. *J. Environ. Chem. Eng.* **9**, 105920 (2021).

62. Mahamadi, C. & Mawere, E. Continuous flow biosorptive removal of methylene blue and crystal violet dyes using alginate-water hyacinth beads. *Cogent Environ. Sci.* **5**, 1594513 (2019).

63. Milojković, J. V. et al. Performance of aquatic weed-waste *Myriophyllum spicatum* immobilized in alginate beads for the removal of Pb(II). *J. Environ. Manage.* **232**, 97–109 (2019).

64. Hosseini, M., Kamani, H., Esrafili, A., Badi, M. Y. & Gholami, M. Removal of malathion by sodium alginate/biosilicate/magnetite nanocomposite as a novel adsorbent: kinetics, isotherms, and thermodynamic. *Study Health Scope* **8**, 11 (2019).

65. Thakre, P. N., Mukherjee, S., Samanta, S., Barman, S. & Halder, G. A mechanistic insight into defluoridation of simulated wastewater applying bio-inspired sodium alginate bead. *Appl. Water Sci.* **10**, 1–12 (2020).

66. Jaafari, J. et al. Effective adsorptive removal of reactive dyes by magnetic chitosan nanoparticles: kinetic, isothermal studies and response surface methodology. *Int. J. Biol. Macromol.* **164**, 344–355 (2020).

67. Han, Z. et al. Recovery of phosphate, magnesium and ammonium from eutrophic water by struvitebiomineralization through free and immobilized Bacillus cereus MRR2. *J. Clean. Prod.* **320**, 128796 (2021).

68. Feng, Q. et al. Simultaneous reclaiming phosphate and ammonium from aqueous solutions by calcium alginate-biochar composite: Sorption performance and governing mechanisms. *Chem. Eng. J.* **429**, 132166 (2022).

69. Mannacharaju, M., Somasundaram, S. & Ganesan, S. Treatment of refractory organics in secondary biological treated post tanning wastewater using bacterial cell immobilized fluidized reactor. *J. Water Process Eng.* **43**, 102213 (2021).

70. Mujtaba, G., Rizwan, M., Kim, G. & Lee, K. Removal of nutrients and COD through co-culturing activated sludge and immobilized *Chlorella vulgaris*. *Chem. Eng. J.* **343**, 155–162 (2018).

71. Garcia-Ochoa, F., Gomez, E., Santos, V. E. & Merchuk, J. C. Oxygen uptake rate in microbial processes: an overview. *Biochem. Eng. J.* **49**, 289–307 (2010).

9 Technological Opportunities for Resource Recovery from Industrial Textile Wastewater

Irwing Ramirez, Ajay Kumar,
Déborah L. Villaseñor-Basulto,
Yolanda G. Garcia-Huante,
Alberto Ordaz, and Lewis S. Rowles

9.1 INTRODUCTION

The textile industry produces around eight tons of fabric per day, consuming nearly two million liters of water [1]. As much as 15% of dyes used to produce these fabrics are released into the environment in wastewater [1,2]. Industrial textile wastewater (ITWW) commonly contains a mixture of chemical compounds, for example, dyes, detergents, waxes, catalytic chemicals, adhesives, surfactants (cationic, anionic, non-ionic), solvents, inorganic salts ($NaCl$, Na_2SO_4), metals, lubricants, dispersing agents, and other organic compounds [3]. ITWW usually has intense coloration with elevated levels of chemical oxygen demand (COD) and low biodegradability (demonstrated by low BOD_5/COD ratios), making traditional biological treatment ineffective [4,5]. For example, COD levels in ITWW range from 5,400 to 865 mg L^{-1} [6–12]. Discharging untreated or partially treated ITWW has been shown to have detrimental impacts on aquatic life and the environment [13]. ITWW has also been shown to have a variety of negative effects on human health, for example, skin rashes, allergies, and potential mutagenic and carcinogenic effects [14].

Currently, much of the ITWW throughout the world goes untreated; thus, identifying a potentially sustainable treatment system for ITWW has become crucial for the textile industry. Ideal treatment systems would be able to treat hazardous wastewater while creating opportunities for water reuse. The main processes that have been used for ITWW treatment include physical (e.g., adsorption and ion exchange), chemical (e.g., chemical coagulation, electrocoagulation), advanced oxidation (e.g., H_2O_2/UV, Fenton, electrochemical Fenton, electro-Fenton ozonation, photocatalysis, photo-electrocatalysis), biological (e.g., aerobic, anaerobic, microalgal-bacterial consortium), and membrane processes (microfiltration nanofiltration, loose nanofiltration, reverse osmosis) [15,16].

DOI: 10.1201/9781003441007-12

When these wastewater treatment processes are operated as single units, removal efficiencies can be inadequate to treat the complex mixture of suspended, colloidal, and dissolved solids in ITWW. For example, employing the nature-based solution of orange seed powder as a bio-adsorbent in textile wastewater achieved a 96% reduction in turbidity, though it had no discernible effect on COD levels post-treatment [17]. Hydrodynamic cavitation with oxidants achieved a 30% reduction in COD from ITWW sourced [18]. A solar-driven photo-Fenton process achieved an impressive 90% discoloration of real wastewater, with a 62.6% COD reduction of only 1 h of equivalent solar radiation [19].

Electron beam irradiation treatment did not improve ITWW biodegradability, likely because of the substantial quantity of organic matter present; however, the processes did enhance the BOD_5/COD ratio after the biological treatment process [20]. Novel materials have also shown remarkable efficiency in treating ITWW. A composite of reduced graphene oxide (rGO)/chalcopyrite ($CuFeS_2$), aided by microwave irradiation and hydrogen peroxide, effectively eliminated 99% of total organic carbon (TOC) in a mere 6 min from ITWW sourced from a dyeing factory in South America [21]. While novel materials and intensive energy techniques may show promise for treating dye-containing wastewater, it is essential to understand the overall sustainability of such methods before scaling them up for real-world industrial applications.

Effective wastewater treatment often necessitates combining multiple techniques, for example, incorporating biological processes and physical–chemical operations. A combination of physical–chemical water treatments have been demonstrated to remove color and COD in industrial wastewater. Badawi and Zaher showed the combination of chemical coagulation-flocculation, adsorption on zero-valent iron, and filtration with micro zeolite removed 97.5% of COD from ITWW in continuous flow (1.5 L/min) [22]. Combining biological–physical–chemical treatments in a pilot plant reduced color, COD, and phenols [23]. While physical–chemical–biological water treatment effectively reduces COD and color, they may not adequately mitigate toxicity in the presence of persistent contaminants [24].

Coupling advanced oxidation processes (AOPs) with chemical coagulation, electrochemical methods, bioprocesses, or membrane processes has shown to be a viable approach for achieving significant mineralization and toxicity reduction [25,26]. AOPs generate highly oxidative and non-selective reactive species, primarily hydroxyl radicals, responsible for organic oxidation. Sequential coagulation with photo-Fenton treatment of ITWW has been shown to increase the BOD_5/COD ratio from 0.21 to 0.74 [7]. Biological degradation (with 96 h retention time) in series with photocatalysis (120 min retention time) using a ZnO/polypyrrole composite resulted in 99.8% TOC degradation in ITWW [27]. While advanced treatment processes have garnered recognition as highly efficient and for being sustainable technologies, it is imperative to understand how they much be appropriately used for the textile water sector. Understanding such opportunities necessitates a delicate balance between water remediation (encompassing pollutant degradation, mineralization, and toxicity reduction) and equally important considerations in applying circular economy principles and management for climate change adaptation.

System-level analysis can be used to help understand how industries can adopt new practices to increase efficiency while meeting sustainability goals. One such analysis was developed by Ezhilarasan and Umakanta where a supply chain model applied circular economy in the textile industry [28]. The optimized model was able to demonstrate how to improve profits while also considering constraints for waste management, novel wastewater treatment technologies, and carbon emissions. Other research on wastewater treatment technologies for industrial effluents, including ITWW, has emphasized technologies capable of effectively removing toxic pollutants and the need to develop water reuse in a circular economy for industries [29]. Others have explored a chemical perspective on the circular economy via biotechnological or chemical processes, where fibers' biotransformation into purified glucose served as a versatile precursor to produce materials and value-added chemicals. This approach suggested more than 70% of the textiles, including a mixture of cotton and polyester, can be recycled [30]. Despite the demonstration of technologies for ITWW treatment and various system-level analyses highlighting their opportunities for the circular economy, the adoption of water treatment technologies by the textile industry has been slow.

A wastewater treatment plant can generate both direct and indirect greenhouse gas (GHG) emissions. Indirect emissions pertain to the carbon footprint associated with power and chemical inputs. On the other hand, direct emissions, also known as fugitive emissions, represent the release of CO_2, CH_4, and N_2O directly from the water surface into the atmosphere [31]. Wastewater treatment plants are widely recognized as significant sources of direct and indirect GHGs, and numerous efforts have been made to estimate the extent of these emissions [32–34]. As an example, estimates for England and Wales suggest that treating the daily generated municipal wastewater requires approximately 2,800 GWh, leading to the emission of around two million metric tons of GHGs [35]. However, studies have concentrated on estimating GHG emissions from municipal wastewater treatment plants, with minor emphasis on industrial wastewater. Yapıcıoglu and Demir estimated both direct and indirect GHG emissions in centralized industrial wastewater treatment plants, suggesting their potential as sources of GHG emissions [36]. In addition, some reports provide estimations of GHG emissions from advanced water treatment processes. Kang et al. conducted a comparative study on direct GHG emissions arising from advanced oxidation methods, specifically ozone, photocatalysis, and H_2O_2/UV, during the degradation of an emerging pollutant. Their findings indicated that the ozonation process accounted for most GHG emissions, primarily attributable to electricity consumption [37].

Mitigating GHG emissions can enhance the sustainability of wastewater treatment plants [31]. It has been identified that electricity consumption during operations such as pumping, mixing, heating, and aeration significantly contributes to GHG emissions [31,38]. Moreover, the choice of technologies employed in a wastewater treatment plant substantially impacts indirect and direct GHG emissions [38]. Reducing GHG emissions from wastewater treatment plants can be achieved by decreasing electricity consumption and operating costs, such as harnessing and utilizing biogas generated within the plant, minimizing new energy consumption, and promoting alternative energy sources such as bioenergy or hydrogen [39].

This book chapter focuses on the practical applications of real ITWW to recovery resources, emphasizing its role in implementing circular economy principles and strategies for climate change adaptation. Scientific literature that focused on the analysis of single dyes or simulated wastewater has been excluded from consideration as the aim is to address the technical and societal challenges intrinsic to achieving sustainability and decarbonization in the textile water sector. The objectives of this chapter are (i) to provide an overview of the role of ITWW in the circular economy, (ii) to examine the contribution of greenhouse gas emissions from ITWW treatment operations, and (iii) to identify existing research that can facilitate the utilization of ITWW as a valuable resource. This book chapter has the potential to provide valuable insight to researchers and industry to chart a path toward sustainable treatment of ITWW.

9.2 CIRCULAR ECONOMY IN THE FASHION INDUSTRY

Clothing holds paramount significance in human life [40]. Its purpose extends beyond protection and warmth by encompassing identity, values, modesty, decency, self-expression, social symbolism, traditional meaning, and psychological well-being [41]. Clothing also serves as a canvas for creative expression, among other functions. The clothing industry is a critical pillar of the global economy, producing essential textile products integral to our daily lives [42]. And it is pivotal in generating employment for millions across various nations. China is the foremost clothing producer worldwide (accounting for 35.6% of global textile exports) followed by Vietnam (5.03%) and Bangladesh (4.82%) [43]. However, the growing textile industry has raised serious concerns about the depletion of natural resources. The adverse effects of the clothing industry have been amplified by the worldwide proliferation of the fast fashion trend, further exacerbating the textile industry's footprint [44]. The rise of fast fashion has correlated with the widespread use of social networks. This trend is characterized by swiftly changing styles and inexpensive clothing, which promotes overconsumption. Consequently, demand leads to heightened garment production, transportation, and disposal levels [45,46]. The culture of massive consumption associated with the rapid proliferation of fast fashion brands, known for their low-cost, trendy clothing, contributes significantly to carbon emissions, solid wastes, and wastewater generation [45].

The awareness of the adverse effects of the clothing industry has inspired consumers worldwide to adopt sustainable practices. For instance, Gen-Z consumers in China, have emerged as a powerful influence in shaping the industry's sustainability agenda due to their preference for sustainable fashion [47]. This change in the consuming habits of the youngest Chinese people may positively impact the textile industry in China, which is the leading cloth producer in the world. Over the past few decades, the clothing industry has significantly contributed to environmental degradation, primarily due to its excessive water consumption, chemical usage, and generation of substantial textile waste. Textile production releases significant volumes of wastewater, often containing organic pollutants, nutrients, dyes, heavy metals, and potentially hazardous chemicals, among other substances [48]. Textile wastewater may also contain chemical additives like detergents, surfactants, solvents, emulsifiers, pH adjusters (such as acids or bases), and stabilizers [49–53]. Particulate matter, including fibers, lint, and other solid residues from textile manufacturing, constitutes the suspended solids of this wastewater [54]. Among the main compounds found in wastewater are the dyes and pigments that were not absorbed by the textiles and subsequently washed off during the dyeing and finishing stages, resulting in colored compounds in the wastewater [55]. There are natural and synthetic dyes whose main differences lie in the source of these compounds, the production scheme, the color range, the fastness, stability, and environmental impact [55].

Synthetic dyes are utilized in the industry for their specific chemical properties, which include stability, fastness, and cost-effectiveness; these attributes provide a much broader spectrum of colors, including vivid and intense hues [56]. Synthetic dyes can also be tailored to attain precise and consistent color results. Synthetic dyes used in the industry are azo, reactive, acid, sulfur, mordant, direct, dispersed, and basic dye [57]. Azo dyes represent the primary category of synthetic dyes, comprising 60% of the total synthetic dye market [56]. Certain textile dyes and chemicals used in textile production may contain heavy metals which have the potential to leach into wastewater [57]. Those heavy metals include lead, cadmium, mercury, chromium, and arsenic [57]. Thus, the ITWW is complex, often containing a variety of pollutants and contaminants [58].

The textile industry also contributes to GHG emissions, primarily because of energy-intensive processes like fiber production, dyeing, finishing, and transportation [59]. Indeed, it is worth noting that synthetic fibers, like polyester, release significant amounts of carbon dioxide (CO_2), contributing to the industry's carbon footprint [43]. The industry's traditional linear "make-use-dispose" model is inherently unsustainable in the long term, as it has led to significant environmental repercussions, including resource depletion and extensive waste generation. Ongoing efforts are to tackle the textile sector's environmental impacts by implementing a circular economy to maximize resource efficiency and minimize waste by assuring resources, raw materials, and final products are used for as long as possible [60].

It has been discussed in several areas that the future of the textile industry lies in its ability to embrace the circular economy concept [28]. Circular economy initiatives within the textile industry have been focused primarily on reducing waste and extending product lifecycles [61]. This objective is achieved through recycling, upcycling, altering attitudes toward fashion trends, and promoting responsible disposal methods [43,45,46]. These measures aim to decrease the demand for new textile production, which helps reduce emissions. This shift towards circular practices is essential for a more sustainable textile industry.

The circular economy application to the clothing industry entails various methodologies. These include designing for durability and longevity, promoting reuse and repair, implementing recycling, and upcycling practices, and prioritizing sustainable materials. Furthermore, responsible production methods are emphasized in minimizing water and energy consumption. This holistic approach contributes to a more sustainable and environmentally conscious clothing industry [43,45,47,60,62,63]. Ta et al. studied customer perceptions regarding using and recycling clothes; the results revealed a remarkably positive response despite the potential stigma surrounding second-hand clothing, indicating a growing acceptance of sustainable fashion practices among consumers [64]. However, reuse and recycling management is not easy, and there are several challenges, such as the expenses tied to human resources, materials, technology, and transportation. Moreover, the process involves significant energy and water consumption, potentially leading to solid waste generation and wastewater production. Addressing the sufficiency of collected resources and adhering to safety regulations are vital components of this endeavor [65].

Policymakers, such as governments, hold a significant role in driving this change by implementing policies and incentives that support circular practices within the industry. Furthermore, consumer education about the benefits of circular fashion in contrast to the environmental impacts of fast fashion is a crucial undertaking that should be integrated into the sustainable agenda of the textile industry. This collective approach can lead to meaningful progress in creating a sustainable fashion sector [63,64].

In the context of top-producing countries like India, Vietnam, and Bangladesh, Saha et al. highlighted crucial steps and challenges of a circular economy in the textile industry:(i) stopping using non-sustainable toxic materials, (ii) using fully recoverable materials, (iii) minimizing the environmental impact of each raw resource, (iv) using of sustainable energy sources, (v) optimizing forward and reverse supply chain, (vi) adopting green marketing strategy and product system services, and (vii) implementing efficient reuse-recycle system [66].

While the literature has extensively documented efforts to apply circular economy principles focused on clothing, there is still a significant gap in global research regarding the recovery processes of valuable materials from pre- and post-consumer industrial streams. For instance, the reutilization of ITWW not only has the potential to achieve a zero-discharge standard but also offers an opportunity to extract valuable materials from this wastewater source. Indeed, significant opportunities exist in the bio or chemical transformation of fibers within post-consumer textiles. This process can potentially yield feedstock to prepare value-added chemicals and materials.

9.3 GHG EMISSIONS FROM MUNICIPAL AND INDUSTRIAL WASTEWATER TREATMENT

In recent years, a significant focus has been on studying and quantifying the GHG generated by wastewater treatment processes. These studies played a central role in comprehending and addressing the environmental impact of wastewater operations. Carbon emissions are commonly quantified as CO_2-equivalents (CO_2-eq). CO_2-eq emissions encompass direct emissions (including-CO_2, CH_4, and N_2O emissions originating directly from the water treatment operations) and indirect emissions (electricity consumption, chemical usage, and other related factors) [67]. Varies protocols that quantify CO_2-eq are used in the literature, for example, the IPCC, bridle model [68] and BSM2G [69].

Estimations of CO_2-eq emissions from conventional wastewater treatment facilities have been calculated in various countries. For instance, it has been estimated wastewater treatment plants in Canada produced approximately 1,600 megagrams per year of methane (CH_4) and 669,100 megagrams per year of carbon dioxide (CO_2) [70]. China has experienced a significant surge, increasing from 326.54 gigagrams of CO_2-eq (Gg CO_2-eq) in 2005 to 1,294.03 Gg CO_2-eq in 2014 [71]. These figures demonstrate that wastewater treatment can be a significant source of GHG emissions.

N_2O emissions are also crucial to GHG emissions with an equivalence of 310 times greater that of CO_2 on global warming potential [72]. N_2O emissions are affected by N-organic content and dissolved oxygen. The concentration of dissolved oxygen plays a crucial role in N_2O emissions. When dissolved oxygen increases, denitrification is incomplete, leading to elevated N_2O emissions and increased energy consumption due to the additional aeration required [73].

Indirect GHG emissions constitute most total emissions and are directly tied to energy consumption [73–76], chemical production [73–76], and transportation [76]. The proportion of indirect GHG emissions varies according to water treatment operation. For example, aerobic biological water treatment can account for the highest indirect CO_2 emission. Biological-activated sludge produces more elevated CO_2 (52%) and CH_4 (20%) emissions than anaerobic digestion [77]. Membrane bioreactor systems (MBR) and sequencing batch reactors (SBR) are known for GHG emissions due to the electricity for their aeration tanks [74,76,78].

Several strategies have been suggested to consistently manage indirect and direct GHG emissions from municipal wastewater treatment plants. These approaches are important for mitigating the environmental impact of wastewater treatment plants. Regarding biological water treatment treatments, anaerobic treatment has been highly recommended for reducing direct GHG emissions [79]. Nonetheless, ensuring the proper utilization of the biogas produced during anaerobic treatment is crucial. Also, transitioning from an aerobic to an anaerobic digester can reduce energy consumption by around 50%, presenting a significant efficiency gain. However, it is important to note that this shift may increase GHG emissions, rising from 0.64 kg CO_2-eq/m^3 to 1.65 kg CO_2-eq/m^3 [80]. Controlling parameters in water treatment, such as fuzzy controllers and dissolved oxygen control, along with integrating innovative treatment sequences and natural bio-based water treatments, have shown remarkable potential in reducing GHG emissions, improving treated water quality, and reducing operational costs [81,82]. Natural bio-based water treatments, for example, constructed wetlands and microalgae cultivation, offer a significant advantage in capturing GHG from the environment [83,84]. Direct GHG emissions of constructed wetlands are affected by the species used. The choice of species influences direct GHG emissions. C. indica stands out for its lower global warming potential among other similar species [85]. Duckweed ponds can fix three times more CO_2 than they emit, confirming their sustainability [86].

In Mexico, a study explored the potential of combining biological treatments for GHG mitigation. The findings revealed that the aerobic-anaerobic process reduced GHG emissions by 27% compared to the aerobic process. This improvement was attributed to the ability to reuse CH_4 and its lower energy consumption [87]. Another action to reduce emissions of CH_4 and N_2O is to reduce the treatment volume [69]. For example, in rural wastewater treatment, an integrated gravitational-flow wastewater treatment system incorporated a sand filter and anaerobic process and constructed wetlands to mitigate GHG emissions effectively [88].

In particular, the textile industry generates around one billion tons of CO_2-eq, nearly 10% of world GHG emissions. Thus, this industry significantly contributes to global warming potential and even exceeds those of the international aviation and shipping sectors, due to extended supply chains and energy-intensive production operations [43]. To mitigate GHG emissions, the textile sector within the fashion industry needs to transition towards decarbonizing its entire supply chain, including wastewater management.

A limited number of studies estimate GHG emissions from the treatment of ITWW. However, an essential strategy for minimizing emissions in wastewater treatment within the textile industry may involve focusing on energy, water efficiency, and pollution reduction. Some promising technologies to reduce consumption in the textile industry include super-critical CO_2 and electrochemical dyeing, digital printing, plasma and microwave technologies, foam technology in textile finishing, and alternative textile auxiliaries [89]. On the other hand, there has been a rising trend in adopting zero liquid discharge technologies [90]. For instance, the solar wastewater treatment system was recommended instead of conventional biological treatments in Algeria, with remarkable effectiveness in treating textile wastewater [91]. Given that most studies on GHG emissions have been conducted in municipal wastewater treatment plants, notable opportunities exist for conducting extensive research on both direct and indirect GHG emissions from treating ITWW. Such work would provide valuable insights for suggesting strategies to decarbonize the industrial textile wastewater sector.

9.4 COMBINED RESOURCE RECOVERY AND GHG MITIGATION THROUGH ADVANCED WATER TREATMENT PROCESSES

Treating wastewater requires substantial energy, chemicals, and materials to meet discharge standards. A paradigm shift has occurred in wastewater treatment which recognizes the valuable resources that can be reclaimed [92]. For instance, within municipal wastewater, organic recovery can significantly reduce CO_2 emissions and sludges, while phosphate recovery can alleviate the depletion of deposits on Earth [93]. Energy recovery can also be done by implementing anaerobic digestion [94]. Similarly, industrial textile wastewater treatment has the potential to enhance overall sustainability by minimizing waste generation, reducing non-renewable resource consumption, and facilitating resource recycling [92]. The recovery of chemicals, nutrients, energy, and water could be achieved by properly selecting a sequence of treatments.

Since the electrocoagulation (EC) process was patented [95], countless studies have demonstrated the effectiveness of this process in removing various contaminants such as suspended solids, colloidal particles, heavy metals, oil droplets, dyes, natural organic matter, and nutrients from different wastewater sources [96]. In recent years, studies on EC have expanded the scope to explore innovative applications beyond its traditional uses. These include the removal of contaminants of emerging concern [97], microplastics [98], and per/polyfluoroalkyl substances [99]. Innovative methods such as electrocoagulation have the potential to fractionate ITWW by facilitating the recovery of valuable resources.

EC is a treatment system that forms hydroxy-metallic coagulants using electrical currents and electrodes. Commonly used sacrificial anodes, such as aluminum and iron electrodes, facilitate the formation of metal hydroxide flocs as a crucial step in this process. These hydroxy-metallic species can neutralize the superficial electrostatic charges on suspended solids, thereby promoting their aggregation and subsequent sedimentation from the aqueous phase. Compared with traditional chemical coagulation-flocculation, EC produces a low amount of sludge [50]. In addition to the in-situ formation of coagulants, the concurrent hydrogen evolution at the cathode provides the opportunity for harvesting it alongside pollutant removal through sedimentation and flotation. For aluminum electrodes, hydrogen gas generation from water reduction at the cathode is shown in Equations (9.1) and (9.2) [100].

At Anode:

$$Al \rightarrow Al^{3+} + 3e^- \rightarrow Al_{(s)} \qquad E^0 = -1.662\,V \tag{9.1}$$

At Cathode:

$$2H_2O + 2e^- \rightarrow H_{2(g)} \uparrow + 2OH^- \quad E^0 = -0.834\,V \tag{9.2}$$

The coupling of EC with other water treatment technologies may facilitate the recovery of valuable resources. Aouni et al. observed that the hybrid EC-nanofiltration process can remove turbidity, COD, sodium, magnesium, potassium, chloride, and sulfate from ITWW [50]. A hybrid EC-nanofiltration process to treat ITWW has shown to be an effective process for dye removal with high and medium molecular weight (reducing 68% of COD), and the nanofiltration fractionated dyes with low molecular weight and salts, allowing the reuse of salts and water [8]. EC can be a pre-treatment method that enhances permeate flux, reduces cake resistance, minimizes cake layer formation, and fractionates high molecular weight chemicals [101].

Hydrogen generation can facilitate the aggregation of lighter particles; thus, enhancing pollutant removal efficiency during the flotation process [100]. Another notable benefit of electrocoagulation is the concurrent production of hydrogen as a valuable by-product. EC can be directly powered by photovoltaic energy without battery storage to degrade dyes while simultaneously producing hydrogen [102]; this method involved dynamically adjusting the flow rate in response to fluctuations in photovoltaic current to ensure the current/flow ratio remained constant. Following the conventional up-flow anaerobic sludge bed design, the hydrogen product was harvested in a three-phase separator [51,102]. Similarly, Deghles and Kurt researched hydrogen harvesting for tannery wastewater treatment, employing aluminum and iron electrodes in highly colored wastewater [103]. This study revealed that the optimal COD removal occurred at pH 6 using a current density of 14 mA cm^{-2}, producing hydrogen equivalent to 16% of the electrical energy demand of the EC process [103].

These examples demonstrate that electrocoagulation holds promise as a practical approach for energy recovery, resource recovery, and wastewater reuse, as illustrated in Figure 9.1. Although electrocoagulation does not require additional chemicals to remove pollutants, drawbacks include anode replacement, cathode passivation, sludge generation, and electricity consumption [10]. Some studies have demonstrated that renewable energy sources like solar panels can power electrocoagulation processes [102]. However, in the context of the industrial textile sector, a gap exists in pilot studies and comprehensive cost analyses of hydrogen harvesting and resource recovery from sludge.

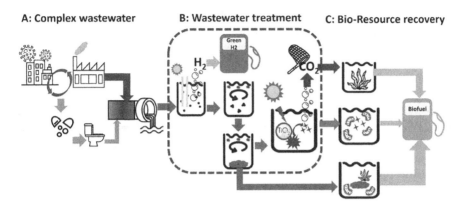

A: Complex wastewater   B: Wastewater treatment   C: Bio-Resource recovery

FIGURE 9.1 EC for resource recovery and greenhouse gas mitigation. Dashed lines note wastewater treatment project boundaries. Solid lines note added value streams. Generation of complex wastewater from combined industrial and domestic sources (a). Treatment opportunities for mitigating GHG emissions and recovery of resources (b). Biotechnological alternatives for further resource recovery (c).

Generating energy from wastewater during the treatment process is considered a novel idea in the circular economy [104]. A fuel cell can turn chemical energy into fuel sources. Ethanol, methanol, hydrogen, and organic matter could be converted into electricity through electrochemical reactions. In recent decades, photocatalytic fuel cells relying on semiconductors have emerged as promising technologies for decomposing organic pollutants while generating electricity and producing hydrogen. Photocatalysis has been extensively studied and applied to degrade recalcitrant pollutants in water [105]. When a semiconductor absorbs a photon with higher energy than its bandgap, photoexcitation of electrons in the valence band promotes them towards the conduction band, creating a hole in the valence band. The hole can react with absorbed hydroxide ions on the surface to form hydroxyl radicals or transfer electrons from absorbed organic molecules serving as electron donors. Photogenerated electrons in the conduction band can either be transferred to electron acceptors under aerobic conditions, bringing reactive oxygen species. In the absence of O_2, electrons can react with hydrogen ions to produce H_2 gas [106]. Indeed, electron-hole pairs may recombine, which typically results in the release of heat energy rather than participating in the photocatalytic reaction for pollutant degradation or hydrogen generation. Some semiconductors that are promising in photocatalysis for organic pollutant degradation include g-C_3N_4, TiO_2, MnO_2, Fe_2O_3, CuO, ZnO, MoS_2, WO_3, $BiWO_6$, $BiVO_4$, and CeO_2 [107].

The photocatalytic processes can be integrated into a fuel cell for simultaneous pollutant removal and energy generation [105,108]. A photocatalytic fuel cell consists of a photoanode (n-type semiconductor) and a photocathode (p-type semiconductor) with higher and lower Fermi level potential, respectively [109]. Upon light activation, the photocatalytic reaction occurs due to the difference between bandgap energies of electrodes and results in photoelectrons that can be transferred through an external circuit from the photoanode to the cathode, leading to an electrical current [109]. Khalik et al. conducted research on energy generation with concurrent degradation of ITWW using ZnO loaded on a carbon plate as the photoanode and CuO loaded on a carbon plate as the photocathode. Their findings indicated that photocurrent generation facilitated the formation of electron-hole pairs, contributing to the degradation of ITWW [11]. Another study treated textile dye effluent from small-scale workshops by employing a solar photocatalytic fuel cell, and the optimal configuration involved a closed circuit with an external resistor of 250 Ω, with 31.9% dye degradation [5].

While numerous effective strategies for enhancing the photocatalytic fuel cell in ITWW have been previously reported, further optimization is necessary to reduce prolonged irradiation times. This optimization is essential to make photocatalytic fuel cells a viable candidate for energy recovery and conversion technology in industrial textile applications.

9.5 RESOURCE RECOVERY USING BIOTECHNOLOGICAL APPROACHES

The textile industry generates vast volumes of wastewater characterized by elevated concentrations of hazardous substances. Azo dyes are the most prevalent, known for their high solubility in water, low degradability, and significant toxicity to aquatic life and humans. Considering the substantial environmental and public health risks linked to azo dyes, it is imperative to implement treatment processes for their targeted removal [110,111].

The complex molecular structures of these dyes pose a challenge to their removal from effluents through sole reliance on physicochemical treatments due to the potential generation of hazardous sludge and incomplete degradation through the formation of by-products [110,112]. Hence, a growing interest exists in exploring innovative and eco-friendly technologies. Biotechnological strategies demonstrate promise for the degradation of dyes. Several bacterial strains have proven effective in treating dye-containing textile effluents. Typically, these bacterial strains are isolated from the soil near textile effluent sludge waste, textile manufacturing plants, and wastewater contaminated with dyes. A diverse range of bacterial genera, including *Bacillus, Pseudomonas, Proteus*, and *Geobacter,* have shown significant potential in effectively degrading dyes found in textile wastewater [113,114]. In various biotechnological approaches, the immobilization of bacterial cells has been employed. This technique enhances their tolerance to high dye concentrations, resulting in a remarkable degradation efficiency compared to free cells (see Table 9.1).

Immobilization can also provide a protective shield to the cells against toxic dye concentrations. It ensures operational stability, rendering this approach cost-effective since the cells can be reused for numerous removal cycles. Natural and synthetic polymers employed for cell immobilization include cellulose, chitosan, alginate, polyurethane, polyvinyl chloride, and polyacrylamide. In addition, microbial consortium cultures have also yielded important results in dye degradation, achieving more than 90% remotion efficiency compared to their corresponding monocultures thus, decreasing the treatment process time. In general, bacterial mechanisms for dye decolorization include the sorption by the microbial cell and biodegradation through different enzyme activities, *e.g.*, azoreductase, peroxidase, laccase, and ligninolytic [115].

TABLE 9.1
Microorganisms Commonly Used to Degrade Dyes in ITWW

Microorganism or Biosystem	Approach Used	Dye	Efficiency	References
Bacillus cereus	Liquid monoculture	Novacron, Super	91%	[116]
Alcaligenes faecalis		black G	90%	
Bacillus spp.			88%	
Bacillus albus DD1	Liquid monoculture	Reactive Black 5	98.5%	[120]
Bacillus sp.	Cells immobilized on polyvinyl alcohol-calcium alginate-activated carbon beads	Reactive Blue 19		[115]
Pseudomonas guariconensis	Calcium alginate	Reactive red 190	91%	[121]
Bacillus flexus TS8	Consortium culture in continuous up-flow packed bed bioreactor	Indanthrene Blue RS	90%	[122]
Proteus mirabilis				
Pseudomonas aeruginosa				
Pseudomonas putida	Consortium culture	CI Reactive blue 40	99%	[123]
Chlorella				
Lactobacillus plantarum				
WGC-D	Bacterial consortium	DR23, DB15, DY12	80%, 70%, 83%, respectively	[124]
Trametes hirsuta D7	Cell immobilized in light-expanded clay aggregate (LECA)	Reactive black (RB5) Acid Blue 113, Acid Orange 7	92%, 97%, 30%, respectively	[125]
Fusarium oxysporum	Monoculture	Crystal Violet	0.93%–15.45%	[126]
		Aniline Blue	10.9%–35.5%	
		Reactive Black 5	9.69%–34.67%	
		Orange II	2.87%–10.98%	
Laccase from Pleurotus ostreatus	Immobilized on LECA	Reactive Black 5	58%	[127]
Vetiver plant and microbial consortium	Sequential anaerobic–aerobic	Methyl red	92%	[119]
Yarrowia sp. SSA1642, Barnettozyma californica SSA1518, Sterigmatomyces halophilus SSA1511	Three yeast consortiums	Red HE3B	>82%	[128]

Biotechnological processes are advantageous because the dyes can be transformed into safe compounds [116].

Similarly, eukaryotic organisms like plants, fungi, and microalgae have been employed for decolorizing textile wastewater (see Table 9.1 for specific examples). These organisms use diverse biological mechanisms to remove dyes from wastewater.

For instance, microalgae undergo metabolic processes to convert chromophores into CO_2 and H_2O molecules. and the algal biomass can adsorb these chromophores [117]. Fungi can decompose organic matter by producing an enzymatic arsenal that allows them to effectively degrade dyes found in textile wastewater. These enzymes include laccase, manganese, and lignin peroxidase [118]. Another notable biosystem involves using the vetiver plant with a microbial consortium, which has proved highly effective in decolorizing textile wastewater through an aerobic–anaerobic process. This approach capitalizes on the vetiver plant's remarkable capacity to remediate various pollutants, *e.g.*, phenol, petroleum hydrocarbons, and heavy metals [119].

Furthermore, using biological systems for the treatment of wastewater creates opportunities to convert toxic compounds into valuable resources such as biofuels, electricity, and pigments. The circular economy supported by biotechnology is an attractive concept for reusing, reducing, and treating industrial wastewater effluents [129]. The nutrients and carbon in wastewater make it ideal for the large-scale cultivation of microalgae [130]. Microalgae have received increasing attention as it is a prominent feedstock for biofuel production. Many algae species exhibit high lipid content within their cells, enabling them to produce and accumulate energy-rich oil [131]. The cultivation of microalgae has found successful application in textile wastewater, effectively removing both organic and inorganic compounds [129].

Pathak et al. found *Chlorella pyrenoidosa* grew in concentrated textile wastewater. The benefits of that were *Chlorella* reduced phosphate, nitrate, and BOD by 87%, 82%, and 63%, respectively [132]. Huy et al. conducted a study on microalgae consortia cultivation in textile wastewater. Interestingly, adding extra nitrogen and phosphorous was able to enhance removal efficiencies with rates of 78.78% for COD, 93.3% for N, and 100% for P [131]. Notably, biodiesel from microalgae grown in a 50% dilution of ITWW using *Chlorella vulgaris* resulted in an 82% COD removal [133]. Furthermore, Marazzi et al. suggested a mixture of a microalgal and bacterial community can maximize valorization and bioremediation of ITWW as some forms of microalgae (e.g., *Scenedesmaceae*) can grow in synergy with dye-degrading bacteria [48].

The carbohydrate content of microalgae is directly correlated with hydrogen production in anaerobic dark fermentation, and the protein and lipid content can be converted into methane and biodiesel. Thus, textile wastewater can serve as an ideal medium for microalgae cultivation. Anaerobic fermentation offers a dual advantage of remediating the textile wastewater while providing a valuable feedstock for further applications.

Microbial fuel cells represent another bacteria-based biotechnology that can be effectively employed for textile wastewater treatment [134]. Simultaneously, this process degrades pollutants and generates bioelectricity, promoting a circular economy and enhancing overall sustainability [113]. Since Porter proposed that the degradation of organic matter by microorganisms liberates electrical energy [135], the development of fuel cells containing microorganisms that use wastewater as a substrate for their reactions has made remarkable progress [136,137]. Microbial fuel cells offer several advantages, including directly converting organic substrates into bioelectricity and reducing sludge generation [138]. A microbial fuel cell consists of an anaerobic anode compartment and a cathode exposed to aerobic solutions, separated by a proton exchange membrane. Protons move from the anode to the cathode through the membrane, while an external circuit allows for the flow of an electrical current [139]. Electrons and protons, generated through substrate oxidation in the anode, combine with oxygen in the cathode to produce water.

Utilizing microalgal biocathodes, Logrono et al. demonstrated the simultaneous reduction of COD by 98% in ITWW alongside bioelectricity generation with a maximum open circuit voltage of 420 mV [140]. Kumar et al. conducted a study demonstrating the economic viability of energy recovery while simultaneously removing dye contaminants from textile industry effluents through stacking microbial fuel cells arranged in a parallel configuration [138]. This study achieved an open circuit voltage of 1.09 V and a COD removal efficiency of 82%. Another experiment showed that the microalgal biocathodes could generate a power density of 123.2 mW m^{-3}, with notable removals of COD (92%–98%), Cr (54%–80%), and Zn (98%) [140].

These studies demonstrate that the microalgal systems could potentially serve as a promising alternative to a Pt cathode and recover nutrients. Integrating microbial fuel cells and microalgae cultivation represents a promising approach. It addresses textile wastewater treatment and offers a potentially economical method for producing value-added products and generating bioelectricity [134]. Thus, microbial fuel cells can be an efficient energy recovery approach from ITWW. Nevertheless, numerous challenges are associated with low overall performance and scaling up this technology to industrial levels, especially considering that most investigations have been conducted at the lab scale.

9.6   DYES RECOVERY FROM TEXTILE WASTEWATER

The industrial wastewater obtained after dye synthesis or dying processing has high salinity (i.e., >6.0% NaCl or >5.0% Na_2SO_4) due to the neutralization reaction and addition of additives [141]. For the production of one ton of dyed products in the textile industry, approximately 200–350 m^3 of pure water is required, resulting in the substantial generation of textile wastewater [142]. The evolving approach of viewing wastewater as a resource rather than a waste recognizes its potential for recycling nutrients, energy generation, resource extraction, and water reclamation.

Reusing dyes obtained from textile wastewater can lead to a substantial reduction in costs and decreasing effluent loads. However, conventional approaches like adsorption, chemical coagulation-flocculation [143], or advanced oxidation process [144] primarily focus on dye degradation. This limitation restricts the potential for dye and salt recovery from ITWW. Therefore, emerging techniques such as enhanced nanofiltration, electrodialysis [145], pressure retarded osmosis [146], forward osmosis [147], and bipolar membrane electrodialysis [148] are gaining prominence as resource recovery operations from ITWW.

Nanofiltration is one of the most effective techniques for separating dyes and salts in textile wastewater due to its high removal efficiency. Nevertheless, commercially available dense nanofiltration membranes tend to exhibit high salt retention, leading to a significant decrease in permeate flux. Growing interest exists in identifying enhanced nanofiltration membranes capable of fractioning dyes and salts from ITWW, facilitating dye recovery and salt recycling. Porous nanofiltration membranes with molecular weight cut-offs from 500 to 1000 Da have gained attention for their effectiveness in separating dyes and salts [149]. The loose surface structure of these nanofiltration membranes enables substantial passage of NaCl while effectively retaining dye molecules, signifying their significant potential in the field of dye-salt separation.

Ye et al. modified the superficial properties of a porous nanofiltration membrane through bioinspired deposition of graphitic carbon nitride nanosheets with a polydopamine/polyethylenimine layer. This modification enabled the membrane to achieve over 99% rejection of reactive dyes while allowing nearly unhindered transmission of salts such as NaCl and Na_2SO_4 [150]. Zheng et al. employed an interfacial polymerization process involving the active monomers 1,3,5-tris(4-aminophenyl) benzene and trimesoyl chloride, to engineer a loose yet highly crosslinked nanofiltration membrane [151]. This system demonstrated exceptional water permeance, low salt rejection, and remarkable efficacy in rejecting various reactive dyes, including Congo red, reactive orange 16, direct red 23, and reactive black 5. Furthermore, the resulting membrane displayed antifouling properties and remarkable stability at up to 72 h of continuous filtration.

Furthermore, Harruddin et al. employed a liquid membrane process to selectively remove and recover black B reactive dye from simulated textile wastewater [152]. In this process, a kerosene–tridodecylamine liquid membrane was supported on a commercial polypropylene membrane. This process achieved an impressive 99% recovery of Black B dye achieved at pH 2 while operating at a flow rate of 150 mL min^{-1}. Ye et al. designed an ultrafiltration-diafiltration membrane process to fractionate an aqueous mixture of dye/Na_2SO_4, with 98% desalination efficiency and over 97% dye recovery from textile wastewater [153]. To assess the efficiency of this membrane process, the retention behavior of four dyes (reactive blue 2, direct red 80, direct red 23, and Congo red) was thoroughly investigated.

Electrodialysis is another promising technique for separating dyes and salts in textile wastewater. When exposed to a steady electric current, the cations, and anions within the highly saline mixture of dyes and salts undergo continuous transport across the ion exchange membranes. This process holds significant potential for efficiently desalinating highly saline mixtures. Berkessa et al. conducted a study on the viability of bipolar membranes for electrodialysis in the desalination of reactive dye/ salt mixtures [154]. They found efficiency is mainly attributed to the formation of a dye cake layer on the anion exchange membrane. This cake layer facilitated the entry of dye into the pores of the anion exchange membranes, where they bound to positively charged groups through electrostatic attraction.

Lin et al. introduced an innovative configuration of loose nanofiltration membranes in electrodialysis to address this limitation as a substitute for anion exchange membranes [155]. This nanofiltration membrane achieved a desalination efficiency of 99% as an anion-conductive membrane in electrodialysis for NaCl solutions. This efficiency was attributed to the porous surface structure, which provided ample nano-channels, enhancing ion transport.

Bipolar membrane electrodialysis is a sustainable and environmentally friendly technique for textile wastewater treatment. The membrane electrodialysis technique integrates elements of both conventional electrodialysis and bipolar membranes. This technique presents opportunities for the clean production of acids or bases, energy generation, synthesis of ionic liquids, and CO_2 extraction [156]. Lin et al. combined the membrane electrodialysis process with a loose nanofiltration membrane (Sepro NF 6) to fractionate dyes and salt. A very low rejection of salt (0.27% in 120 g L^{-1} NaCl solution) and up to 99.9% rejection of reactive dyes was observed [148]. The dissociation of salts into acids and bases through the membrane electro-dialysis technique introduces new directions for raw material recycling and achieving zero liquid discharge. Like many membrane processes, the efficiency of this technique is hindered by fouling.

9.7 CONCLUSIONS

Clothing is an integral aspect of our society tied to the textile industry. The textile industry is a generator of polluted wastewater. Treating industrial textile wastewater should encompass more than mere dye or contaminant removal. It should achieve sustainability by using renewable resources, minimizing waste generation, enabling resource recycling, and mitigating greenhouse emissions. The success of implementing those sustainable strategies hinges on the active involvement of industry, the government, and society. Although the literature on direct GHG mitigation in textile wastewater treatment plants is limited, strategies previously implemented in municipal wastewater treatment may be employed in textile wastewater regarding the combination of water treatments or adopting hybrid processes that have multifunctional treatment necessary for ITWW and can mitigate GHG emissions.

This book chapter shows that electro-separation technologies are gaining attention for their crucial role in achieving contaminant removal, resource recovery, and potentially reducing GHG emissions. However, notable opportunities exist to increase the selectivity of recovery, optimize operating conditions, and scale up to realistic industrial scenarios. Several electro-separation systems are worth continued study in textile wastewater treatment, including electro-membrane processes, peroxy-coagulation, electron-precipitation, electro-deposition, electro-sorption, and electro-filtration. Furthermore, membrane fractionation or bio-based methods could significantly enhance their efficiency. Resource recovery in the textile industry can potentially significantly decrease overall environmental impact, but there has been limited research on an industrial scale, necessitating the outline of comprehensive implementation strategies.

ACKNOWLEDGMENTS

This book chapter received support from the Open University and the Waste2Fresh grant under the Horizon 2020 program (reference number 9584912). The authors express their gratitude to the Open University for providing the necessary financial resources, access to facilities, and valuable training that enabled the successful completion of this project.

REFERENCES

1. Jorge, A. M. S., Athira, K. K., Alves, M. B., Gardas, R. L. & Pereira, J. F. B. Textile dyes effluents: A current scenario and the use of aqueous biphasic systems for the recovery of dyes. *J. Water Process Eng.* **55**, 104125 (2023).
2. Falini, G. et al. Natural calcium phosphates from circular economy as adsorbent phases for the remediation of textile industry waste-waters. *Ceram. Int.* **49**, 243–252 (2023).
3. Paździor, K., Bilińska, L. & Ledakowicz, S. A review of the existing and emerging technologies in the combination of AOPs and biological processes in industrial textile wastewater treatment. *Chem. Eng. J.* **376**, 120597 (2019).
4. Asgari, G. et al. Mineralization and biodegradability improvement of textile wastewater using persulfate/dithionite process. *Biomass Convers. Biorefin.* (2023) doi:10.1007/s13399-023-04128-6.
5. Khalik, W. F. et al. Enhancement of simultaneous batik wastewater treatment and electricity generation in photocatalytic fuel cell. *Environ. Sci. Pollut. Res.* **25**, 35164–35175 (2018).
6. Jorfi, S. et al. Enhanced coagulation-photocatalytic treatment of Acid red 73 dye and real textile wastewater using UVA/synthesized MgO nanoparticles. *J. Environ. Manage.* **177**, 111–118 (2016).
7. GilPavas, E., Dobrosz-Gómez, I. & Gómez-García, M. Á. Coagulation-flocculation sequential with Fenton or photo-Fenton processes as an alternative for the industrial textile wastewater treatment. *J. Environ. Manage.* **191**, 189–197 (2017).
8. Tavangar, T., Jalali, K., Alaei Shahmirzadi, M. A. & Karimi, M. Toward real textile wastewater treatment: Membrane fouling control and effective fractionation of dyes/inorganic salts using a hybrid electrocoagulation–nanofiltration process. *Sep. Purif. Technol.* **216**, 115–125 (2019).
9. Bener, S., Atalay, S. & Ersöz, G. The hybrid process with eco-friendly materials for the treatment of the real textile industry wastewater. *Ecol. Eng.* **148**, 105789 (2020).
10. Bulca, Ö., Palas, B., Atalay, S. & Ersöz, G. Performance investigation of the hybrid methods of adsorption or catalytic wet air oxidation subsequent to electrocoagulation in treatment of real textile wastewater and kinetic modelling. *J. Water Process Eng.* **40**, 101821 (2021).
11. Khalik, W. F. et al. Converting synthetic azo dye and real textile wastewater into clean energy by using synthesized CuO/C as photocathode in dual-photoelectrode photocatalytic fuel cell. *Environ. Sci. Pollut. Res.* **30**, 58516–58526 (2023).
12. Sathya, U., Keerthi, Nithya, M. & Balasubramanian, N. Evaluation of advanced oxidation processes (AOPs) integrated membrane bioreactor (MBR) for the real textile wastewater treatment. *J. Environ. Manage.* **246**, 768–775 (2019).
13. Alves, P. A. et al. Photo-assisted electrochemical degradation of real textile wastewater. *Water Sci. Technol.* **61**, 491–498 (2010).
14. Lellis, B., Fávaro-Polonio, C. Z., Pamphile, J. A. & Polonio, J. C. Effects of textile dyes on health and the environment and bioremediation potential of living organisms. *Biotechnol. Res. Innov.* **3**, 275–290 (2019).
15. Ahsan, A. et al. Wastewater from the textile industry: Review of the technologies for wastewater treatment and reuse. *Korean J. Chem. Eng.* **40**, 2060–2081 (2023).
16. Holkar, C. R., Jadhav, A. J., Pinjari, D. V, Mahamuni, N. M. & Pandit, A. B. A critical review on textile wastewater treatments: Possible approaches. *J. Environ. Manage.* **182**, 351–366 (2016).
17. Flores Alarcón, M. A. D. et al. Efficient dye removal from real textile wastewater using orange seed powder as suitable bio-adsorbent and membrane technology. *Water (Basel)* **14**, 4104, (2022).
18. Khajeh, M., Taheri, E., Amin, M. M., Fatehizadeh, A. & Bedia, J. Combination of hydrodynamic cavitation with oxidants for efficient treatment of synthetic and real textile wastewater. *J. Water Process Eng.* **49**, 103143 (2022).

19. Bandala, E. R., Peláez, M. A., García-López, A. J., Salgado, M. de J. & Moeller, G. Photocatalytic decolourisation of synthetic and real textile wastewater containing benzidine-based azo dyes. *Chem. Eng. Process.: Process Intensif.* **47**, 169–176 (2008).

20. He, S. et al. Enhancement of biodegradability of real textile and dyeing wastewater by electron beam irradiation. *Radiat. Phys. Chem.* **124**, 203–207 (2016).

21. Vieira, Y., Ceretta, M. B., Foletto, E. L., Wolski, E. A. & Silvestri, S. Application of a novel rGO-CuFeS2 composite catalyst conjugated to microwave irradiation for ultra-fast real textile wastewater treatment. *J. Water Process Eng.* **36**, 101397 (2020).

22. Badawi, A. K. & Zaher, K. Hybrid treatment system for real textile wastewater remediation based on coagulation/flocculation, adsorption and filtration processes: Performance and economic evaluation. *J. Water Process Eng.* **40**, 101963 (2021).

23. Ayed, L., Ksibi, I. El, Charef, A. & Mzoughi, R. El. Hybrid coagulation-flocculation and anaerobic-aerobic biological treatment for industrial textile wastewater: Pilot case study. *J. Text. Inst.* **112**, 200–206 (2021).

24. Kaur, P., Kushwaha, J. P. & Sangal, V. K. Electrocatalytic oxidative treatment of real textile wastewater in continuous reactor: Degradation pathway and disposability study. *J. Hazard. Mater.* **346**, 242–252 (2018).

25. Oller, I., Malato, S. & Sánchez-Pérez, J. A. Combination of advanced oxidation processes and biological treatments for wastewater decontamination–A review. *Sci. Total Environ.* **409**, 4141–4166 (2011).

26. Titchou, F. E. et al. Removal of organic pollutants from wastewater by advanced oxidation processes and its combination with membrane processes. *Chem. Eng. Process.–Process Intensif.* **169**, 108631 (2021).

27. Ceretta, M. B., Vieira, Y., Wolski, E. A., Foletto, E. L. & Silvestri, S. Biological degradation coupled to photocatalysis by ZnO/polypyrrole composite for the treatment of real textile wastewater. *J. Water Process Eng.* **35**, 101230 (2020).

28. Peter John, E. & Mishra, U. A sustainable three-layer circular economic model with controllable waste, emission, and wastewater from the textile and fashion industry. *J. Clean. Prod.* **388**, 135642 (2023).

29. Ahmed, M. et al. Recent developments in hazardous pollutants removal from wastewater and water reuse within a circular economy. *NPJ Clean Water* **5**, 12 (2022).

30. To, M. H., Uisan, K., Ok, Y. S., Pleissner, D. & Lin, C. S. K. Recent trends in green and sustainable chemistry: Rethinking textile waste in a circular economy. *Curr. Opin. Green Sustain. Chem.* **20**, 1–10 (2019).

31. Zhan, X. Greenhouse gas emissions from wastewater treatment facilities. In *Greenhouse Gas Emission and Mitigation in Municipal Wastewater Treatment Plants* 17–28 (International Water Association, 2018). doi:10.2166/9781780406312_017.

32. He, Y. et al. Net-zero greenhouse gas emission from wastewater treatment: Mechanisms, opportunities and perspectives. *Renew. Sustain. Energy Rev.* **184**, 113547 (2023).

33. Mannina, G. et al. Greenhouse gases from wastewater treatment –A review of modelling tools. *Sci. Total Environ.* **551–552**, 254–270 (2016).

34. Gulhan, H., Cosenza, A. & Mannina, G. Modelling greenhouse gas emissions from biological wastewater treatment by GPS-X: The full-scale case study of Corleone (Italy). *Sci. Total Environ.* 167327, **905** (2023) doi:10.1016/j.scitotenv.2023.167327.

35. Environment Agency. *Transforming Wastewater Treatment to Reduce Carbon Emissions.* (U.K. Environmental Agency, 2009). https://assets.publishing.service.gov.uk/media/5a7c6e9540f0b62aff6c1a0e/scho1209brnz-e-e.pdf

36. Yapıcıoğlu, P. & Demir, Ö. Minimizing greenhouse gas emissions of an industrial wastewater treatment plant in terms of water-energy nexus. *Appl. Water Sci.* **11**, 180 (2021).

37. Kang, Y.-M., Kim, T.-K., Kim, M.-K. & Zoh, K.-D. Greenhouse gas emissions from advanced oxidation processes in the degradation of bisphenol A: A comparative study of the H2O2/UV, TiO2/UV, and ozonation processes. *Environ. Sci. Pollut. Res.* **27**, 12227–12236 (2020).

38. Finnegan, W., Wu, G. & Zhan, X. Life cycle assessment of a wastewater treatment plant. In *Greenhouse Gas Emission and Mitigation in Municipal Wastewater Treatment Plants* 127–148 (International Water Association, 2018). doi:10.2166/9781780406312_127.

39. Hu, W., Tian, J. & Chen, L. Greenhouse gas emission by centralized wastewater treatment plants in Chinese industrial parks: Inventory and mitigation measures. *J. Clean. Prod.* **225**, 883–897 (2019).

40. Harsanto, B., Primiana, I., Sarasi, V. & Satyakti, Y. Sustainability innovation in the textile industry: A systematic review. *Sustainability* **15**, 1549, (2023).

41. Huang, M. Application of behavioral psychology in clothing design from the perspective of big data. *Appl. Artif. Intell.* **37**, 2194118 (2023).

42. Millward-Hopkins, J., Purnell, P. & Baurley, S. A material flow analysis of the UK clothing economy. *J. Clean. Prod.* **407**, 137158 (2023).

43. Leal Filho, W. et al. An overview of the contribution of the textiles sector to climate change. *Front. Environ. Sci.* vol. 10, 2 doi:10.3389/fenvs.2022.973102 (2022).

44. Palamutcu, S. 2–Energy footprints in the textile industry. In *Handbook of Life Cycle Assessment (LCA) of Textiles and Clothing* (ed. Muthu, S. S.) 31–61 (Woodhead Publishing, 2015). doi:10.1016/B978-0-08-100169-1.00002-2.

45. Hageman, E., Kumar, V., Duong, L., Kumari, A. & McAuliffe, E. Do fast fashion sustainable business strategies influence attitude, awareness and behaviours of female consumers? *Bus. Strategy Environ.* (2023) doi:10.1002/bse.3545.

46. Simurina, J. & Mustac, N. *Impact of textile industry on the environment as a consequence of the development of social networks. Proceedings of FEB Zagreb 10th International Odyssey Conference on Economics and Business* vol. 1 (University of Zagreb, Croatia 2019).

47. Zhang, Y., Liu, C. & Lyu, Y. Profiling consumers: Examination of Chinese Gen Z consumers'sustainable fashion consumption. *Sustainability (Switzerland)* **15**, 8447, (2023).

48. Marazzi, F. et al. Wastewater from textile digital printing as a substrate for microalgal growth and valorization. *Bioresour. Technol.* **375**, 128828 (2023).

49. Balkan, M., Ozturk, E. & Kitis, M. Economic and cross-media effect analyses of best available techniques for caustic recovery from mercerization textile wastewater. *Clean. Technol. Environ. Policy* **25**, 1043–1058 (2023).

50. Aouni, A., Fersi, C., Ben Sik Ali, M. & Dhahbi, M. Treatment of textile wastewater by a hybrid electrocoagulation/nanofiltration process. *J. Hazard. Mater.* **168**, 868–874 (2009).

51. Phalakornkule, C., Sukkasem, P. & Mutchimsattha, C. Hydrogen recovery from the electrocoagulation treatment of dye-containing wastewater. *Int. J. Hydrogen Energy* **35**, 10934–10943 (2010).

52. Kaushik, A., Mona, S. & Kaushik, C. P. Integrating photobiological hydrogen production with dye-metal bioremoval from simulated textile wastewater. *Bioresour. Technol.* **102**, 9957–9964 (2011).

53. Pachwarya, R. B. & Meena, R. C. Degradation of Azo dyes ponceau S, S-IV from the wastewater of textile industries in a new photocatalytic reactor with high efficiency using recently developed photocatalyst MBIRD-11. *Energy Sour., Part A: Recov., Utiliz., Environ. Effects* **33**, 1651–1660 (2011).

54. Hutagalung, S. S. et al. Combination of ozone-based advanced oxidation process and nanobubbles generation toward textile wastewater recovery. *Front. Environ. Sci.* **11**, 1154739, (2023).

55. Slama, H.Ben et al. Diversity of synthetic dyes from textile industries, discharge impacts and treatment methods. *Appl. Sci.* **11**, 6255, (2021).

56. Ardila-Leal, L. D., Poutou-Piñales, R. A., Pedroza-Rodríguez, A. M. & Quevedo-Hidalgo, B. E. A brief history of colour, the environmental impact of synthetic dyes and removal by using laccases. *Molecules* vol. 26 Preprint at doi:10.3390/molecules26133813 (2021).

57. Al-Tohamy, R. et al. A critical review on the treatment of dye-containing wastewater: Ecotoxicological and health concerns of textile dyes and possible remediation approaches for environmental safety. *Ecotoxicol. Environ. Saf.* vol. 231 Preprint at doi:10.1016/j.ecoenv.2021.113160 (2022).

58. Correia, V. M., Stephenson, T. & Judd, S. J. Characterisation of textile wastewaters –A review. *Environ. Technol.* **15**, 917–929 (1994).

59. Farhana, K., Kadirgama, K., Mahamude, A. S. F. & Mica, M. T. Energy consumption, environmental impact, and implementation of renewable energy resources in global textile industries: An overview towards circularity and sustainability. *Mater. Circ. Econ.* **4**, 15 (2022).

60. Koszewska, M. Circular economy –Challenges for the textile and clothing industry. *Autex Research Journal.* **18(4)**, 337–347 (2018).

61. Jia, F., Yin, S., Chen, L. & Chen, X. The circular economy in the textile and apparel industry: A systematic literature review. *J. Clean. Prod.* **259**, 120728 (2020).

62. Wysokińska, Z. Innovative textiles industry and its future within the concept of circular economy –From the global to regional perspective. *J. Vasyl Stefanyk Precarpathian National Univ.* 6(3–4), 67–76, (2019).

63. Chen, X., Memon, H. A., Wang, Y., Marriam, I. & Tebyetekerwa, M. Circular economy and sustainability of the clothing and textile industry. *Mater. Circ. Econ.* **3**, 12 (2021).

64. Ta, A. H., Aarikka-Stenroos, L. & Litovuo, L. Customer experience in circular economy: Experiential dimensions among consumers of reused and recycled clothes. *Sustainability (Switzerland)* **14**, 509, (2022).

65. dos Santos, P. S. & Campos, L. M. S. Practices for garment industry's post-consumer textile waste management in the circular economy context: An analysis on literature. *Braz. J. Oper. Prod. Manag.* **18**, 1–17 (2021).

66. Saha, K., Dey, P. K. & Papagiannaki, E. Implementing circular economy in the textile and clothing industry. *Bus. Strategy Environ.* **30**, 1497–1530 (2021).

67. Bao, Z., Sun, S. & Sun, D. Assessment of greenhouse gas emission from A/O and SBR wastewater treatment plants in Beijing, China. *Int. Biodeterior. Biodegradation* **108**, 108–114 (2016).

68. Pratama, M. A. & Setiarini, J. Application of bridle model in estimating greenhouse gases emissions from three wastewater treatment plants in Fukushima Prefecture, Japan. *IOP Conf. Ser. Earth Environ. Sci.* **724**, 012061 (2021).

69. Corominas, L., Flores-Alsina, X., Snip, L. & Vanrolleghem, P. A. Comparison of different modeling approaches to better evaluate greenhouse gas emissions from whole wastewater treatment plants. *Biotechnol. Bioeng.* **109**, 2854–2863 (2012).

70. Sahely, H. R., MacLean, H. L., Monteith, H. D. & Bagley, D. M. Comparison of on-site and upstream greenhouse gas emissions from Canadian municipal wastewater treatment facilities. *J. Environ. Eng. Sci.* **5**, 405–415 (2006).

71. Yan, X. et al. Spatial and temporal distribution of Greenhouse gas emissions from municipal wastewater treatment plants in China from 2005 to 2014. *Earths Future* **7**, 340–350 (2019).

72. Préndez, M. & Lara-González, S. Application of strategies for sanitation management in wastewater treatment plants in order to control/reduce greenhouse gas emissions. *J. Environ. Manage.* **88**, 658–664 (2008).

73. Flores-Alsina, X., Corominas, L., Snip, L. & Vanrolleghem, P. A. Including greenhouse gas emissions during benchmarking of wastewater treatment plant control strategies. *Water Res.* **45**, 4700–4710 (2011).

74. Chen, Y.-C. Estimation of greenhouse gas emissions from a wastewater treatment plant using membrane bioreactor technology. *Water Environ. Res.* **91**, 111–118 (2019).

75. Kang, Y.-M., Kim, T.-K., Kim, M.-K. & Zoh, K.-D. Greenhouse gas emissions from advanced oxidation processes in the degradation of bisphenol A: A comparative study of the H2O2/UV, TiO2/UV, and ozonation processes. *Environ. Sci. Pollut. Res.* **27**, 12227–12236 (2020).

76. Kyung, D., Kim, M., Chang, J. & Lee, W. Estimation of greenhouse gas emissions from a hybrid wastewater treatment plant. *J. Clean. Prod.* **95**, 117–123 (2015).

77. Bai, R. L., Jin, L., Sun, S. R., Cheng, Y. & Wei, Y. Quantification of greenhouse gas emission from wastewater treatment plants. *Greenhouse Gases: Sci. Technol.* **12**, 587–601 (2022).

78. Mamais, D., Noutsopoulos, C., Dimopoulou, A., Stasinakis, A. & Lekkas, T. D. Wastewater treatment process impact on energy savings and greenhouse gas emissions. *Water Sci. Technol.* **71**, 303–308 (2014).

79. Campos, J. L. et al. Greenhouse gases emissions from wastewater treatment plants: Minimization, treatment, and prevention. *J. Chem.* **2016**, 3796352 (2016).

80. Aghabalaei, V., Nayeb, H., Mardani, S., Tabeshnia, M. & Baghdadi, M. Minimizing greenhouse gases emissions and energy consumption from wastewater treatment plants via rational design and engineering strategies: A case study in Mashhad, Iran. *Energy Rep.* **9**, 2310–2320 (2023).

81. Santín, I., Barbu, M., Pedret, C. & Vilanova, R. Fuzzy logic for plant-wide control of biological wastewater treatment process including greenhouse gas emissions. *ISA Trans.* **77**, 146–166 (2018).

82. Vieira, A. et al. The impact of the art-ICA control technology on the performance, energy consumption and greenhouse gas emissions of full-scale wastewater treatment plants. *J. Clean. Prod.* **213**, 680–687 (2019).

83. Maity, J. P., Bundschuh, J., Chen, C.-Y. & Bhattacharya, P. Microalgae for third generation biofuel production, mitigation of greenhouse gas emissions and wastewater treatment: Present and future perspectives –A mini review. *Energy* **78**, 104–113 (2014).

84. Molinos-Senante, M., Hernández-Sancho, F., Sala-Garrido, R. & Cirelli, G. Economic feasibility study for intensive and extensive wastewater treatment considering greenhouse gases emissions. *J. Environ. Manage.* **123**, 98–104 (2013).

85. Chen, X. et al. Greenhouse gas emissions and wastewater treatment performance by three plant species in subsurface flow constructed wetland mesocosms. *Chemosphere* **239**, 124795 (2020).

86. Mohedano, R. A., Tonon, G., Costa, R. H. R., Pelissari, C. & Belli Filho, P. Does duckweed ponds used for wastewater treatment emit or sequester greenhouse gases? *Sci. Total Environ.* **691**, 1043–1050 (2019).

87. Noyola, A., Paredes, M. G., Morgan-Sagastume, J. M. & Güereca, L. P. Reduction of greenhouse gas emissions from municipal wastewater treatment in Mexico based on technology selection. *Clean (Weinh.)* **44**, 1091–1098 (2016).

88. Song, P. et al. Performance analysis and life cycle greenhouse gas emission assessment of an integrated gravitational-flow wastewater treatment system for rural areas. *Environ. Sci. Pollut. Res.* **26**, 25883–25897 (2019).

89. Hasanbeigi, A. & Price, L. A technical review of emerging technologies for energy and water efficiency and pollution reduction in the textile industry. *J. Clean. Prod.* **95**, 30–44 (2015).

90. Mohan, S., Oke, N. & Gokul, D. Conventional and zero liquid discharge treatment plants for textile wastewater through the lens of carbon footprint analysis. *J. Water Clim. Change* **12**, 1392–1403 (2021).

91. Igoud, S. et al. Solar wastewater treatment: Advantages and efficiency for reuse in agriculture and industry. In *2019 7th International Renewable and Sustainable Energy Conference (IRSEC)* 1–5 (IEEE, 2019). doi:10.1109/IRSEC48032.2019.9078228.

92. Mo, W. & Zhang, Q. Energy-nutrients-water nexus: Integrated resource recovery in municipal wastewater treatment plants. *J. Environ. Manage.* **127**, 255–267 (2013).
93. Hao, X., Furumai, H. & Chen, G. Resource recovery: Efficient approaches to sustainable water and wastewater treatment. *Water Res.* **86**, 83–84 (2015).
94. Cornejo, P. K., Zhang, Q. & Mihelcic, J. R. How does scale of implementation impact the environmental sustainability of wastewater treatment integrated with resource recovery? *Environ. Sci. Technol.* **50**, 6680–6689 (2016).
95. Vik, E. A., Carlson, D. A., Eikum, A. S. & Gjessing, E. T. Electrocoagulation of potable water. *Water Res.* **18**, 1355–1360 (1984).
96. Mao, Y., Zhao, Y. & Cotterill, S. Examining current and future applications of electrocoagulation in wastewater treatment. *Water (Basel)* **15**, (2023).
97. Hajalifard, Z. et al. The efficacious of AOP-based processes in concert with electrocoagulation in abatement of CECs from water/wastewater. *NPJ Clean Water* **6**, 30 (2023).
98. Liu, F. et al. A systematic review of electrocoagulation technology applied for microplastics removal in aquatic environment. *Chem. Eng. J.* **456**, 141078 (2023).
99. Sivagami, K. et al. Electrochemical-based approaches for the treatment of forever chemicals: Removal of perfluoroalkyl and polyfluoroalkyl substances (PFAS) from wastewater. *Sci. Total Environ.* **861**, 160440 (2023).
100. Nippatla, N. & Philip, L. Electrocoagulation-floatation assisted pulsed power plasma technology for the complete mineralization of potentially toxic dyes and real textile wastewater. *Process Saf. Environ. Prot.* **125**, 143–156 (2019).
101. Güneş, E. & Gönder, Z. B. Evaluation of the hybrid system combining electrocoagulation, nanofiltration and reverse osmosis for biologically treated textile effluent: Treatment efficiency and membrane fouling. *J. Environ. Manage.* **294**, 113042 (2021).
102. Phalakornkule, C., Suandokmai, T. & Petchakan, S. A solar powered direct current electrocoagulation system with hydrogen recovery for wastewater treatment. *Sep. Sci. Technol.* **55**, 2353–2361 (2020).
103. Deghles, A. & Kurt, U. Hydrogen gas production from Tannery wastewater by electrocoagulation of a continuous mode with simultaneous pollutants removal. *IOSR J. Appl. Chem.* **10**, 40–50 (2017).
104. Smol, M. Circular economy in wastewater treatment plant – Water, energy and raw materials recovery. *Energies (Basel)* **16**, 3911, (2023).
105. Mishra, S. & Sundaram, B. A review of the photocatalysis process used for wastewater treatment. *Mater. Today Proc.* (2023) doi:10.1016/j.matpr.2023.07.147.
106. Ramírez-Sánchez, I. M., Máynez-Navarro, O. D. & Bandala, E. R. Degradation of emerging contaminants using Fe-doped TiO2under UV and visible radiation BT – Advanced research in nanosciences for water technology. In *Advanced Research in Nanosciences for Water Technology* (eds. Prasad, R. & Karchiyappan, T.) 263–285 (Springer International Publishing, 2019). doi:10.1007/978-3-030-02381-2_12.
107. Villaseñor-Basulto, D. L., Bandala, E. R., Ramirez, I. & Rodriguez-Narvaez, O. M. Synthesis and photocatalytic applications of CuxO/ZnO in environmental remediation. In *Sustainable Nanotechnology for Environmental Remediation* (eds. Koduru, J. R., Karri, R. R., Mubarak, N. M. & Bandala, E. R. 397–433 (Elsevier, 2022). doi:10.1016/B978-0-12-824547-7.00026-6.
108. Ni, J. et al. Light-driven simultaneous water purification and green energy production by photocatalytic fuel cell: A comprehensive review on current status, challenges, and perspectives. *Chem. Eng. J.* **473**, 145162 (2023).
109. He, Y. et al. Photocatalytic fuel cell –A review. *Chem. Eng. J.* **428**, 131074 (2022).
110. Sen, S. K., Patra, P., Das, C. R., Raut, S. & Raut, S. Pilot-scale evaluation of biodecolorization and biodegradation of reactive textile wastewater: An impact on its use in irrigation of wheat crop. *Water Resour. Ind.* **21**, 100106 (2019).

111. Mulinari, J. et al. Enhanced textile wastewater treatment by a novel biofilm carrier with adsorbed nutrients. *Biocatal. Agric. Biotechnol.* **24**, 101527 (2020).

112. Khehra, M. S., Saini, H. S., Sharma, D. K., Chadha, B. S. & Chimni, S. S.Comparative studies on potential of consortium and constituent pure bacterial isolates to decolorize azo dyes. *Water Res.* **39**, 5135–5141 (2005).

113. Moyo, S., Makhanya, B. P. & Zwane, P. E. Use of bacterial isolates in the treatment of textile dye wastewater: A review. *Heliyon* **8**, e09632, (2022).

114. El Bouraie, M. & El Din, W. S. Biodegradation of reactive black 5 by Aeromonas hydrophila strain isolated from dye-contaminated textile wastewater. *Sustain. Environ. Res.* **26**, 209–216 (2016).

115. Cai, J. et al. Effective decolorization of anthraquinone dye reactive blue19 using immobilized *Bacillus* sp. JF4 isolated by resuscitation-promoting factor strategy. *Water Sci. Technol.* **81**, 1159–1169 (2020).

116. Hossen, Md. Z. et al. Biodegradation of reactive textile dye Novacron Super Black G by free cells of newly isolated *Alcaligenes faecalis* AZ26 and *Bacillus* spp obtained from textile effluents. *Heliyon* **5**, e02068, (2019).

117. Khan, R., Bhawana, P. & Fulekar, M. H. Microbial decolorization and degradation of synthetic dyes: A review. *Rev. Environ. Sci. Biotechnol.* **12**, 75–97 (2013).

118. Zafar, S., Bukhari, D. A. & Rehman, A. Azo dyes degradation by microorganisms –An efficient and sustainable approach. *Saudi J. Biol. Sci.* **29**, 103437 (2022).

119. Jayapal, M., Jagadeesan, H., Shanmugam, M., DanishaJ, P. & Murugesan, S. Sequential anaerobic-aerobic treatment using plant microbe integrated system for degradation of azo dyes and their aromatic amines by-products. *J. Hazard. Mater.* **354**, 231–243 (2018).

120. Srivastava, A., Dangi, L. K., Kumar, S. & Rani, R. Microbial decolorization of Reactive Black 5 dye by *Bacillus albus* DD1 isolated from textile water effluent: Kinetic, thermodynamics & decolorization mechanism. *Heliyon* **8**, e08834, (2022).

121. Reddy, S. & Osborne, J. W. Biodegradation and biosorption of Reactive Red 120 dye by immobilized *Pseudomonas guariconensis*: Kinetic and toxicity study. *Water Environ. Res.* **92**, 1230–1241 (2020).

122. Mohanty, S. S. & Kumar, A. Biodegradation of Indanthrene Blue RS dye in immobilized continuous upflow packed bed bioreactor using corncob biochar. *Sci. Rep.* **11**, 13390 (2021).

123. Ayed, L., Ladhari, N., El Mzoughi, R. & Chaieb, K. Decolorization and phytotoxicity reduction of reactive blue40 dye in real textile wastewater by active consortium: Anaerobic/aerobic algal-bacterial-probiotic bioreactor. *J. Microbiol. Methods* **181**, 106129 (2021).

124. Thiruppathi, K., Rangasamy, K., Ramasamy, M. & Muthu, D. Evaluation of textile dye degrading potential of ligninolytic bacterial consortia. *Environ. Challenges* **4**, 100078 (2021).

125. Alam, R. et al. Understanding the biodegradation pathways of azo dyes by immobilized white-rot fungus, Trametes hirsuta D7, using UPLC-PDA-FTICR MS supported by in silico simulations and toxicity assessment. *Chemosphere* **313**, 137505 (2023).

126. Thoa, L. T. K. et al. Microbial biodegradation of recalcitrant synthetic dyes from textile-enriched wastewater by Fusarium oxysporum. *Chemosphere* **325**, 138392 (2023).

127. George, J. et al. Efficient decolorization and detoxification of triarylmethane and azo dyes by porous-cross-linked enzyme aggregates of Pleurotus ostreatus laccase. *Chemosphere* **313**, 137612 (2023).

128. Ali, S. S. et al. Valorizing lignin-like dyes and textile dyeing wastewater by a newly constructed lipid-producing and lignin modifying oleaginous yeast consortium valued for biodiesel and bioremediation. *J. Hazard. Mater.* **403**, 123575 (2021).

129. Ummalyma, S. B., Sahoo, D. & Pandey, A. Resource recovery through bioremediation of wastewaters and waste carbon by microalgae: A circular bioeconomy approach. *Environ. Sci. Pollut. Res.* **28**, 58837–58856 (2021).
130. Goveas, L. C., Nayak, S., Vinayagam, R., Loke Show, P. & Selvaraj, R. Microalgal remediation and valorisation of polluted wastewaters for zero-carbon circular bioeconomy. *Bioresour. Technol.* **365**, 128169 (2022).
131. Huy, M., Kumar, G., Kim, H.-W. & Kim, S.-H. Photoautotrophic cultivation of mixed microalgae consortia using various organic waste streams towards remediation and resource recovery. *Bioresour. Technol.* **247**, 576–581 (2018).
132. Pathak, V. V, Kothari, R., Chopra, A. K. & Singh, D. P. Experimental and kinetic studies for phycoremediation and dye removal by *Chlorella pyrenoidosa* from textile wastewater. *J. Environ. Manage.* **163**, 270–277 (2015).
133. Javed, F. et al. Real textile industrial wastewater treatment and biodiesel production using microalgae. *Biomass Bioenergy* **165**, 106559 (2022).
134. Deka, R. et al. A techno-economic approach for eliminating dye pollutants from industrial effluent employing microalgae through microbial fuel cells: Barriers and perspectives. *Environ. Res.* **212**, 113454 (2022).
135. Potter, M. C. & Waller, A. D. Electrical effects accompanying the decomposition of organic compounds. *Proc. R. Soc. Lond. Ser. B, Containing Pap. Biol. Char.* **84**, 260–276 (1911).
136. Bazina, N., Ahmed, T. G., Almdaaf, M., Jibia, S. & Sarker, M. Power generation from wastewater using microbial fuel cells: A review. *J. Biotechnol.* **374**, 17–30 (2023).
137. Arun, J. et al. New insights into microbial electrolysis cells (MEC) and microbial fuel cells (MFC) for simultaneous wastewater treatment and green fuel (hydrogen) generation. *Fuel* **355**, 129530 (2024).
138. Sonu, K., Syed, Z. & Sogani, M. Up-scaling microbial fuel cell systems for the treatment of real textile dye wastewater and bioelectricity recovery. *Int. J. Environ. Stud.* **77**, 692–702 (2020).
139. Kouam Ida, T. & Mandal, B. Microbial fuel cell design, application and performance: A review. *Mater. Today Proc.* **76**, 88–94 (2023).
140. Logroño, W. et al. Single chamber microbial fuel cell (SCMFC) with a cathodic microalgal biofilm: A preliminary assessment of the generation of bioelectricity and biodegradation of real dye textile wastewater. *Chemosphere* **176**, 378–388 (2017).
141. Ye, W. et al. Enhanced fractionation of dye/salt mixtures by tight ultrafiltration membranes via fast bio-inspired co-deposition for sustainable textile wastewater management. *Chem. Eng. J.* **379**, 122321 (2020).
142. Liang, C.-Z., Sun, S.-P., Li, F.-Y., Ong, Y.-K. & Chung, T.-S. Treatment of highly concentrated wastewater containing multiple synthetic dyes by a combined process of coagulation/flocculation and nanofiltration. *J. Memb. Sci.* **469**, 306–315 (2014).
143. Zhao, C. et al. Application of coagulation/flocculation in oily wastewater treatment: A review. *Sci. Total Environ.* **765**, 142795 (2021).
144. Ganiyu, S. O., Martínez-Huitle, C. A. & Oturan, M. A. Electrochemical advanced oxidation processes for wastewater treatment: Advances in formation and detection of reactive species and mechanisms. *Curr. Opin. Electrochem.* **27**, 100678 (2021).
145. Othman, N. H., Kabay, N. & Guler, E. Principles of reverse electrodialysis and development of integrated-based system for power generation and water treatment: A review. *Rev. Chem. Eng.* **38**, 921–958 (2022).
146. Shi, Y. et al. Recent development of pressure retarded osmosis membranes for water and energy sustainability: A critical review. *Water Res.* **189**, 116666 (2021).
147. Wu, X., Zhang, X., Wang, H. & Xie, Z. Smart utilisation of reverse solute diffusion in forward osmosis for water treatment: A mini review. *Sci. Total Environ.* **873**, 162430 (2023).

148. Lin, J. et al. Toward resource recovery from textile wastewater: Dye extraction, water and base/acid regeneration using a hybrid NF-BMED process. *ACS Sustain. Chem. Eng.* **3**, 1993–2001 (2015).

149. Ye, W. et al. Advanced desalination of dye/NaCl mixtures by a loose nanofiltration membrane for digital ink-jet printing. *Sep. Purif. Technol.* **197**, 27–35 (2018).

150. Ye, W. et al. High-flux nanofiltration membranes tailored by bio-inspired co-deposition of hydrophilic g-C3N4 nanosheets for enhanced selectivity towards organics and salts. *Environ. Sci. Nano* **6**, 2958–2967 (2019).

151. Liu, Y. et al. Facile and novel fabrication of high-performance loose nanofiltration membranes for textile wastewater recovery. *Sep. Purif. Technol.* **308**, 122867 (2023).

152. Harruddin, N., Othman, N., Lim Ee Sin, A. & Raja Sulaiman, R. N. Selective removal and recovery of Black B reactive dye from simulated textile wastewater using the supported liquid membrane process. *Environ. Technol.* **36**, 271–280 (2015).

153. Lin, J. et al. Tight ultrafiltration membranes for enhanced separation of dyes and Na2SO4 during textile wastewater treatment. *J. Memb. Sci.* **514**, 217–228 (2016).

154. Berkessa, Y. W. et al. Anion exchange membrane organic fouling and mitigation in salt valorization process from high salinity textile wastewater by bipolar membrane electrodialysis. *Desalination* **465**, 94–103 (2019).

155. Ye, W. et al. Loose nanofiltration-based electrodialysis for highly efficient textile wastewater treatment. *J. Memb. Sci.* **608**, 118182 (2020).

156. Liang, X. & Zhang, Y. Controllable recovery and regeneration of bio-derived ionic liquid choline acetate for biomass processing via bipolar membrane electrodialysis-based methodology. *Sep. Purif. Technol.* **297**, 121455 (2022).

10 Biogranulation as an Option for Wastewater Treatment Improvement

M.A. Gomez-Gallegos, Ernestina Moreno Rodriguez, Erick R. Bandala, and J.L. Sanchez-Salas

10.1 INTRODUCTION

In modern times, water availability for different uses (0.1%–0.3% of fresh water) and wastewater (industry, agriculture, urban) effluents released to the environment with poor or zero treatment are among the most pressing problems faced by the world population. Therefore, establishing a policy to treat all wastewater is urgent and its reclamation has become paramount for reducing water scarcity. Considering wastewater treatment in general, biological treatments are the most commonly used for wastewater treatments (WWT) [1]. The most used biological treatment in WWT plants is activated sludge which has been identified with several drawbacks including the considerable operation area required [2–4], slow settling velocity, low nutrient (e.g., nitrogen and phosphorous) removal, and presence of a great variety of microorganisms in the system that led to washing phenomenon [5,6]. These factors represent a challenge for contaminant removal in wastewater. Hence, new environmental bioprocesses for WWT have been introduced for the removal of organic matter, and nutrients (nitrogen and phosphorus), considered the main parameters needed to reduce wastewater effluents [7]. Among these new technologies, the biogranulation process has been the most explored in recent years.

Biogranulation has been suggested as a very competitive choice, transcendental, and with significant achievements for WWT [8–11], mainly for total phosphorus, chemical oxygen demand (COD), and total nitrogen (NT) removal [8,12,13]. Other benefits of biogranulation include low investment [8,14], attractive settling characteristics [13], energy cost reduction [3], and high biomass retention [15].

Several physical–chemical–biological phenomena are involved in biogranules formation, such as cellular interaction involving trophic bacteria. In these microbiological interactions, different microbial molecules such as carbohydrates, proteins, and ions (e.g., cations or anions) are involved, conferring different charges that will depend on the pH and temperature of the system [16], resulting in packed unions and immobilized cells components [17]. These packed elements are, in fact, a microbial consortium with particular features that make it unique, favoring permeability, size distribution, physical resistance, microbial activity (Figure 10.1), mechanical stability, rheology, porosity, surface adhesion, and surrounded by extracellular polymeric

DOI: 10.1201/9781003441007-13

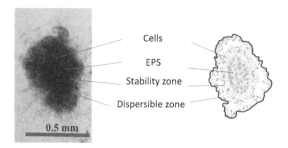

FIGURE 10.1   General biogranule structure [18].

substances giving hydrophobicity surface and thermodynamics, which in turn improve the settling rate, and pollutants removal [16].

Different strategies have been tested for biogranules development, the most commonly accepted is using sequential batch reactors (SBR) [19], reported as the most convenient method when compared with experiments carried out using sequential batch reactors [2,18,20]. Nevertheless, the challenge remains on how to maintain adequate biogranule structural stability and identify appropriate operating conditions to secure biogranule stability in the long term.

10.2   CHARACTERISTICS OF AEROBIC GRANULES

The following characteristics have been identified particularly for granular sludges [7,17,21]: (i) dense and strong structure, (ii) round, regular, smooth shape, and transparent outer surface, (iii) remains as independent entities in the mixing and settling phases, (iv) high biomass retention and excellent settling properties, and (v) withstands high flow rates. Some of the specific biogranule features are explained in more detail next.

Morphology. Biogranule morphology is usually confirmed by microscopic examination showing spherical pellets with a clear surface [22] and multilayers through multicellular interactions [23,24]. Size is very important for biogranule characterization. Hydrodynamic shear forces are significant in suspended biosolids size with much larger, spherical biogranules produced at low shear [7,17]. Therefore, hydrodynamic shear forces must be adjusted with the oxygen supply to have a convenient biogranule size. Microbial diversity and structure were affected by the hydrodynamic shear force [25]. There is no established condition to form and maintain a particular size and structure with any wastewater quality and with the proper microorganism (aerobic, microaerophilic, anaerobic) that could use not only organic matter but also nitrogen, phosphorous, or persistent or emergent pollutants. Then the challenge is to find the best protocol to use according to the wastewater source.

Settleability. The sedimentation characteristics of biogranules will influence solids and liquids separation in WWT processes. It has been observed that greater biomass retention through aerobic granulation positively affects reactor performance producing faster pollutant degradation [7,16,17]. Density and strength. Biogranules are formed with high physical strength, tolerate high abrasion and shear, and, generally, with specific gravity values between the 1.004 and 1.065 range [22] generating differences with flocs in terms of density, microbial aggregation, and size [7,16].

Cell surface hydrophobicity. Hydrophobicity is very important for cell aggregation processes. In granular sludge, hydrophobicity has been suggested twice than conventional bio-flocs [26] as bacteria become hydrophobic during feast-famine periods [7] allowing granulation [16]. Mature granules usually are found with increased hydrophobicity and decreased zeta potential [21,27].

Specific oxygen utilization rate (SOUR). Defined as the milligram of oxygen consumed per gram of volatile suspended solids (VSS) per hour, SOUR is a property used to characterize aerobic granules formed in industrial WWT. SOUR increases when hydrodynamic shear force is increased as a function of air surface velocity [7,17]. However, care must be taken because when SOUR increases growing the air surface velocity, therefore, rise the share force and it could break the biogranules reducing microbial community structure (aerobic, microaerophilic, and anaerobic).

Storage stability. Storage temperature effect on biogranules is related to stability and activity decrease due to prolonged inactivity period. Nevertheless, low storage temperatures are preferred because, when high storage temperature occurs, it can cause biogranules disintegration related to substrate decrease [17,28].

10.3 BIOGRANULE TYPES

The biological granulation process is classified, based on WWT operating conditions, as anaerobic or aerobic systems [16]. During WWT processes, microbial consortia consume organic matter and transform it into biomass, independently of their affinity or presence of oxygen. Anaerobic and aerobic biogranulation conditions are detailed below:

10.3.1 ANAEROBIC BIOGRANULATION

Anaerobic granular sludge includes millions of microorganisms per gram of mass and it is sometimes considered a biofilm type [17]. The significance of this biofilm is related to consortia protection from harmful/toxic environments allowing cells to survive [29]. In recent years, interest has increased in developing physicochemical models to explain and interpret anaerobic granulation formation. During anaerobic biogranulation, adherence of different microorganisms (e.g., bacteria, protozoa-fungal-micro nematodes) to different surfaces occurs. Based on thermodynamics, some physicochemical models for anaerobic pelleting have been developed, the inert nuclei model, pressure selection model, extracellular polymer model, and surface tension model are among the most interesting [17,30] and detailed below.

Inert nuclei model. This model suggests that miniature nuclei or biocarriers for bacterial attachment are the first stages in anaerobic biogranulation. In up-flow anaerobic sludge blanket (UASB) reactors, anaerobic bacteria adhere to inert particles' surfaces to create an initial biofilm called "embryonic particle". Additional mature granules will be formed through the growth of adherent bacteria under specific operating conditions. The initial granulation phase refers to the presence of micronuclei for bacterial aggregation. Inert particles such as zeolites, sand, and water-absorbing polymers have been studied for this purpose [31]. However, this kind of approach has not been implemented for WWT and therefore it is an opportunity for further study and to identify drawbacks that must be addressed before successful implementation for water treatment. Must be considered that for the formation of anaerobic biogranules in UASB reactors, two to eight months are required, which is a disadvantage of starting these types of reactors [30] and can produce bad odors, corrosion due to sulfur byproducts and cannot achieve a direct surface water discharge without a posttreatment.

Selection pressure model. This model proposes that microbial incorporation into UASB reactors is an efficient protective strategy against elevated selection pressures resulting from an upward flow pattern. High pressure chosen at high flow rates is considered beneficial for anaerobic biogranules formation and production. Using this model, the best microbial granules are selected by washing off light, dispersed particles [32]. However, even with its possible advantage, so far, there has not been any report of its application in any wastewater treatment.

Extracellular polymers (ECP) model. The ECP model mediates cell cohesion and adhesion while maintaining the matrix structural integration. One of its disadvantages, however, is limiting microbial absorption due to exopolysaccharide formation [33,34]. Some studies suggest that ECP could modify bacteria's surface charge to negative and, therefore, change their capacity for attaching to nearby cells as well as inert particles [32,34].

Surface tension model. Thermodynamic theory mentions that, in microbial biogranulation, a new interface is formed between biogranules and the liquid phase preventing bacteria and liquid from interfacing individually. Molecular contact between the surfaces of two attached bacteria is also observed. Increased binding of hydrophilic cells at low liquid surface tension, and the opposite for hydrophobic cells, has been suggested for UASB reactors [35]. Biogranulation is favored by bacteria with lower surface energy [32,36]. In anaerobic systems, in general, biomass is separated within the reactor because its production is limited and associated with support.

10.3.2 AEROBIC BIOGRANULATION

The aerobic granulation process consists primarily of microbial self-adhesion by electrostatic forces and interactions [17]. In general terms, a large microorganism's conglomerate is formed and supported by individual microorganisms (bacteria, protozoa-fungal-micronematodes) kept inside and better adapted to oxygen presence. Aerobic biogranules are formed through two decisive operating conditions called "famine and feast" conditions and taking short settling periods. Under these conditions, the formation of stable biogranules is allowed and, in parallel, the removal of organic matter and nutrients will occur simultaneously [37].

In aerobic systems, biomass is removed through secondary settlers by gravity. Different experiments have been carried out for bigger particle formation using continuous flow reactors but these particles are removed faster than those formed in sequential batch reactors (SBR) [38]. Therefore, regular aerobic-activated sludge

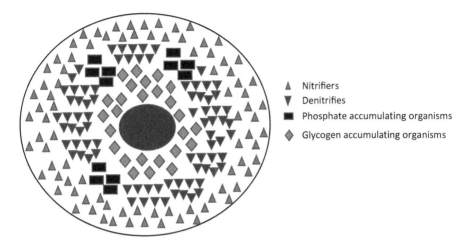

FIGURE 10.2   Microorganism distribution in biogranules developed under aerobic conditions. (Adapted from [39].)

characteristics depend on sludge settling properties [5]. However, aerobic biogranules can develop different zones that are able to harbor different aerobic, microaerophilic, and anaerobic microorganisms capable of metabolic pathways related to C, N, and P removal, as shown in Figure 10.2 [39].

10.4   FACTORS AFFECTING BIOGRANULATION

As shown in Figure 10.2, biogranules will be more efficient using only one single bioreactor instead of several bioreactors with different aerobic/anaerobic conditions. Therefore, several factors are involved in producing the proper biogranules for different WWT systems which will be briefly discussed below.

For biogranules to be structured, different conditions that influence their formation are required [17] such as substrate composition, organic load rate, hydrodynamic shear force, settling time, hydraulic retention time (HRT), aerobic starvation, presence of calcium ion in feed, intermittent feeding strategy, dissolved oxygen, pH and temperature, seed sludge, reactor configuration, and inhibition to aerobic granulation. All these factors are briefly explored next.

Substrate composition. Substrate composition is very important because of its influence on microbial variety and aggregation process [40]. A wide variety of substrates (e.g., glucose, acetate, ethanol, phenol, and synthetic wastewater) have been used in aerobic pellet formation [17] and biogranules type developed has been found depending on substrate composition quality. *Organic loading rate.* Organic load is considered a determining factor limiting aerobic granulation. The degree of microorganism starvation is measured using the organic substrate loading rate (OLR), and an increase in OLR means rapid microbial growth. Particularly, anaerobic granulation is highly sensitive to OLR [17] because heterotrophic bacteria are the main organic matter consumers [41] responsible for contaminants and nutrient removal [42]. Some recent studies suggest that higher OLR would increase biomass growth rate [18]. *Hydrodynamic shear force*. Aerobic granule formation is favored by relatively high hydrodynamic shear force values and stability is due to structured granules [22]. Also, the granule-cutting force has been suggested as proportionally related to its density and resistance [43] and related to extracellular polysaccharide production because it offers stability to the formed granules [22,44]. Extracellular polysaccharide production is also related to high shear to obtain aerobic granules with compact structures [17].

Settling time. Settling time is related to the existing microbial community in wastewater [17]. A short settling time favors fast-settling bacteria growth, while those microorganisms showing poor settling are washed off. Biogranules with short settling times have been observed to form granules with larger diameters [32,45–47]. *Hydraulic retention time (HRT)*. In aerobic granulation, dispersed low-density sludge is washed off and structured granules are kept within the reactor. Short HRT should be to reduce suspended growth, but necessarily long HRT values are needed for microbial growth and accumulation. Short HRT promotes microbial activity as well as cellular polysaccharide production and increases cellular hydrophobicity. Extracellular polymeric substances play a significant role in aerobic granule formation because are considered as a glue [17]. *Aerobic starvation*. Bacteria have been suggested to increase hydrophobicity under starvation conditions, which enhanced aggregation. Increasing hydrophobicity is probably a strategy against starvation, modifying surface properties under this condition [22] and highly significant in granule conformation providing strength and density [17]. *Presence of calcium*. Ca^{2+} ions stabilize agglomerates by reducing negatively charged groups on bacteria surfaces and extracellular polymeric substances [48]. *Intermittent feeding strategy*. Cell hydrophobicity is affected during starvation periods, which plays a very important role in aerobic granulation. Therefore, the formation of compact aerobic granules is favored by operating strategies in the feeding stage [17]. *Dissolved oxygen (DO), pH, and temperature*. DO, pH, and temperature, are usually not considered determinant variables in aerobic granulation. However, these parameters must be adjusted depending on wastewater type or quality [17]. Stability and granular formation are, in fact, affected by pH in the liquid phase [7,49]. No granulation restrictions have been identified using low (e.g., 0.7–1.0 mg/L) or high DO values [7,50]. *Seed sludge*. Seed sludge quality is determined by its macroscopic characteristics, settling, surface properties, hydrophobicity, density, and microbial activity to obtain dense granules [17]. Reactor configuration. Different types of reactors have been reported including column-type up-flow reactors and full mixed tank reactors with different hydrodynamic characteristics according to flow and microbial aggregates relationship. In SBR reactors, the height/diameter (H/D) ratio must be taken into account to improve granule formation [17]. *Inhibition of aerobic granulation*. Various strategies have been proposed for granular inhibition, among them, free ammonia has been shown to inhibit energy metabolism [51], but other strategies have been used affecting feeding time and filamentous growth on aerobic granular systems [52].

10.5 BIOGRANULE COMPONENTS

10.5.1 BACTERIA

Microbial structure and diversity in aerobic granules are quite simple. Information is obtained through microscopy studies showing, for example, that aerobic ammonium-oxidizing bacteria *Nitrosomonas* spp. locates to a depth of 70–100 µm from the granule surface and anaerobic bacteria *Bacteroides* spp. was detected at a depth of 800–900 µm from the granule surface. The nitrifying bacteria population was located mainly at a depth of 70–100 µm from the granule surface. This microbial diversity is connected to culture media structure [17].

In activated sludge, the vast majority of bacterial genera are Gram-negative. Most genera identified correspond to *Arthrobacter, Comamonas, Lophomonas, Zoogloea, Sphaerotilus, Azotobacter, Chromobacterium, Achromobacter, Flavobacterium, Bacillus,* and *Nocardia*. At some point, *Zoogloea* was considered helping to bind activated sludge flocs, but this is no longer the case because various types of bacteria have been identified to generate polymers that allow union between flocs [41]. Several studies have suggested been Proteobacteria dominance in aerobic biogranules by exhibiting 21 phylotypes and related clones grouped into Alpha-, Beta-, and Gamma-proteobacteria [53].

10.5.2 NUTRIENTS (NITROGEN, PHOSPHORUS) AND ORGANIC MATTER REMOVAL

Total nitrogen, total phosphorous, and organic matter have been successfully removed from wastewater by biogranulation. However, removal efficacy depends on the presence of different microorganisms such as ammonia-oxidizing bacteria and heterocyclic denitrifiers that allow complete elimination within the system [39,54]. *Nitrogen removal.* Ammonium is oxidized during the nitrite pathway, by ammonia-oxidizing bacteria, to nitrates and then it is denitrified by heterocyclic denitrifies to nitrogen (gas) [39,55,56]. Phosphorus removal. The process is based on alternating anaerobic and aerobic conditions with substrates limited to the anaerobic stage, focusing on biomass culture suspensions that require large reactor volumes [17]. During the anaerobic phase, activated sludge produces phosphorus which accumulates as polyphosphates and is stored as polyhydroxy alkanoate substances (PHAs). In the aerobic phase, polyphosphates storing microorganisms also allow nitrification and, at the same time, organic substrate oxidation [54–57]. *Fungal species.* Activated sludge contains a wide variety of microorganisms. Fungi are rarely important members of the community [41] and, therefore, they are not usually mentioned for WWT using aerobic granular sludge. Some studies have been considered fungal granule technology [7]. Fungi can be dominant in aerobic granular reactors, growing more easily at low pH values. The topic, however, remains relatively unknown and there is a significant need for differences between fungal and bacterial granules in terms of formation, mechanisms, and structure [58]. Despite this, at the operational level, an excess of filamentous fungi is usually considered a significant problem that prevents the application of granulation processes. Filamentous organisms belonging to the

Geotrichum genera and phylum *Ascomycota* have been reported to require less nutrients compared to bacteria and, therefore, affect nutrient removal [59]. The lack of information about fungal granules, their formation, and performance for WWT is a knowledge gap worth exploring.

Protozoa. There are different indicators related to the health of a wastewater system, the presence of protozoa is one of them meaning wastewater is free of toxic chemicals [41]. Predators' microorganisms cause the reduction of suspended bacterial communities, helping to remove undesirable pollutants [60]. Ciliated protozoa have been suggested to play a significant role in granulation because they have been found on biogranule surfaces and proposed to promoting aerobic granule formation, nevertheless, the real role of protozoa in aerobic biogranulation is unclear, which is considered an avenue for future research worth exploring [61]. *Algae.* Algae are key microorganisms able to improve pollutants removal at low-cost, generating high quality biomass for other purposes. The main goal using microalgae for wastewater treatment is transformation of pollutants used as nutrients into biomass, to improve the quality of the effluent driven by solar radiation [62]. This treatment option is proper when no harmful interaction occur between bacteria and algae, taking advantage that bacteria consume oxygen and produce carbon dioxide, while algae will use the carbon dioxide produced by the bacteria and produce oxygen after photosynthesis. Microalgae such as *Chlorella vulgaris* has been tested for nutrient removal showing high ammonium-nitrogen removal capacity within 48 h and with potential for biofuels production [63]. *Chlorella sp.*, was used for nitrogen, phosphorus, COD and metallic ions removal showing between 70% and 90% efficacy after 216 h [64]. *Cladophora fascicularis* was found capable to adsorb/desorb Pb(II) in wastewater depending on pH, ionic strength, temperature, and initial Pb(II) concentration [65]. *Chlorella miniate* was found able to reduce Cr(VI) to Cr(III) through carboxylate groups of intracellular proteins and to adsorb Cr(VI) by interaction with the amino group of proteins found in this alga at pH 3 [66]. Biomass of different families of microalgae such as *Chlorophyceae (Botryococcus, Chlamydomonas, Chlorella, Gonium, Pandorina, Scenedesmus, Spirogyra, Volvox), Cyanophyceae (Oscillatoria, Spirulina) and Euglenophyceae (Phacus)*, can adsorb As(III) and As(V) from freshwater. Arsenic-tolerant microalgae were chosen for biosorption studies, obtaining a dry biome with a higher percentage of biosorption [67].

10.6 BIOGRANULATION APPLICATIONS IN WWTS

10.6.1 MUNICIPAL WWT

Aerobic granulation is considered a promising alternative for simultaneous removal of nutrients and organic matter [68]. Studies to verify the technology applications using domestic wastewater are available but lack information on actual performance for low-medium strength wastewater, which is considered a knowledge gap worth exploration [42]. A study looking at aerobic granulation efficiency for domestic wastewater treatment using SBR reactor found removal efficiency as high as 49% for total COD, 11% for NH_4^+-N and total N, and 33% for phosphate [69]. Another study using domestic wastewater found high total COD (85%–95%) and NH_4^+-N (90%–99%) removal [29].

Application of aerobic granular sludge for domestic wastewater treatment using SBR generated removal percentages as high as total COD = 92%, total N = 81%, total P = 85% [70]. Municipal water treatment was treated using SBR (98 L) obtaining 75% COD removal; 73% NH_4^+-N removal [71]. Using two 9,600 m^3SBR, municipal wastewater treatment was carried out using biogranulation generating 88% COD removal; 96% BOD removal; 97% NH_4^+-N removal; 86% total N removal; and 87% total P removal [72]. Using three SBR reactors, septic wastewater treatment was carried out obtaining 94% COD removal; 99% NH_4^+-N removal; 83.5% TP removal; 98% SS removal [73]. The challenge on these WWT, is keep the biogranules structure for any WW quality or consider unify the WW quality. Both conditions are difficult, however, can develop biogranules with a microorganism's association adapted to different WW quality.

10.6.2 INDUSTRIAL WWT

Industrial wastewater (synthetic and real) were tested for biogranulation treatment obtaining solved COD removal 78%–80%, NH_4^+-N removal 98% [74]. Biogranulation technology is applied for WWT in different industries (Feed industry 48%, pharmaceutical industry 32%, production of polymer and polypropylene 7.2%, production of biofuel 5.5%, production of fertilizer 2.2%, food industry 2.0%, chemical industry 1.4%, manufacture of metal 1.2%, others 0.5%) [75]. Biogranulation was used to treat rubber industry wastewater in a 0.6 L SBR, obtaining 96.5% COD removal and 95% ammonia removal [76]. Palm oil mill effluents were treated in a 3 L SBR reactor using granulation obtaining 91.1% COD removal and 97.6% ammonia removal [77]. The treatment of explosive industry wastewater was carried out in SBR, generating 93%–97% COD removal and >80% total N removal [78]. Using a 4 L SBR, livestock wastewater treatment was carried out obtaining 74% COD removal; 73% total N removal; and 70% total P removal [79].

Different organic load conditions and nutrient concentrations have been tested to find the best conditions for biogranulation and reduce organic matter and micronutrients using a batch aerobic reactor. Results suggested using 1,100 mg/L COD, 16 mg/L total N, and 11 mg/L of total P, biogranulation removed 99% COD, 99.4% total N, and 45% total P. Settling velocity in the process was improved almost seven times compared with non-biogranulation conditions [18], indicating the importance of organic matter and micronutrient load which must be established for a particular wastewater system.

The optimization of biogranulation process will develop new microorganism association and at the same time will necessary to find the proper strategy to reduce possible biological risks or the way to discard the biomass obtained during the process. However, also, can be used to obtain other kind of energy using that kind of waste. For example, use that biomass for composting, natural gas production, or oil extraction from algae cells.

10.7 CONCLUSIONS

Turning wastewater into good quality water for reuse in different applications, including human consumption is a significant need to tackle current water shortages all around the world. However, wastewater treatment for reclamation is mandatory only in developed countries. Biological wastewater treatment involves significant investment including structures for aerobic, anaerobic, and micro-aerophilic bioreactors that is not considered mandatory in some places and, particularly, mainly in developing countries. This infrastructure investment could be significantly reduced if the designed biological system requires only one cost-efficient bioreactor to eliminate organic matter, micronutrients (e.g., nitrogen, phosphorous) and other pollutants regularly present in wastewater effluents from different sources. The use of biogranulation is a cost-effective alternative for different regions because of its easy implementation to generate biogranules conformed of a mix of different microbial populations. Depending on wastewater quality, biogranules will include microorganisms with different metabolic characteristics and different oxygen requirements (e.g., aerobic, anaerobic, and micro-aerophilic) all in one. Biogranulation has been identified with a high potential use in wastewater reclamation and pollution reduction with the additional potential of using the residual biomass to generate energy.

REFERENCES

1. Abdel-Raouf, N., Al-Homaidan, A. A., & Ibraheem, I. B. (2012). Microalgae and wastewater treatment. *Saudi Journal of Biological Sciences*, 19(3), 257–275.
2. Del Rio, A. V., Figueroa, M., Arrojo, B., Mosquera-Corral, A., Campos, J. L., García-Torriello, G., & Méndez, R. (2012). Aerobic granular SBR systems applied to the treatment of industrial effluents. *Journal of Environmental Management*, 95, S88–S92.
3. Świątczak, P., & Cydzik-Kwiatkowska, A. (2018). Performance and microbial characteristics of biomass in a full-scale aerobic granular sludge wastewater treatment plant. *Environmental Science and Pollution Research*, 25(2), 1655–1669.
4. Cydzik-Kwiatkowska, A., Bernat, K., Zielińska, M., Bułkowska, K., & Wojnowska-Baryła, I. (2017). Aerobic granular sludge for bisphenol A (BPA) removal from wastewater. *International Biodeterioration & Biodegradation*, 122, 1–11.
5. Arrojo, B., Mosquera-Corral, A., Garrido, J., & Méndez, R. (2004). Aerobic granulation with industrial wastewater in sequencing batch reactors. Water Research, *38*(14–15), 3389–3399.12
6. Liu, Y., & Liu, Q. S. (2006). Causes and control of filamentous growth in aerobic granular sludge sequencing batch reactors. *Biotechnology Advances*, 24(1), 115–127.
7. Winkler, M. K. H., Meunier, C., Henriet, O., Mahillon, J., Súarea-Ojeda, M. E., Del Moro, G., ... Weissbrodt, D. G. (2018). An integrative review of granular sludge for the biological removal of nutrients and recalcitrant organic matter from wastewater. *Chemical Engineering Journal*, 336, 489–502.
8. Czarnota, J., Masłoń, A., & Zdeb, M. (2018). Powdered keramsite as unconventional method of AGS technology support in GSBR reactor with minimum-optimum OLR. *E3S Web of Conferences*, 44, 00024.
9. Cai, W., Huang, W., Lei, Z., Zhang, Z., Lee, D. J., & Adachi, Y. (2019). Granulation of activated sludge using butyrate and valerate as additional carbon source and granular phosphorus removal capacity during wastewater treatment. *Bioresource Technology*, 282, 269–274.
10. Tiwari, S. S., Iorhemen, O. T., & Tay, J. H. (2018). Semi-continuous treatment of naphthenic acids using aerobic granular sludge. *Bioresource Technology Reports*, 3, 191–199. doi:10.1016/j.biteb.2018.08.007
11. Pronk, M., Abbas, B., Al-Zuhairy, S. K., Kraan, R., Kleerebezem, R., & Van Loosdrecht, M. M. (2015). Effect and behaviour of different substrates in relation to the formation of aerobic granular sludge. *Applied Microbiology and Biotechnology*, 99(12), 5257–5268.

12. Bassin, J. P., Tavares, D. C., Borges, R. C., & Dezotti, M. (2019). Development of aerobic granular sludge under tropical climate conditions: The key role of inoculum adaptation under reduced sludge washout for stable granulation. *Journal of Environmental Management*, 230, 168–182.
13. Seow, T. W., Lim, C. K., Nor, M. M., Mubarak, M. M., Lam, C. Y., Yahya, A., & Ibrahim, Z. (2016). Review on wastewater treatment technologies. *International Journal of Applied Environmental Science*, 11, 111–126.
14. De Sousa Rollemberg, S. L., Barros, A. R., Firmino, P. I., & dos Santos, A. B. (2018). Aerobic granular sludge: Cultivation parameters and removal mechanisms. *Bioresource Technology*, 270, 678–680.
15. Gao, D., Liu, L., Liang, H., & Wu, W. M. (2011). Aerobic granular sludge: characterization, mechanism of granulation and application to wastewater treatment. *Critical Reviews in Biotechnology*, 31(2), 137–152.
16. Liu, X. W., Sheng, G. P., & Yu, H. Q. (2009). Physicochemical characteristics of microbial granules. *Biotechnology Advances*, 27(6), 1061–1070.
17. Liu, Y., & Tay, J. H. (2004). State of the art of biogranulation technology for wastewater treatment. *Biotechnology Advances*, 22(7), 533–563.
18. Gomez-Gallegos, M. A., Reyes-Mazzoco, R., Flores-Cervantes, D. X., Jarayathne, A., Goonetilleke, A., Bandala, E. R., & Sanchez-Salas, J. L. (2021). Role of organic matter, nitrogen and phosphorous on granulation and settling velocity in wastewater treatment. *Journal of Water Process Engineering*, 40, 101967.
19. Ionescu, I. A., Bumbac, C., & Cornea, P. (2015). Formation of aerobic granules in sequencing batch reactor SBR treating dairy industry wastewater. *Scientific Bulletin. Series F. Biotechnologies*, 19, 235–238.
20. Bumbac, C., Ionescu, I. A., Tiron, O., & Badescu, V. R. (2015). Continuous flow aerobic granular sludge reactor for dairy wastewater treatment. *Water Science and Technology*, 71(3), 440–445.
21. Zheng, Y. M., Yu, H. Q., & Sheng, G. P. (2005). Physical and chemical characteristics of granular activated sludge from a sequencing batch airlift reactor. *Process Biochemistry*, 40, 645–650.
22. Tay, J. H., Liu, Q. S., & Liu, Y. (2001). Microscopic observation of aerobic granulation in sequential aerobic sludge blanket reactor. *Journal of Applied Microbiology*, 91, 168–175.
23. Cui, F., Park, S., Kim, M., (2014). Characteristics of aerobic granulation at mesophilic temperatures in wastewater treatment. *Bioresource Technology* 151, 78–84. https://doi.org/10.1016/j.biortech.2013.10.025
24. Oliveira, A.S., Amorim, C.L., Ramos, M.A., Mesquita, D.P., Inocêncio, P., Ferreira, E.C., van Loosdrecht, M., Castro, P.M.L., (2020). Variability in the composition of extracellular polymeric substances from a full-scale aerobic granular sludge reactor treating urban wastewater. *Journal of Environmental Chemical Engineering* 8, 104156. https://doi.org/10.1016/j.jece.2020.104156
25. He, Q., Chen, L., Zhang, S., Chen, R., & Wang, H. (2019). Hydrodynamic shear force shaped the microbial community and function in the aerobic granular sequencing batch reactors for low carbon to nitrogen (C/N) municipal wastewater treatment. *Bioresource Technology*, 271, 48–58.
26. Liu, Y., Yang, S. F., Liu, Q. S., & Tay, J. H. (2003b). The role of cell hydrophobicity in the formation of aerobic granules. *Current Microbiology*, 46, 270–274.
27. Su, B., Qu, Z., Song, Y., Jia, L., & Zhu, J. (2014). Investigation of measurement methods and characterization of zeta potential for aerobic granular sludge. *Journal of Environmental Chemical Engineering*, 2(2), 1142–1147.
28. Xu, H. C., He, P. J., Wang, G. Z., Yu, G. H., & Shao, L. M. (2010). Enhanced storage stability of aerobic granules seeded with pellets. *Bioresource Technology*, 101(21), 8031–8037.

29. Ni, B., Xie, W., Liu, S., Yu, H., Wang, Y., Wang, G., & et al. (2009). Granulation of activated sludge in a pilot-scale sequencing batch reactor for the treatment of low-strength municipal wastewater. *Water Research, 43*(3), 751–761.

30. Liu, Y., Xu, H. L., Yang, S. F., & Tay, J. H. (2003). Mechanisms and models for anaerobic granulation in upflow anaerobic sludge blanket reactor. *Water Research,* 37(3), 661–673.

31. Nuntakumjorn, B., Khumsalud, W., Vetsavas, N., Sujjaviriyasup, T., & Phalakornkule, C. (2008). Comparison of sludge granule and UASB performance by adding chitosan in different forms. *Chiang Mai Journal of Science,* 35(1), 95–102.

32. Milferstedt, K., Hamelin, J., Park, C., Jung, J., Hwang, Y., Cho, S. K., … Kim, D. H. (2017). Biogranules applied in environmental engineering. *International Journal of Hydrogen Energy,* 42(45), 27801–27811.

33. Rudd, T., Sterritt, R. M., & Lester, J. N. (1984). Complexation of heavy metals by extracellular polymers in the activated sludge process. *Journal of the Water Pollution Control Federation,* 56(12), 1260–1268.

34. Schmidt, J. E. T., & Ahring, B. K. (1994). Extracellular polymers in granular sludge from different upflow anaerobic sludge blanket (UASB) reactor. *Applied Microbiology and Biotechnology,* 42, 457–462.

35. Thaveesri, J., Daffonchio, D., Liessens, B., Vandermeren, P., & Verstraete, W. (1995). Granulation and sludge bed stability in upflow anaerobic sludge bed reactors in relation to surface thermodynamics. *Applied and Environmental Microbiology,* 61(10), 3681.

36. Rouxhet, P. G., & Mozes, N. (1990). Physical chemistry of the interface between attached micro-organisms and their support. *Water Science and Technology,* 22((1–2)), 1–16.

37. Stamatelatou, K., & Tsagarakis, K. P. (2015). *Sewage Treatment Plants: Economic Evaluation of Innovative Technologies for Energy Efficiency.* United Kingdom: IWA Publishing.

38. Show, K. Y., & Tay, J. H. (2014). Aerobic sludge granulation: Current perspectives, advances and the way forward. *Environmental Science and Technology,* 1, 4.

39. Nancharaiah, Y. V., & Reddy, G. K. (2018). Aerobic granular sludge technology: mechanisms of granulation and biotechnological applications. *Bioresource Technology,* 247, 1128–1143.

40. Lee, D. J., Chen, Y. Y., Show, K. Y., Whiteley, C. G., & Tay, J. H. (2010). Advances in aerobic granule formation and granule stability in the course of storage and reactor operation. *Biotechnology Advances,* 28(6), 919–934. doi: https://doi.org/10.1016/j.bi

41. Rittmann, B. E., & McCarty, P. L. (2012). *Environmental Biotechnology: Principles and Applications.* Tata McGraw-Hill Education.

42. We, A. C. E., Aris, A., & Zain, N. A. (2020). A review of the treatment of low-medium strength domestic wastewater using aerobic granulation technology. *Environmental Science: Water Research & Technology,* 6(3), 464–490.

43. Tay, J. H., Liu, Q. S., & Liu, Y. (2003). *Shear Force Influences the Structure of Aerobic Granules Cultivated in Sequencing Batch Reactor.* Cape Town, South Africa.

44. Tay, J. H., Liu, Q. S., & Liu, Y. (2001b). The role of cellular polysaccharides in the formation and stability of aerobic granules. *Letters in Applied Microbiology,* 33, 222–226.

45. Adav, S. S., Lee, D. J., & Lai, J. Y. (2009). Aerobic granulation in sequencing batch reactors at different settling times. *Bioresource Technology,* 100(21), 5359–5361.

46. Qin, L., Liu, Y., & Tay, J. H. (2004). Effect of settling time on aerobic granulation in sequencing batch reactor. *Biochemical Engineering Journal,* 21(1), 47–52.

47. McSwain, B. S., Irvine, R. L., & Wilderer, P. A. (2004). The influence of settling time on the formation of aerobic granules. *Water Science and Technology,* 50(10), 195–202.

48. Jiang, H. L., Tay, J. H., Liu, Y., & Tay, S. T. L. (2003). Ca2+ augmentation for enhancement of aerobically grown microbial granules in sludge blanket reactors. *Biotechnology Letters,* 25, 95–99.

49. Beun, J. J., Van Loosdrecht, M. C.M., & Heijnen, J. J. (2002). Aerobic granulation in a sequencing batch airlift reactor. *Water Research*, 36(3), 702–712.
50. Khan, M. Z., Mondal, P. K., & Sabir, S. (2013). Aerobic granulation for wastewater bioremediation: A review. *The Canadian Journal of Chemical Engineering*, 91(6), 1045–1058.
51. Yang, S. F., Tay, J. H., & Liu, Y. (2004). Inhibition of free ammonia to the formation of aerobic granules. *Biochemical Engineering Journal*, 17, 41–48.
52. Liu, L., Gao, D. W., Zhang, M., & Fu, Y. (2010). Comparison of Ca^{2+} and Mg^{2+} enhancing aerobic granulation in SBR. *Journal of Hazardous Materials*, 181(1–3), 382–387.
53. Adav, S. S., Lee, D. J., & Lai, J. Y. (2010). Microbial community of acetate utilizing denitrifiers in aerobic granules. *Applied Microbiology and Biotechnology*, 85, 753–762.
54. Wang, F., Bai, Y., Yang, F., Zhu, Q., Zhao, Q., Zhang, X., ..., Liao, H. (2021). Degradation of nitrogen, phosphorus, and organic matter in urban river sediments by adding microorganisms. *Sustainability*, 13(5), 2580.
55. Lopes, R. B., Ribeiro, J. S., Neves, S. C. B., Lameira, L. F., de Moura, L. S., de Santana, M. B., & Taube, P. S. (2022). Dissolved oxygen, organic matter and nutrients in fish systems combined with bio-addition of friendly microorganisms. *Research, Society and Development*, 11(4), e26111427382–e26111427382.
56. Baranu, Barisiale, S., & Lawrence, E. (2022). Nitrogen bacteria associated with bioremediation soil and their ability to degrade hydrocarbon. *International Journal of Current Microbiology and Applied Sciences*, 11(2), 306–314.
57. Pérez, M. L., Dautant, R., Contreras, A., & González, H. (2002). Remoción de fósforo y nitrógeno en aguas residuales utilizando un reactor discontinuo secuencial (SBR). *Congreso Interamericano de Ingeniería Sanitaria y Ambiental*. 28, págs. 1–8. FEMISCA.
58. Yang, S. F., Li, X. Y., & Yu, H. Q. (2008). Formation and characterisation of fungal and bacterial granules under different feeding alkalinity and pH conditions. *Process Biochemistry*, 43(1), 8–14.
59. Mille-Lindblom, C., Fischer, H., & J. Tranvik, L. (2006). Antagonism between bacteria and fungi: substrate competition and a possible tradeoff between fungal growth and tolerance towards bacteria. *Oikos*, 113(2), 233–242.
60. Li, J., Ma, L., Wei, S., & Horn, H. (2013). Aerobic granules dwelling vorticella and rotifers in an SBR fed with domestic wastewater. *Separation and Purification Technology*, 110, 127–131.
61. Chan, S. H., Ismail, M. H., Tan, C. H., Rice, S. A., & McDougald, D. (2021). Microbial predation accelerates granulation and modulates microbial community composition. *BMC Microbiology*, 21(1), 1–18.
62. Salazar, M. (2006). Aplicación e importancia de las microalgas en el tratamiento de aguas residuales. *Contactos*, 59, 64–70.
63. Kim, J., Lingaraju, B. P., Rheaume, R., Lee, J. Y., & Siddiqui, K. F. (2010). Removal of ammonia from wastewater effluent by Chlorella vulgaris. *Tsinghua Science and Technology*, 15(4), 391–396.
64. Wang, L., Min, M., Li, Y., Chen, P., Chen, Y., Liu, Y., & Ruan, R. (2010). Cultivation of green algae *Chlorella* sp. in different wastewaters from municipal wastewater treatment plant. *Applied Biochemistry and Biotechnology*, 162, 174–1186.
65. Deng, L., Su, Y., Su, H., Wang, X., & Zhu, X. (2007). Sorption and desorption of lead (II) from wastewater by green algae *Cladophora fascicularis*. *Journal of Hazardous Materials*, 143(1–2), 220–225.
66. Han, X., Wong, Y. S., Wong, M. H., & Tam, N. F. (2007). Biosorption and bioreduction of Cr (VI) by a microalgal isolate, *Chlorella miniata*. *Journal of Hazardous Materials*, 146(1–2), 65–72.

67. Sibi, G. (2014). Biosorption of arsenic by living and dried biomass of fresh water microalgae-potentials and equilibrium studies. *Journal of Bioremediation & Biodegredation*, 5(6), 1.

68. Sepúlveda-Mardones, M., Campos, J. L., Magrí, A., & Vidal, G. (2019). Moving forward in the use of aerobic granular sludge for municipal wastewater treatment: an overview. *Reviews in Environmental Science and Bio/Technology*, 18, 741–769.

69. De Kreuk, M. K., & Van Loosdrecht, M. C. (2006). Formation of aerobic granules with domestic sewage. *Journal of Environmental Engineering*, 132, 694–697.

70. Su, B., Cui, X., & Zhu, J. (2012). Optimal cultivation and characteristics of aerobic granules with typical domestic sewage in an alternating anaerobic/aerobic sequencing batch reactor. *Bioresource Technology*, *110*, 125–129.

71. Guimarães, L.B., Mezzari, M.P., Daudt, G.C., da Costa, R.H.R. (2017). Microbial pathways of nitrogen removal in aerobic granular sludge treating domestic wastewater. *Journal of Chemical Technology & Biotechnology* DOI: 10.1002/jctb.5176.

72. Pronk, M., de Kreuk, M.K., de Bruin, B., Kamminga, P., Kleerebezem, R., van Loosdrecht, M.C., (2015b). Full scale performance of the aerobic granular sludge process for sewage treatment. *Water Research* 84, 207–217

73. Giesen, A., de Bruin, L., Niermans, R., van der Roest, H., (2013). Advancements in the application of aerobic granular biomass technology for sustainable treatment of wastewater. *Water Practice and Technology* 8 (1), 47–54

74. Liu, Y., Moy, B. Y., & Tay, J. (2007). COD removal and nitrification of low-strength domestic wastewater in aerobic granular sludge sequencing batch reactors. *Enzyme and Microbial Technology*, 42, 23–28

75. Yegorov, B. V., Batievskaya, N. O., & Fedoryaka, V. P. (2017). Application of granulation technology in various industries. *Зернові продукти і комбікорми*, (17), 33–38.

76. Rosman, N.H., Nor Anuar, A., Othman, I., Harun, H., Sulong Abdul Razak, M.Z., Elias, S.H., Mat Hassan, M.A., Chelliapan, S., Ujang, Z., (2013). Cultivation of aerobic granular sludge for rubber wastewater treatment. *Bioresource Technology* 129, 620–623.

77. Abdullah, N., Ujang, Z., Yahya, A., (2011). Aerobic granular sludge formation for high strength agro-based wastewater treatment. *Bioresource Technology* 102, 6778–6781

78. Zhang, J. H., Wang, M. H., & Zhu, X. M. (2013). Treatment of HMX-production wastewater in an aerobic granular reactor. *Water Environment Research, 85*(4), 301–307.

79. Othman, I., Anuar, A.N., Ujang, Z., Rosman, N.H., Harun, H., Chelliapan, S., (2013). Livestock wastewater treatment using aerobic granular sludge. *Bioresource Technology* 133, 630–634.

11 Alternatives for Urban Stormwater Treatment

Buddhi Wijesiri

11.1 INTRODUCTION

Stormwater is the largest non-point source of pollution in urban waterways. Cities around the world discharge thousands of liters of runoff carrying pollutants such as nutrients, heavy metals, and hydrocarbons. For example, cities in Australia are estimated to produce 3,000 GL of runoff annually [1]. The toxicity of stormwater pollutants poses direct risks to aquatic fauna and flora (e.g., salmon die-off in the U.S. Pacific Northwest [2]), and potential risks to human health via non-potable water use.

Urban areas typically consist of a range of land uses, among which residential, commercial, and industrial lands largely contribute to pollutants in stormwater. The common pollutants include particulate solids, nutrients, heavy metals, and hydrocarbons, and their impacts can range from eutrophication of waterways to carcinogenic effects. Based on the research conducted over the past few decades, the typical pollutant concentrations are reported in Table 11.1 below.

It is important to note that the current stormwater monitoring systems have limited capability of detecting multiple sources of pollution. The traditional monitoring systems mostly aim at above-ground pollutant deposition and wash-off, while sub-surface sources such as illicit discharges go undetected. Hence, the reported pollutant concertation levels are likely associated with high variability.

The impacts of polluted stormwater are being exacerbated due to rapid urbanization and climate change. The human populations continue to migrate to cities as such arable lands are transformed into impervious settlements. This results in an increase in pollutant loads (via human activities) and their dispersion (via roads, roofs, and parking areas) across the urban landscape. On the other hand, global warming alters wet and dry weather patterns, such that the variability in regional warming could worsen these changes in weather patterns. For example, high intensity-short duration rainfall events are predicted to occur more frequently in Australia over the coming decades [4,5], leading to washing off large quantities of pollutants accumulated on urban surfaces into receiving waters. Critically, interdisciplinary research (water engineering, hydrology, and climate sciences) is necessary to create new knowledge on stormwater pollution (pollution processes and site-specific pollution levels) under the above circumstances, without which the design of effective measures to improve stormwater quality would not be possible in the coming decades.

DOI: 10.1201/9781003441007-14

TABLE 11.1

Average Concentrations of Pollutants in Urban Stormwater Runoff

Pollutant Type		Concentration Based on Land Use (mg/L)		
		Residential	Commercial	Industrial
Total suspended solids (TSS)		173.2 ± 256.0	164.2 ± 125.3	163.3 ± 201.3
Nutrients	Total nitrogen (TN)	2.5 ± 1.3	3.4 ± 2.1	2.7 ± 1.6
	Total phosphorous (TP)	0.5 ± 0.5	0.4 ± 0.3	0.6 ± 0.8
Heavy metals	Cd	0.001 ± 0.002	0.005 ± 0.01	0.001 ± 0.002
	Cr	0.009 ± 0.01	0.02 ± 0.01	0.02 ± 0.02
	Cu	0.04 ± 0.06	0.04 ± 0.05	0.03 ± 0.02
	Ni	0.009 ± 0.01	0.02 ± 0.02	0.02 ± 0.009
	Pb	0.1 ± 0.3	0.09 ± 0.1	0.05 ± 0.06
	Zn	0.2 ± 0.3	0.3 ± 0.3	0.2 ± 0.4

Source: Adapted from Simpson et al. [3].

Despite risks associated with stormwater, its discharge into urban waters such as rivers and creeks is not adequately regulated compared to the discharge of wastewater. Furthermore, unlike wastewater, stormwater is considered a nuisance primarily because of flooding. Hence, its potential opportunities for reuse as a non-potable and even potable (if drinking water standards are met) alternative to costly and high carbon footprint sources such as desalination and recycling, are largely ignored. Governments invest in developing advanced technologies and infrastructure to treat wastewater and explore its other benefits such as resource recovery. However, for stormwater management, the annual water services budget only allocates for large networks of stormwater collection and disposal without adequate treatment [6]. Therefore, it is imperative to develop novel sustainable technologies to improve stormwater quality and potentially transform stormwater into a safe-to-use alternative water resource.

11.2 URBAN STORMWATER MANAGEMENT— STORMWATER QUALITY

Stormwater runoff is generated by urban impervious surfaces during rainfall events. The management of this runoff is typically meant to minimize disturbance to public activities by removing the runoff via subsurface drainage networks which discharge polluted water into receiving waters. Yet, most urban drainage systems get overwhelmed during heavy rainfall events, triggering floods. It is common practice to attenuate the volume of runoff before being let into the drainage network either by delaying the discharge (detention) or through infiltration into the ground (retention). While detention/retention systems can vary in size, city-wide large-scale systems also exist, for example, the underground flood water diversion facility in Japan. In addition to volume reduction, detention/retention systems reduce particulate matter (sediments), which may also reduce a fraction of particle-bound toxic pollutants. However, since most toxicants are attached to fine particle fractions [7], a large fraction of these pollutants are still allowed to be discharged into receiving waters or infiltrated into soil and groundwater. In response, more sophisticated pre-control measures have been implemented, such as constructed wetlands, swales, rain gardens, and bioretention systems. These are nature-based systems, but commonly termed as Best Management Practices (BMP), Low Impact Design (LID) and Water Sensitive Urban Design (WSUD), and more recently Sponge City. Despite their high potential to treat pollutants in stormwater runoff, they are traditionally being used to control runoff volume while partially targeting stormwater quality [8].

Since stormwater carries a range of different pollutants that have different physico-chemical characteristics and different levels of bioavailability, it is a challenging task to engineer nature-based systems to achieve the set standards of water quality. It is even more difficult to transform stormwater for different purposes (e.g., recreational, drinking, and direct discharge into waterways), which require different levels of water quality. The treatment systems are expected to accommodate multiple physical, chemical, and biological processes that can remove or reduce pollutants from stormwater under changing environmental conditions. It is crucial to ensure that stormwater treatment systems are designed such that they are resilient against changing weather patterns, which the variability is likely to be exacerbated at a regional scale due to climate change.

11.3 NATURE-BASED SOLUTIONS FOR STORMWATER TREATMENT

Nature-based solutions (NBS) are engineered systems that mimic physical, chemical, and biological processes within natural environments. For example, natural wetlands (Figure11.1a) are highly effective in holding large volumes of water, allowing slow discharge and in turn preventing flooding, as well as improving water quality using soil–water–vegetation interactions. Such natural systems have long been integrated into urban planning in the form of constructed wetlands (Figure11.1b) in cities such as Melbourne, Australia [9] and cities around China as part of the Sponge City agenda [10] and around the world [11]. Moreover, NBS have also been included as an integrated component of infrastructure design, for example, vegetated rooftop gardens (Figure11.1c) and green walls (Figure11.1d).

The discussion below briefly delves into the stormwater treatment potential of typical NBS (constructed wetlands, green walls, bioswales, green roofs, and porous pavements). This section also discusses in detail a specific type of NBS, namely, biofiltration systems that have the potential to treat stormwater within a range of scenarios such as stormwater treatment, intermittent stormwater and wastewater treatment, and sewer overflow (raw sewage diluted by stormwater during heavy rainfall events) treatment.

11.3.1 Constructed Wetlands

Constructed wetlands are ecosystems engineered to replicate the natural hydrologic/hydraulic and pollutant removal processes of wetlands. They are a sustainable

FIGURE 11.1   (a) Natural wetland; (b) Constructed wetland; (c) Industrial building top covered with vegetation; (d) Green wall; (e) Bioswale; (f) Residential green roof.

approach to treating polluted waters and ecological restoration in urban environments. Constructed wetlands utilize vegetation, soil media, and microorganisms to enable natural physical, chemical, and biological processes to remove pollutants from water [12].

The polluted water is fed into carefully designed basins or channels (optimally sized) that contain a mix of wetland plants, such as common reeds (*Phragmites australis*) and cattails (*Typha*). The pollutants are removed as the water flows through the wetland (filtration and sedimentation) horizontally or vertically (depending on the design) and comes into contact with the plants' roots (plant uptake) and the various microorganisms (biodegradation into less harmful forms) [13]. Constructed wetlands are versatile systems, such that they can be designed to manage different volumes and pollution levels of industrial wastewater, municipal sewage, and stormwater.

These systems can perform consistently despite water flows over the soil surface (surface flow constructed wetlands) or water flows through the soil media (subsurface flow constructed wetlands).

In addition to treating polluted waters, constructed wetlands provide a range of other benefits to improve the livability of urban spaces. They can be designed and integrated into urban settings to offer habitats for a diverse group of plants and animals, preserving urban biodiversity. This not only allows native species to thrive in local environments but also attracts other wildlife to cities.

It is important to note the role of constructed wetlands as a flood control measure during heavy storm events. These wetlands have a great capacity to adsorb and retain large volumes of stormwater and slowly discharge into receiving waters, reducing potential damages to infrastructure and ecosystems. Nevertheless, constructed wetlands' ability to mitigate floods may have shifted the attention to their ability to treat polluted water, in particular stormwater. The natural treatment processes can be harnessed to create a cost-effective, environmentally sustainable, and aesthetically pleasing alternative to typically expensive water treatment methods.

11.3.2 GREENWALLS

Green walls, living walls, or vertical gardens, are innovative features incorporated with an array of living plants. They offer a range of environmental and aesthetic benefits as an integral part of modern sustainable architecture and urban design. Primarily, green walls are designed to introduce greenery into limited spaces in urban settings, making use of vertical surfaces such as building facades, interior walls, and freestanding structures. The walls consist of modular panels or planting systems, and specific plants are selected based on their adaptability to the environmental conditions of the location [14].

Even though green walls are more popular in urban areas to improve air quality, and aesthetics and to increase energy efficiency of buildings, there is potential for in-situ treatment of stormwater, which is yet to be fully harnessed. The building spaces themselves account for a large portion of impervious surfaces that generate stormwater (in the form of roof water which could be flushed into the stormwater drainage). By collecting this stormwater and diverting it to a green wall, not only the volume of stormwater that otherwise would be flushed out can be reduced, but also be treated.

While the initial installation and maintenance of green walls require careful planning and specialized expertise, they offer significant long-term benefits. These living ecosystems represent a harmonious union between urban living spaces and the natural world, promoting sustainability, well-being, and a greener future for cities.

11.3.3 BIOSWALES

Vegetated swales or biofiltration swales (bioswales) are stormwater management systems designed to capture, treat, and infiltrate runoff. These sustainable green infrastructure elements are commonly incorporated into roadside areas and serve as a natural solution to mediate fast-flowing runoff that picks up pollutants, debris, and excess nutrients along the way [15,16].

Bioswales intercept and manage stormwater runoff before it reaches the storm drainage systems and waterways. They are typically constructed as shallow and elongated channels, which are engineered to attenuate and direct runoff. These swales provide a natural filtration system, allowing the water to percolate into the ground through vegetation and engineered soils, retaining pollutants within the soil and plants. The diverse root systems of the plants within the swales trap pollutants and uptake toxicants, and provide a favorable environment for microorganisms to thrive, promoting biological processes that transform pollutants into less harmful compounds. Bioswales, in the meantime, also play a role in recharging groundwater supplies as the water is gradually released, minimizing the impact on conventional drainage systems.

Utilizing Bioswales as a way of improving stormwater quality could be easier compared to other alternatives because they can be incorporated into urban planning to bring in additional benefits. These systems contribute to mitigating the urban heat island effect, where cities experience higher temperatures than surrounding vegetated areas due to human activities and impervious surfaces. The vegetation in bioswales contributes to evaporative cooling, reducing the overall temperature and improving the livability of urban environments. Additionally, bioswales enhance the aesthetic appeal of urban landscapes. With their lush vegetation and naturalistic design, they create pockets of greenery amidst concrete jungles, attracting wildlife and offering pleasant green spaces for pedestrians and residents. Bioswales also act as buffers against flooding during heavy rainfall, reducing the burden on traditional stormwater drainage systems.

11.3.4   GreenRoof

Green roofs, living roofs, or vegetated roofs, are innovative architectural features that transform conventional rooftops into thriving nature-based systems. These sustainable nature-based systems typically consist of a layer of vegetation, a growing medium, and a waterproofing membrane. They can characteristically host a diverse range of vegetation such as sedums, vibrant wildflowers, and even small trees. Like other vegetated nature-based systems, it is critical to choose plants that can thrive in local climates and environmental conditions [14].

Green roofs play a crucial role in stormwater management. As traditional rooftops are impermeable, they cause rainwater to flow rapidly and overwhelm stormwater drainage systems. Green roofs act as natural sponges, absorbing rainfall by retaining a significant amount of water within their vegetation. Consequently, the discharge of stormwater runoff is delayed, reducing the burden on drainage systems, and minimizing the risk of flooding during heavy rain events.

In addition to environmental benefits, green roofs contribute to energy efficiency in buildings. They act as an additional insulation layer to reduce heat transfer, keeping buildings cooler in the summer and warmer in the winter. This insulation effect leads to energy savings and reduced reliance on heating and cooling systems, lowering greenhouse gas emissions and overall energy consumption, which significantly contributes to a lower carbon footprint [14].

Green roofs also foster biodiversity in urban settings. These elevated nature-based systems create habitats for a variety of flora and fauna that may have otherwise struggled to find space in densely populated urban areas. In addition to supporting native species, green roofs contribute to biodiversity conservation in urban settings by attracting beneficial insects and pollinators.

Like all other nature-based systems, green roofs offer aesthetically pleasing and calming spaces for relaxation, recreation, and social interactions. These are popular features in most regions of the world, which can be seen in the form of rooftop gardens, accessible terraces, or green workspaces.

Like green walls, the initial cost of installing green roofs can be higher than traditional roofs, they offer substantial benefits in terms of building energy savings, an opportunity to control stormwater (untreated roof water that flows into stormwater drains), and improving air quality, which altogether can increase property value, making green roofs an economically and environmentally sustainable urban development.

11.4 POROUS PAVEMENTS

As a sustainable alternative to traditional impervious pavements, porous/permeable/pervious pavements are designed to allow water infiltration into the soil. The aim is to reduce stormwater runoff volume that is discharged into receiving waters, which can potentially minimize the amount of pollutants that can be discharged into receiving waters [17].

Conventional pavements are vast impervious areas where pollutants accumulate in large quantities during dry periods and large volumes of runoff are generated during rainfall, overwhelming the drainage system. Since porous pavements are made with interconnected voids using permeable concrete, pervious asphalt, permeable pavers, and gravel-filled surfaces, they mimic the natural infiltration of water similar to a previous soil surface [17]. In addition to effective stormwater management, porous pavements also contribute to groundwater replenishment as the water infiltrates through underlying soil, enabling aquifer recharge. This is a sustainable approach to conserving limited water resources in urban areas, particularly in the areas of groundwater depletion.

Porous pavements also satisfy the aesthetic aspects of NBS, by adding visually appealing green features to public spaces, parking lots, and roadways. Further, they can be best implemented in the urban landscape when integrated with other NBS such as bioswales and rain gardens, maximizing the benefits of individual NBS systems.

Nevertheless, porous pavements may require substantial maintenance as they are subject to direct impacts of dry and wet weather. In the long term, sediment build-up can reduce infiltration capacity; hence regular cleaning is essential to ensure satisfactory efficiency of these systems.

In the context of stormwater treatment, porous pavements may have limited potential compared to other NBS, because the treatment largely depends on the underneath soil. However, porous pavements can be reinvented to incorporate some key design features of other soil-based filtration systems. This can be done by introducing multi-layered soils (even including carbon sources for enhancing treatment processes) rather than simply constructing a porous pavement on the existing soil column. Yet, it is important to note that this could be feasible and effective for limited purposes such as parking lots given the need for maintenance (engineered soil layers may require replacement in the long term).

11.5   BIOFILTRATION SYSTEMS

Biofilters are soil and plant-based passive water treatment technology. They are typically used to control runoff volume (reduce peak flow) and to improve runoff quality before discharge into urban waterways [18]. Biofilters can accommodate physical, chemical, and biological treatment processes simultaneously and their soil media, plants, hydrological, and hydraulic conditions can be varied to form different configurations (e.g., unsaturated, and saturated zones enable specific conditions for specific treatment processes, see Figure 11.2). As such, biofilters have the potential to treat common stormwater pollutants such as sediment, nutrients, and heavy metals [19–21]. Nevertheless, challenges to harness the full potential of biofilters persist, including vegetation dying off during extended dry periods [22,23], resulting in poor long-term performance, which can be worsened in regions where longer dry periods are becoming common due to climate change [24,25]. Additionally, there is a significant opportunity to increase the efficiency of existing treatment processes as well as incorporate novel treatment processes within biofilters, so that a range of pollutants at a range of concentrations can be treated.

In the context of maintaining a consistent long-term performance of biofilters, it is important to manage the intermittent nature of stormwater occurrence. In response,

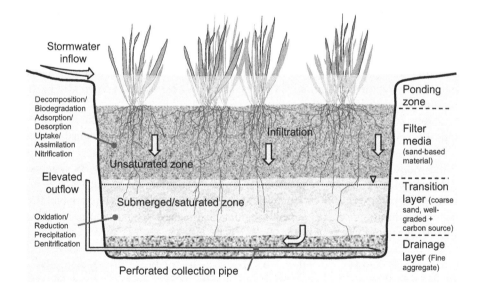

FIGURE 11.2   Typical biofilter configuration and key pollutant removal processes.

research has recognized the potential of dual-mode biofilters, where the biofilter is fed with stormwater during wet periods while wastewater is used to maintain vegetation and microbiological activity during dry periods [26–29]. This dual-mode approach would not only increase the efficiency of stormwater treatment and reduce costs of long-term maintenance but also minimize the burden of end-of-pipe treatment of wastewater [30].

Considering the limited space in urban areas and the need for reducing maintenance costs, biofilters are best implemented with alternating water sources (stormwater–groundwater, and stormwater–wastewater), which is shown to be effective in recent studies [28,31]. However, since different water sources have different pollutant concentrations, it is important to modify typical treatment processes and develop new processes to enable the effective design of biofilters. Table 11.2 summarizes the characteristics of key design features of biofilters.

Biofilters create complex physical, chemical, and biological interactions among plants, filter media, and microorganisms to remove pollutants. The operational conditions within biofilters are highly dynamic and difficult to optimize to create favorable conditions for pollutant treatment processes [32]. Two of the key processes within biofilters are nitrification (which transforms NH_4^+ into NO_2 and NO_3) and denitrification (which converts NO_2 and NO_3 into N_2 gas). Since nitrification and denitrification are incomplete processes (partial conversion of pollutants into other chemical forms), current biofilters require much improvement in pollutant removal efficiency, particularly under field conditions [26,33–36].

In addition to typical treatment processes, biofilters also host more efficient pollutant transformation processes, although their mechanisms are not well-understood and are not fully integrated into biofilter designs. For example, anaerobic ammonium oxidation (Anammox) directly converts NH_4^+ into N_2, which is shown to be highly efficient (>85%) in treating nitrogen in wastewater within controlled bioreactors [37].

Taking a step further, recent research shows the potential of combining multiple treatment processes as an attempt to complete the removal of pollutants. For example, the Simultaneous Nitrification, Anammox, and Denitrification (SNAD) process aims to optimum utilization of resources within the treatment system such as media, plants, and microorganisms to remove nitrogen from polluted waters. Even though a few studies on constructed wetlands treating wastewater have shown some intriguing results [38–40], more research is necessary to find ways to facilitate highly demanding operational conditions that are required to support each individual process.

11.6 KNOWLEDGE GAPS

11.6.1 Government Actions and Community Perception

The United Nations Sustainable Development Goals (SDGs) specifically focus on ensuring access to safe and adequate water for all (SDG 6). Currently, the expected outcomes of these goals seem to have become too ambitious given the progress of goals is already lagging the target timeline of 2030. Regarding SDG 6, one of the main reasons for the lack of progress is limited effort and investment in alternative water resources despite existing resources continuing to diminish both in quantity

TABLE 11.2

Characteristics of Biofilter Design Features

Design Feature	Wastewater Biofilters	Stormwater Biofilters	Dual-Mode Biofilters (Stormwater + Greywater)	References
	Typical arrangement includes top to bottom: filter media, saturated/submerged zone (SZ), and drainage layer.			
	• External carbon source is not necessarily due to the abundance of biodegradable organic matter (can be reduced in the filter media).	• Transition layer is placed between SZ and drain layer.	• Laboratory columns are found to have included a transition layer between the SZ and drain layer.	[41] [33]
		• SZ of at least 350 mm, ideally 450–500 mm 450–500 mm.	• However, field-based filters have placed the transition later between filter media and SZ, which is appropriate.	[18]
Media		• External carbon sources (e.g., mulch, woodchip, wheat straw, maize cobs) need to be introduced in the SZ because stormwater lacks the organic carbon required for denitrification.		[42] [20] [43] [28]
		• Novel hydrochar has been tested as not only a carbon source but also as an ammonia adsorbent.		[26] [27]

(Continued)

TABLE 11.2 (Continued)
Characteristics of Biofilter Design Features

Design Feature	Wastewater Biofilters	Stormwater Biofilters	Dual-Mode Biofilters (Stormwater + Greywater)	References
Vegetation		Plants need to be established for a certain period before transferring into a biofilter (15–24 weeks).		[29] [22]
	• Ornamental species contribute to pollutant removal.	• An extensive and fine root system with high biomass is preferred to facilitate nitrogen removal and to withstand evaporation losses during dry periods.	• Should thrive under greywater conditions of regular watering, nutrient-rich inflows, and salinity.	
	Should withstand water-logged conditions, high nutrient loads, and elevated salinity.	• Inclusion of a saturated zone would support during dry spells.	• Alternating water sources have minimal impact on treatment performance if plants are selected carefully.	[44] [26] [27]
		• Intermittent wetting may be required to prevent plants from dying		
Infiltration		• Satisfactory rates: 100–600 mm/h		[29] [28] [26] [27]
		• Higher rates (i.e., >300 mm/h) may cause plant stress (which can be overcome with dual-mode systems) and limit treatment time (which can be balanced with a SZ).		
		• Higher infiltration rates for stormwater inflow can be expected due to the larger driving head of stormwater flow and greywater not being able to cause ponding across the entire surface of the system.		
		• As the system ages, the infiltration may reduce due to sediment build-up, which can be scraped out, and dense vegetation can maintain long-term hydraulic conductivity of the system (by creating preferential flow paths).		

and quality. As the last underutilized source of water, stormwater needs much attention from both the public and the government. To transform stormwater as a fit-for-purpose resource, governments must make a continuous flow of long-term investment in research and in changing the perception of the public.

11.6.2 OPTIMIZATION OF TREATMENT PROCESSES

Removal of pollutants in NBS is a combination of physical, chemical, and biological processes. These processes are facilitated by soil media, plants, and microorganisms. While most solids are filtered through or settled in during treatment, micropollutants such as nutrients and heavy metals are not efficiently removed. For example, nitrogen removal using nitrification and denitrification are at satisfactory levels, but there is much opportunity to enhance their performance by optimizing the design (e.g., optimizing depth and filter media configurations including a submerged zone). Further, phosphorous removal remains at low levels compared to nitrogen removal, as such adsorption and plant intake need substantial improvement by introducing novel adsorptive materials and plant species with high intake.

Regarding highly toxic pollutants such as heavy metals which are common in stormwater, still need much attention in understanding their removal processes. Even though treatment processes such as adsorption by organic carbon-rich materials can be effective, one of the challenges is to prevent the leaching of toxicants. Overcoming these challenges is necessary to design NBS that can treat a range of pollutants simultaneously.

11.6.3 NOVEL TREATMENT PROCESSES

In addition to classical treatment processes, NBS needs to be incorporated with new and more efficient processes. This is due to the need to minimize associated costs and treat emerging pollutants. One of the key problems in nitrogen removal in NBS is incomplete nitrification and denitrification processes. To tackle this issue, Anammox (direct ammonia oxidation to nitrogen gas) and Simultaneous Nitrification, Anammox, and Denitrification (SNAD) have been proposed. While the potential of these new processes is long realized, their integration within NBS needs research, particularly to establish a knowledge base on operational and environmental conditions favorable to microorganisms, particularly those involved with Anammox bacteria.

11.6.4 STORMWATER QUALITY MODELLING

Modeling stormwater pollution processes is an essential step in designing treatment systems. There are a number of commercially available modeling tools with a range of capabilities (e.g., conceptual, physically-based, and statistical) and limitations (e.g., simplified/lumped representation of pollution processes and limitation of uncertainty assessment). However, a stand-alone model/modeling tool is yet to be developed, such that catchment surface and sub-surface pollution sources are accounted for; catchment-specific pollution levels can be estimated; and uncertainties in data and model can be quantified. Such a modeling approach is a significant step forward in transforming stormwater as an alternative water resource under the circumstances of rapid urbanization and climate change (i.e., considering expected variability arising from the effects of changes in urban landscape and wet and dry weather patterns).

11.6.5 STORMWATER MONITORING

One of the biggest challenges to stormwater quality modeling and design of treatment systems is the lack of reliable field measurements. Current field monitoring does not provide accurate and adequate data on ever-changing stormwater pollution levels. Nevertheless, a major advancement in this area has emerged recently, where real-time stormwater monitoring is done using sensors. Currently, such sensor systems can measure a limited number of parameters such as flow depth, pH, and electrical conductivity, while active research is necessary to improve sensor capacity to measure turbidity, nutrients (ammonia and nitrates), and even heavy metals, which are essential data to design effective treatment systems.

11.6.6 RESOURCE RECOVERY

Resource recovery is a priority area in wastewater treatment research. This could be due to high concentrations of materials with potential benefits. Even though stormwater only contains relatively low concentrations of such materials, it is generated in large volumes, highlighting the opportunities for recovering reusable materials such as nutrients. Critically, the type of materials transported via stormwater varies by their origin (e.g., residential lands produce more organic content and nutrients while industrial and commercial lands produce heavy metals, hydrocarbons, and emerging contaminants such as pesticides and pharmaceuticals). Therefore, fit-for-purpose treatment systems (specifically optimized to treat nutrients and heavy metals) can be designed to recover useful materials.

11.7 CONCLUSIONS

Stormwater runoff is the largest contributor to polluting urban waterways, carrying pollutants ranging from suspended solids to toxicants such as heavy metals, hydrocarbons and even emerging pollutants including pharmaceuticals and pesticides. However, managing stormwater has long been centered on controlling the runoff volume due to the significant effects of flooding. Therefore, much work is necessary, from the governments, industry, and research community, to improve the quality of stormwater. Current efforts such as retention/detention systems and constructed wetlands have shown success in reducing pollution levels prior to discharge into receiving waters, but their primary aim has always been mitigating the effects of the sheer quantity of runoff. Cost-effective and low carbon footprint alternatives for stormwater treatment such as Nature-based Solutions (NBS) need to be developed and implemented at a broad scale, focusing on removing pollutants from stormwater. To make this happen, substantial work is necessary in the areas of accelerating government actions and improving community perception towards stormwater as an alternative water resource, optimization of treatment processes within novel technologies such as NBS, introducing efficient treatment processes, and improving the credibility of stormwater quality predictions, developing new tools for reliable real-time stormwater monitoring, and enabling resource recovery from treatment systems.

REFERENCES

1. ECRC. *Stormwater Management in Australia.* (Commonwealth of Australia 2015, Canberra, Australia, 2015).
2. Tian, Z. et al. A ubiquitous tire rubber-derived chemical induces acute mortality in coho salmon. *Science* **371**, 185–189, doi:10.1126/science.abd6951 (2021).
3. Simpson, I. M., Winston, R. J. & Brooker, M. R. Effects of land use, climate, and imperviousness on urban stormwater quality: A meta-analysis. *Science of the Total Environment* **809**, 152206, doi:10.1016/j.scitotenv.2021.152206 (2022).
4. King, A. D., Karoly, D. J. & Henley, B. J. Australian climate extremes at 1.5 C and 2 C of global warming. *Nature Climate Change* **7**, 412, doi:10.1038/NCLIMATE3296 (2017).
5. Iturbide, M. et al. An update of IPCC climate reference regions for subcontinental analysis of climate model data: definition and aggregated datasets. *Earth System Science Data* **12**, 2959–2970 (2020).
6. ITA. *Australia – Water and Wastewater Treatment,* https://www.trade.gov/country-commercial-guides/australia-water-and-wastewater-treatment (2022).
7. Wijesiri, B., Egodawatta, P., McGree, J. & Goonetilleke, A. Understanding the uncertainty associated with particle-bound pollutant build-up and wash-off: a critical review. *Water Research* **101**, 582–596, doi:10.1016/j.watres.2016.06.013 (2016).
8. Fletcher, T. D. et al. SUDS, LID, BMPs, WSUD and more – The evolution and application of terminology surrounding urban drainage. *Urban Water Journal* **12**, 525–542, doi:10.1080/1573062X.2014.916314 (2015).
9. City_of_Melbourne. *Trin Warren Tam-Boore Wetland – Royal Park,* https://urbanwater.melbourne.vic.gov.au/projects/wetlands/wetlands-sample-project/
10. Cai, Y. et al. Utilization of constructed wetland technology in China's sponge city scheme under carbon neutral vision. *Journal of Water Process Engineering* **53**, 103828, doi:10.1016/j.jwpe.2023.103828 (2023).
11. Cui, M., Ferreira, F., Fung, T. K. & Matos, J. S. Tale of two cities: how nature-based solutions help create adaptive and resilient urban water management practices in Singapore and Lisbon. *Sustainability.* **13**, 10427 (2021).
12. Stefanakis, A. I. J. S. The role of constructed wetlands as green infrastructure for sustainable urban water management. *Sustainability.* **11**, 6981 (2019).
13. Parde, D. et al. A review of constructed wetland on type, treatment and technology of wastewater. *Environmental Technology & Innovation* **21**, 101261, doi:10.1016/j.eti.2020.101261 (2021).
14. Manso, M., Teotónio, I., Silva, C. M. & Cruz, C. O. Green roof and green wall benefits and costs: A review of the quantitative evidence. *Renewable and Sustainable Energy Reviews* **135**, 110111, doi:10.1016/j.rser.2020.110111 (2021).
15. Anderson, B. S. et al. Bioswales reduce contaminants associated with toxicity in urban storm water. *Environmental toxicology and chemistry.* **35**, 3124–3134 (2016).
16. Everett, G., Lamond, J. E., Morzillo, A. T., Matsler, A. M. & Chan, F. K. S. Delivering green streets: An exploration of changing perceptions and behaviours over time around bioswales in Portland, *Oregon. Journal of Flood Risk Management.* **11**, S973–S985 (2018).
17. Li, H., Harvey, J. T., Holland, T. & Kayhanian, M. J. E. R. L. The use of reflective and permeable pavements as a potential practice for heat island mitigation and stormwater management. *Environmental Research Letter.* **8**, 015023 (2013).

18. Payne, E. et al. *Adoption Guidelines for Stormwater Biofiltration Systems.* (Cooperative Research Centre for Water Sensitive Cities, Melbourne, 2015).

19. Bratieres, K., Fletcher, T. D., Deletic, A. & Zinger, Y. Nutrient and sediment removal by stormwater biofilters: A large-scale design optimisation study. *Water Research* **42**, 3930–3940, doi:10.1016/j.watres.2008.06.009 (2008).

20. Nabiul Afrooz, A. R. M. & Boehm, A. B. Effects of submerged zone, media aging, and antecedent dry period on the performance of biochar-amended biofilters in removing fecal indicators and nutrients from natural stormwater. *Ecological Engineering* **102**, 320–330, doi:10.1016/j.ecoleng.2017.02.053 (2017).

21. Feng, W., Hatt, B. E., McCarthy, D. T., Fletcher, T. D. & Deletic, A. Biofilters for stormwater harvesting: understanding the treatment performance of key metals that pose a risk for water use. *Environmental Science & Technology* **46**, 5100–5108 (2012).

22. Payne, E. et al. Which species? A decision-support tool to guide plant selection in stormwater biofilters. *Advances in Water Resources* **113**, 86–99 (2018).

23. McDowell, N. et al. Mechanisms of plant survival and mortality during drought: why do some plants survive while others succumb to drought? *New Phytologist* **178**, 719–739 (2008).

24. Zhang, K., Manuelpillai, D., Raut, B., Deletic, A. & Bach, P. M. Evaluating the reliability of stormwater treatment systems under various future climate conditions. *Journal of Hydrology* **568**, 57–66, doi:10.1016/j.jhydrol.2018.10.056 (2019).

25. Steffen, W., Hughes, L., Alexander, D. & Rice, M. Cranking up the intensity: climate change and extreme weather events. *Climate Council of Australia.* (2017). https://research-management.mq.edu.au/ws/portalfiles/portal/72586064/72585882.pdf

26. Barron, N. J. et al. Dual-mode stormwater-greywater biofilters: The impact of alternating water sources on treatment performance. *Water Research* **159**, 521–537, doi:10.1016/j.watres.2019.04.020 (2019).

27. Barron, N. J., Hatt, B., Jung, J., Chen, Y. & Deletic, A. Seasonal operation of dual-mode biofilters: The influence of plant species on stormwater and greywater treatment. *Science of the Total Environment* **715**, 136680, doi:10.1016/j.scitotenv.2020.136680 (2020).

28. Zhang, K. et al. Pollutant removal performance of field scale dual-mode biofilters for stormwater, greywater, and groundwater treatment. *Ecological Engineering* **163**, 106192, doi:10.1016/j.ecoleng.2021.106192 (2021).

29. Fowdar, H. S., Hatt, B. E., Breen, P., Cook, P. L. M. & Deletic, A. Designing living walls for greywater treatment. *Water Research* **110**, 218–232, doi:10.1016/j.watres.2016.12.018 (2017).

30. Grant, S. B. et al. Taking the "waste" out of "wastewater" for human water security and ecosystem sustainability. *Science* **337**, 681–686 (2012).

31. Aloni, A. & Brenner, A. Use of cotton as a carbon source for denitrification in biofilters for groundwater remediation. *Water* **9**, 714 (2017).

32. Payne, E. et al. Biofilter design for effective nitrogen removal from stormwater – influence of plant species, inflow hydrology and use of a saturated zone. *Water Science and Technology* **69**, 1312–1319, doi:10.2166/wst.2014.013 (2014).

33. Zinger, Y., Prodanovic, V., Zhang, K., Fletcher, T. D. & Deletic, A. The effect of intermittent drying and wetting stormwater cycles on the nutrient removal performances of two vegetated biofiltration designs. *Chemosphere* **267**, 129294, doi:10.1016/j.chemosphere.2020.129294 (2021).

34. Prodanovic, V., McCarthy, D., Hatt, B. & Deletic, A. Designing green walls for greywater treatment: The role of plants and operational factors on nutrient removal. *Ecological Engineering* **130**, 184–195 (2019).

35. Ye, J., Liu, J., Ye, M., Ma, X. & Li, Y.-Y. Towards advanced nitrogen removal and optimal energy recovery from leachate: A critical review of anammox-based processes. *Critical Reviews in Environmental Science and Technology* **50**, 612–653 (2020).

36. Zhang, K. et al. The impact of stormwater biofilter design and operational variables on nutrient removal-a statistical modelling approach. *Water Research* **188**, 116486 (2021).
37. Ji, J. et al.Achieving advanced nitrogen removal from low C/N wastewater by combining endogenous partial denitrification with anammox in mainstream treatment. *Bioresource Technology* **270**, 570–579 (2018).
38. Paranychianakis, N. V., Tsiknia, M. & Kalogerakis, N. Pathways regulating the removal of nitrogen in planted and unplanted subsurface flow constructed wetlands. *Water Research* **102**, 321–329 (2016).
39. Chen, D. et al. Denitrification- and anammox-dominant simultaneous nitrification, anammox and denitrification (SNAD) process in subsurface flow constructed wetlands. *Bioresource Technology* **271**, 298–305 (2019).
40. Li, H. & Tao, W. Efficient ammonia removal in recirculating vertical flow constructed wetlands: complementary roles of anammox and denitrification in simultaneous nitritation, anammox and denitrification process. *Chemical Engineering Journal* **317**, 972–979 (2017).
41. Fowdar, H. S., Hatt, B. E., Breen, P., Cook, P. L. & Deletic, A. Evaluation of sustainable electron donors for nitrate removal in different water media. *Water Research* **85**, 487–496 (2015).
42. Audet, J. et al. Nitrogen removal and nitrous oxide emissions from woodchip bioreactors treating agricultural drainage waters. *Ecological Engineering* **169**, 106328, doi:10.1016/j.ecoleng.2021.106328 (2021).
43. Wang, F., Wang, H., Sun, C. & Yan, Z. Conventional bioretention column with Fe-hydrochar for stormwater treatment: Nitrogen removal, nitrogen behaviour and microbial community analysis. *Bioresource Technology* **334**, 125252 (2021).
44. Read, J., Fletcher, T. D., Wevill, T. & Deletic, A. Plant traits that enhance pollutant removal from stormwater in biofiltration systems. *International Journal of Phytoremediation* **12**, 34–53 (2009).

12 Fluoride Removal from Rural Groundwater

A Systematic Review of Health Effects, Technologies, and Socio-Economic Factors

Palistha Shrestha and Erick R. Bandala

12.1 INTRODUCTION

Developing global water supply to improve health and socioeconomic well-being while simultaneously managing water resources sustainably has continued to be one of the twenty-first century's biggest challenges. Lack of access to safe water, sanitation, and hygiene (WASH) facilities continues to be a global concern, directly affecting the survival and well-being of billions around the world. Recognizing WASH's critical role in preventing diseases and improving the quality of life, the United Nations (UN) declared universal access to safe water and sanitation as part of the Sustainable Development Goals (SDGs) to achieve by 2030. However, despite significant efforts, the world seems far behind in reaching the global goal with 2 billion people still lacking access to safely managed drinking water services along with 3.6 billion people still lacking access to safe sanitation facilities [1]. Accelerating progress from all sectors is critical to achieve the goal by 2030.

In addition to increasing domestic water demand due to population growth, increasing demand for food and energy has snowballed the need for irrigation and water extraction with human water consumption increasing from roughly 500 to around 4,000 km^3 per year over the past century [2]. The availability of safe and reliable water has a longstanding impact on the fragile socioeconomic framework of human civilization. Combined with the effects of climate change, longstanding droughts have increased water insecurity issues and have raised concerns over growing reasons for social unrest around the world. For five consecutive years, from 2015 to 2020, the World Economic Forum declared water crisis as one of the top five Global Risks for livelihood with the 2023 Global Risk Report identifying Natural Resource crisis as one of the top 10 risks for the short term impact in 2 years as well as for the long term impact in the next 10 years [3,4]. Already, 2.4 billion people are living in water-stressed countries [5].

DOI: 10.1201/9781003441007-15

With most freshwater stored in the ground, global importance and dependence on groundwater has significantly increased in the past few decades with groundwater currently supplying around half of the world's domestic water needs [6]. Global water trends show that groundwater abstraction has consistently increased to meet the shortages in surface water availability due to increasing droughts [2]. With increasing variability in available surface water resources, groundwater dependence will only continue to rise. Additionally, for small-scale water supply systems, groundwater is generally the primary water source due to its superior quality to surface water and accessibility [7]. Hence, ensuring access to safe and reliable groundwater, especially for drinking water purposes, is critical in achieving SDG Goal 6 and contributing to a sustainable and safe future for all. This study undertakes a review of the risk posed by one specific, very frequent groundwater contaminant – fluoride ion, its presence, health effects, and available technology for removal to identify knowledge gaps and suggest avenues for future research that may sustainably help address the variety of problems caused by the presence of fluoride in groundwater sources.

12.2 FLUORIDE

12.2.1 FLUORIDE IN GROUNDWATER

The variability in surface water availability has forced millions of people to depend on groundwater for their livelihood. In Africa alone, more than 400 million people depend on groundwater as their main or only source of water supply [8]. And while access to groundwater is a welcomed relief in times of droughts, it can contain a wide range of biological, chemical, and physical contaminants due to both natural and anthropogenic sources.

Arsenic, fluoride, and iodine are some of the main chemical components in groundwater that affect the health of around 450 million people around the world [9]. Among them, fluoride (F^-) is considered the 13th most prevalent element found in the earth's crust in minerals such as fluorite, apatite, or biotite [8,10]. Fluoride can occur naturally in groundwater due to weathering, dissolution, and precipitation of fluoride-bearing minerals from water-rock interactions, ion exchange, and sorption processes [11,12]. The presence of F^- in groundwater depends on geological formations, mineral type found in the rock, groundwater residence time, and geochemical reactions that occur along its path [13]. While the majority of F^- contamination in groundwater is associated with natural sources, it can also occur in groundwater through anthropogenic sources such as phosphate-based fertilizers, coal power plants, ceramic and glass industrial waste, and copper and nickel smelting processes [12,14,15].

The hydrochemistry of groundwater directly affects the release of F^- ions from geological structures. Factors such as high alkalinity, sodium, and bicarbonate (HCO_3^-) directly increase F^- concentration while high levels of total hardness, calcium, and magnesium negatively correlate to F^- concentration in groundwater [12,16]. Fluoride solubility levels in groundwater decrease at high calcium concentrations as it forms insoluble mineral fluorite (CaF_2) [15].

12.2.2 FLUORIDE HEALTH EFFECTS

Fluoride intake possesses both beneficial and harmful effects. Fluoride enters the human body primarily through contaminated drinking water, especially groundwater [12]. The human body is known to readily absorb F⁻ with minimal excretion [17]. Although F⁻ concentrations in drinking water ranging from 0.5 to 1 mg/L are considered beneficial to prevent dental caries, the World Health Organization (WHO) has set the guideline F⁻ drinking water concentration limit to be 1.5 mg/L for a person who consumes 2 L of water per day [18]. The WHO F⁻ guideline should be adjusted based on daily water intake, especially in areas with warm climates, where water consumption is higher as the health effects of fluoride are directly proportional to the amount of fluoride ingested daily [10]. In areas with high fluoride toxicity, fluoride also enters the body through food consumption and can cause stunted growth in children and hormone imbalances [19].

Very little is known about the effects of low F⁻ doses on the cardiovascular system, nervous system, hepatic and renal functions, reproductive system, thyroid function, blood glucose homeostasis, or the immune system [20]. This lack of information is identified as a significant knowledge gap worth attention to as it may be the cause of long-term side effects on human well-being and can become an impediment to sustainable development.

Consumption of high levels of fluoride can lead to several debilitating diseases. In the human body, fluoride replaces hydroxide ions in hydroxyapatite and forms fluorapatite in tooth enamel and bones, which causes skeletal and dental fluorosis [21–23]. Long-term exposure to high levels of F⁻ can lead to conditions such as osteoporosis, arthritis, cancer, miscarriage, and birth defects as well as harmful effects on the respiratory organs, the digestive system, the endocrine system, cardiovascular organs, and immune responses [12,15,19]. Fluoride is also known to damage the neural network [24]. Fluoride in the human body showed several toxicity phenotypes such as skeletal tissue and soft tissue damage, organelle damage, protein inhibition, and oxidative stress [25]. High F⁻ ingestion poses the highest health risk to children due to the accumulation and deposition of F⁻ in their growing bones and joints [26,27].

Although more than 25 countries around the world have reported endemic skeletal fluorosis, it is especially prevalent in rural communities in China, India, and the East African Rift Valley, where people predominately rely on untreated groundwater for drinking water purposes with more than 260 million population affected worldwide [12,26]. Some studies have reported F⁻ concentrations of up to 23.5 mg/L in Kenya Rift Valley, 13.3 mg/L in China, and 17.5 mg/L in India [11,27,28].

12.3 FLUORIDE REMOVAL TECHNOLOGIES

Because of its undesirable effect on human health, the removal of F⁻ in water, a process known as defluoridation, is essential in preventing F⁻ toxicity. However, even after decades of research towards defluoridation, it continues to be a global challenge as the high reactivity of F⁻ makes it difficult to separate it from water, especially in rural and community settings [19]. Several different defluoridation processes have been developed. Some of the widely used defluoridation processes along with novel approaches are discussed below.

12.3.1 ADSORPTION

Adsorption methods have been reported as a promising solution for defluoridation as they have higher removal efficiency, simple application, and the use of easily accessible adsorbents [29]. The use of metal oxide or hydroxides as adsorbents has been continuously tested as cost-effective, innovative adsorbents with easy access. Treating activated alumina with ozone has shown an increase in the defluoridation efficiency from 77.4% to 98.3% [30]. These materials have been reported for a large specific surface area (>200 m^2/g) and high point of zero charges (pzc), which is related to a positively charged surface, capable of attracting negatively charged F$^-$ ions by electrostatic attraction [31]. Minerals such as bauxite have also been explored to remove F$^-$ with high defluoridation efficiency. However, bauxite ores from different geographical locations have been reported to have variable fluoride adsorption rates, which can cause issues with its widespread application [26].

Rare earth metals (e.g. cerium, titanium, lanthanum) have also been found with significant affinity for F$^-$ because of the presence of empty orbitals available for F$^-$ions. Electrostatic attraction between metal oxides/hydroxides and F$^-$, however, may lead to a desorption process with changes in pH, ionic strength, or other conditions that may release F$^-$ back into the environment. Very few studies are devoted to assessing the fate of F$^-$ ions after their adsorption in metal oxides/hydroxides and the risk of recontamination. This is identified as a significant knowledge gap worth attention to prevent further contamination of groundwater or avoid mismanagement of exhausted materials that may lead to further health effects among the water treatment operators.

The use of biopolymers (e.g., sodium alginate, pectin, chitosan) for F$^-$ adsorption has been gaining popularity because of their high efficiency due to the presence of hydroxyl and carboxyl groups, which are capable of coordinating with negatively charged F$^-$ ions to form irreversible hydrogel-like microbeads [32]. A combination of biopolymers and metals is also reported as substrates with interesting adsorption capacity for application in defluoridation. For example, sodium alginate with metals generates co-mingling and cross-linking to produce a gel structure that reduces agglomeration and metal oxide leaching, maximizing F$^-$ adsorption properties by increasing metal active sites and combining organic and inorganic properties [33]. Biopolymers have also been tested as part of some agricultural waste materials showing interesting adsorption capability. For example, studies with exhausted coffee grounds and iron sludge reported F$^-$ removal efficiency to be 56.7% and 62.9% respectively, at optimal experimental conditions [34].

Another interesting method is the use of waste materials (or some of their components) for water defluoridation. However, it is important to consider the environmental impact generated using these materials, which is identified as an interesting avenue for future research. Using, for example, life cycle analysis to properly assess the environmental impact of applying waste biomass to water defluoridation is not frequently reported, creating an information gap worth exploring to prevent undesirable side effects of these practices.

Carbon-based materials have also been reported to have interesting applications as biosorbents for water defluoridation. For example, biochar, a carbon-based material made of waste biomass from a wide range of sources, has been reported to have interesting adsorption capacity for F$^-$ because of its high content of $CaCO_3$, which generated outer-sphere complexation with F$^-$ and enhanced its removal from water [35]. Biochar made of animal bones, usually referred to as bone char, has been reported to have interesting water defluoridation potential. Hydroxyapatite, one of the major components of bone char, is considered to have F$^-$ adsorption capabilities. The mechanism of fluoride adsorption by hydroxyapatite mainly consists of electrostatic attraction by the hydroxyapatite surface to F$^-$, anion exchange between OH$^-$ or PO_4^{2-} and F$^-$, complexation reaction of Ca^{2+} with F$^-$ ligates to generate surface precipitation, and hydrogen bonds between F$^-$ and/or OH$^-$ in the hydroxyapatite lattice [36]. Enhancing biochar's adsorption capability by the addition of metals is another option explored recently for water defluoridation [37]. For example, in his study, La has suggested enhancing anion exchange capacity in biochar, modifying its pzc, and increasing defluoridation potential up to 164.2 mg/g within 30 min of contact time [38]. Other rare earth metals (e.g., Ce, Zr) are reported with a potential increase of biochar's adsorption capability for water defluoridation because of their relatively low ionic potential and a strong tendency to dissociate hydroxyl groups into ions, but very few available studies are devoted to testing that hypothesis. This lack of information is another very interesting avenue for future research because it may generate new materials with enhanced potential for water defluoridation.

12.3.2 COAGULATION

Coagulation is another cost-effective method for water defluoridation, in which charged particles are neutralized and agglomerated to settle down. Different chemicals (e.g. ferric chloride, ferrous sulfate, lime, potash alum, sodium bicarbonates) are commonly used as coagulants with alum ($Al_2(SO_4)_3 \cdot 18H_2O$) and lime ($Ca(OH)_2$) identified as the most extensively used for water defluoridation but hardly achieving WHO guidelines, depending on initial F$^-$ concentration [39]. Alternative methods including using plant-based coagulants and/or electrocoagulation have attracted attention as a cost-effective approach for water in contrast with conventional coagulants in which operational cost is high and the process usually generates a significant amount of sludge [39–41].

Electrocoagulation (EC) is a nonconventional water technology using sacrificial electrodes to release coagulant precursors into the solution with significant advantages (e.g. higher treatment efficiency, lower sludge production, avoiding chemical storage) compared with conventional chemical coagulation [42]. EC has been tested for water defluoridation with encouraging results when aluminum anodes are used suggesting that the process is efficient [43]. In a study from 2019, Laney et al. used *Moringa oleifera* (MO) extract coupled with EC, using aluminum electrodes at different current density values and different electrode separations in batch and recirculation experiments for water defluoridation [42]. Control experiments using MO extract and EC individually achieved 5% and 54% water defluoridation, respectively. Best experimental batch conditions were achieved using 12.5 mL of MO extract coupled with EC (3.3 mA/cm² and 1.0 cm electrode separation), producing >90% fluoride removal. Recirculation experiments with the EC reactor were performed with DI water and tap water using 1.0 cm electrode separation, 12.5 mL of MO extract, and different current densities. More than 90% fluoride removal was achieved with the EC/MO process, using 3.3 mA/cm², in both DI and tap water after 30 and 60 min, respectively. An energy consumption index (ECI) was developed, which showed that 1.51 and 0.67 W/h/mg were achieved for batch experiments of EC alone and EC/MO extract, respectively. For EC/MO extract, recirculation experiments with tap and DI water resulted in 0.35 and 0.22 W/h/mg, respectively. A cost analysis showed that $0.18 will be needed to treat one cubic meter of water. Operating variables (e.g., flow rate, current intensity, electrode configuration) were found to have a significant effect on F⁻ removal. Other operational issues, such as electrode passivation and treated water alkalization, have been observed but little is known about their effect on defluoridation efficacy. This lack of information is identified as another significant knowledge gap worth exploring to identify scale-up parameters and enhance the technology readiness level of the technology.

12.3.3 BIOREMEDIATION

Constructed wetlands have also been suggested with the potential for water defluoridation through the absorption of contaminants. Unlike adsorption, absorption assimilates through the bulk sorbent volume instead of through its surface and highly depends on biological processes within living biomass to eliminate contaminants [44]. Wetland efficiency has been linked to macrophyte composition, which, in the case of *Azolla pinnata* var. *imbricata* was found capable of absorbing fluoride and heavy metals [45]. Other aquatic macrophytes (e.g., *Pistia stratiotes*, *Eichhornia crassipes*, *Spirodela polyrhiza*, *Hydrilla verticillata*) have also been reported with some defluoridation capability [46]. These authors, however, found that fluoride concentration higher than 40 mg/L induces oxidative stress and metabolic imbalance degrading the plant physiology. Some terrestrial plants identified as fluoride-resistant and fluoride accumulators have also been explored for preventing fluoride groundwater contamination, including *Salix* willow, *Platanus* sp., and *Salix nigra* reported with relatively high resistance to chlorosis and necrosis by fluoride.

Bacteria has also gained interest in water defluoridation. Some bacterial strains (i.e., *Micrococcus luteus*, *Aeromonas hydrophilla*, *Micrococcus varians*, *Pseudomonas aeruginosa*) have been suggested to be fluoride resistant with few studies reporting interesting defluoridation efficacy by some bacterial strains [47,48]. The significance of using bacteria or plants to remove fluoride from water effluents is high because constructed wetlands are highly cost-effective and widely spread technology for multiple applications. In this regard, for generating information on the kinetics of F⁻ removal by the different plants and bacteria species, the effect of detention time or operation conditions is basic to secure the appropriate scaling up of the processes. To our best knowledge, this information is scarce and constitutes a highly interesting knowledge gap urging attention because the generated information will fill a significant gap that can generate safe water, especially in rural communities.

12.4 SOCIOECONOMICS OF DEFLUORIDATION

Most defluoridation projects have failed to produce sustainable widescale implementation, particularly those in rural locations, usually because of the lack of economic resources, the complexity of the defluoridation process, and/or the need for a highly skilled workforce to run and maintain [44]. When developing water treatment processes, it is critical to understand major barriers that can exist against effective technology adoption. Fluoride removal processes such as Reverse Osmosis (RO), some co-precipitation EC processes, and membrane processes are expensive, which creates insurmountable barriers, especially for rural impoverished communities [29]. In areas where communities do not have access to a main water supply system, such as most rural areas in developing countries, groundwater is the primary water source [53]. Hence, in terms of defluoridation technologies, there is a desperate need to focus on small-scale, local solutions for rural areas [54]. Table 12.1 shows some of the water defluoridation alternatives tested in the past with the potential for use in rural areas of developing countries. As shown, adsorption is the most applied, using modified materials that can be easily obtained and/or developed for application in rural communities. Nevertheless, the lack of information on other technologies with the potential for use as a point-of-use or point-of-entry treatment for water defluoridation in small, rural communities is considered another knowledge gap worth attention to generating alternative ways for people in different locations to get rid of the health threat posed by high fluoride concentration in their water sources.

Detection is the first step towards tackling endemic fluorosis. As fluoride is tasteless, odorless, and appears clear in water, its presence in groundwater is easily concealed without specific examination and hence can be easily overlooked in rural communities [19]. Fluoride detection is generally an expensive process and requires skilled expertise for accurate quantification, such as using ion selective electrodes (ISE), molecular absorption spectrometry (MAS), ion chromatography, and colorimetric tests [19]. Even recent academic studies in developing countries have had to ship groundwater samples overseas for ion chromatography testing [21]. This further increases the difficulty and creates barriers to quantifiable fluoride detection in rural settings. Combined with limited available literature about regional and global

TABLE 12.1
Various Water Defluoridation Technologies for Rural Communities

Process Type	Water Source	Defluoridation Efficacy	References
Adsorption on coffee grounds	Groundwater	57% F⁻ removal	[49]
Adsorption on porous starch	Drinking water	$Q_e{}^a = 25.4$ mg/g	[50]
Adsorption on modified alumina	Aqueous phase	$Q_e = 26.4$ mg/g	[33]
Adsorption on modified diatomite	Groundwater	89% F⁻ removal	[51]
Adsorption on modified diatomaceous Earth	Aqueous phase	93% F⁻ removal	[52]

$^a Q_e = Adsorption\ capacity.$

fluoride levels, there is an area for further data-driven studies to identify and monitor environmental areas of concern for public health risks [44].

Although adsorption and other methods have had successes in laboratory settings, defluoridation processes have been known to fail during field tests [55]. Moreover, most experiments are not even tested in the field, which further restricts advancement toward community public health enhancement [8].

Hence, in addition to focusing on novel defluoridation processes, this review also makes a case for holistically looking at the community project application for long-term sustainability. Beyond lab experiments on defluoridation efficiency, community awareness is ultimately the key towards effective water treatment implementation and the final fruition of scientific research to solve a public health concern. Due to the complexities in detection, it is difficult for communities to accept and recognize the presence of high fluoride concentration in their water. As most of the rural population lives in poverty, there is a lack of access to easily available information regarding environmental and health hazards [8]. Studies show that there is an association between household water treatment practice adoption and education status, awareness of water treatment practices, and household income in rural areas [56]. In Malawi, where water sources have fluoride concentrations up to 27.2 mg/L, and where 82.7% of the children showed signs of dental fluorosis, 100% of the parents showed low or no awareness about the causes and the health risk of fluoride contamination [57]. Strong public support is key to widely adopted fluoride removal strategies, especially in remote rural areas [19]. Several community factors such as willingness to adapt, perception, pro-environmental behavior, user-friendliness, altruism, sense of ownership, and environmental commitment along with various financial services such as soft loans, tax incentives, and subsidies are critical towards successful technology adoption, especially in rural areas in developing countries [58].

This further indicates the need to incorporate community integration studies with technical/scientific research to improve long-term defluoridation sustainability.

Although WHO has set the fluoride guideline at 1.5 mg/L for a person who drinks 2 L of water per day, different countries have established their own national standards [18]. For instance, Malawi's fluoride guideline of 6 mg/L is well above the WHO guideline [21]. Hence, there is a need to increase awareness and advocacy on governmental as well as community issues.

Widescale defluoridation can also be an issue due to the need to treat large volumes of water. In areas that do not have access to alternate water sources, it is important to consider defluoridation not only for drinking water but also for household water use as well since cooking techniques such as boiling food in fluoride-rich water can significantly increase the fluoride concentration in the food [7]. In areas where communities utilize multiple water sources, it is additionally difficult to monitor and treat water from different ground and surface waters collectively.

Past studies have reported a very low rate of acceptance of defluoridation technologies stressing that more research is required to understand behavioral factors influencing people's consumption of safer water to enhance public health and household drinking water treatment technologies adoption. These factors can be targeted at rural communities to trigger the most effective behavioral change interventions [59].

It is also important to consider religious beliefs while choosing which defluoridation technology to implement. For example, bone char from cow bones is difficult to accept among Hindus while bone char from pigs is difficult to accept among Muslims [29].

In some studies, the influence of psychological components on the acceptance of fluoride-free services has been tested, collecting information on perceived vulnerability and severity of contracting fluorosis diseases [60]. These studies usually consider knowledge factor that encompasses individual fluorosis awareness as well as actions that can be taken for health protection. Despite awareness, behavior change is still tricky to implement. Even when a majority of the rural population was aware of the direct link between contaminated drinking water and fluorosis, many still believed that the water was safe for consumption [61]. Attitudinal factors in the community are connected to individual belief in their efforts and health consequences of safer drinking water access. Public attitude includes effective components related to feelings of performing a specific behavior with perceived social behavior influencing community member's decision on gathering information about descriptive, injunctive, and personal norms [61]. Descriptive norms explain individual behavior in certain ways related to other people behaving similarly [62]. Injunctive norms relate to behavior supported by relatives, friends, or neighbors. Personal norms include personal feelings such as moral obligations. Self-regulation factors relate to perceived actions linked to continue and/or maintain a specific behavior [63].

Very few studies are available where all these attitudinal considerations have been used to analyze the population's willingness to pay for defluorinated water services. In one available study, 330 respondents were interviewed to estimate the willingness to pay for serving 2 m^3 of fluoride-free water at home compared to the service available at the nearest public water source using an open-ended format and a Tobit regression [64]. The authors found that participants were willing to pay $0.134 for 20 L of fluoride-safe water at home and $0.07 for 20 L of water available from the nearest public tap. The rate suggested was found able to cover defluoridation water cost services, which was not found for other water treatment services for different contaminants (e.g., arsenic [65]) where the willingness to pay for these services was found not sufficient to introduce these devices on a large scale without the use of subsidies. We consider that the lack of information regarding attitudinal factors affecting the adoption of defluoridation technologies and/or the willingness to pay by community members is a very significant knowledge gap urging attention to clearly generate a sense of the major barriers to face after developing cost-effective methodologies for water defluoridation or other water treatment technologies. This can be considered as an interesting avenue for future research, worth exploring where expertise from more than one field is required.

12.5 CONCLUSIONS

Addressing gaps in removing excess fluoride from water, especially in rural communities, requires both extensive scientific research as well as community engagement studies to solidify sustainability depending on the needs of the community. The following are the main findings of this study:

- The presence of F⁻ in drinking water sources is a worldwide problem demanding attention to prevent undesirable health effects in communities without alternative water sources. The required solutions, however, need to be sustainable and cost-effective to increase their adoption by community members and secure successful implementation and operation.
- The significance of health effects generated by consuming water with F⁻ concentration above the international guidelines and without the proper safeguards may generate a significant threat to accomplishing not only SDG 6 but other U.N. goals set for accomplishing in 2030.
- Alternative defluoridation technologies have emerged as potential solutions to address the problem, but many of them remain with low technology readiness levels or pending scientific development to clearly evaluate their feasibility for full-scale application.
- As with several other technologies, socioeconomic conditions may be a limiting factor governing the feasibility of water defluoridation technologies. Behavioral trends impeding technology adoption among community members or lack of willingness to pay for the water services by users were found significant threats to generating affordable, fit-for-purpose water in rural communities. The lack of information regarding all these items has been identified as a significant knowledge gap.
- Water defluoridation is a major challenge and should remain a high priority for water professionals and scholars to generate safeguards for vulnerable community members. It is only when working together and coordinating that all the challenges faced in accomplishing SDG 6 can be addressed.

REFERENCES

1. WHO/UNICEF. *Progress on Household Drinking Water, Sanitation and Hygiene 2000–2020: Five Years into the SDGs.* Word Health Organization (WHO). (2021).
2. Bierkens, M. F. P. & Wada, Y. Non-renewable groundwater use and groundwater depletion: a review. *Environmental Research Letters* **14**, 063002 (2019).
3. World Water Forum. *The Global Risks Report.* (The World Economic Forum, 2020). https://www.weforum.org/publications/the-global-risks-report-2020/
4. World Water Forum. *The Global Risk Report 2023.* (The World Economic Forum, 2023). https://www.weforum.org/publications/global-risks-report-2023/
5. UN DESA. *The Sustainable Development Goals Report 2023: Special Edition.* https://unstats.un.org/sdgs/report/2023/ (The United Nations, 2023).
6. Rodell, M. et al. Emerging trends in global freshwater availability. *Nature* **557**, 651–659 (2018).
7. Sawangjang, B. & Takizawa, S. Re-evaluating fluoride intake from food and drinking water: Effect of boiling and fluoride adsorption on food. *Journal of Hazardous Materials* **443**, 130162 (2023).
8. Onipe, T., Edokpayi, J. N. & Odiyo, J. O. A review on the potential sources and health implications of fluoride in groundwater of Sub-Saharan Africa. *Journal of Environmental Science and Health, Part A* 55, 1078–1093 (2020).
9. Wang, Y. et al. Groundwater quality and health: making the invisible visible. *Environmental Science & Technology* **57**, 5125–5136 (2023).

10. Kumar, P. J. S. Groundwater fluoride contamination in Coimbatore district: a geo-chemical characterization, multivariate analysis, and human health risk perspective. *Environmental Earth Sciences* **80**, 232 (2021).
11. Su, H., Wang, J. & Liu, J. Geochemical factors controlling the occurrence of high-fluoride groundwater in the western region of the Ordos basin, northwestern China. *Environmental Pollution* **252**, 1154–1162 (2019).
12. Noor, S., Rashid, A., Javed, A., Khattak, J. A. & Farooqi, A. Hydrogeological proper-ties, sources provenance, and health risk exposure of fluoride in the groundwater of Batkhela, Pakistan. *Environmental Technology & Innovation* **25**, 102239 (2022).
13. Zhang, B. et al. Hydrochemical characteristics of groundwater and dominant water–rock interactions in the Delingha Area, Qaidam Basin, Northwest China. *Water* **12**, 836 (2020).
14. Podgorski, J. E., Labhasetwar, P., Saha, D. & Berg, M. Prediction modeling and map-ping of groundwater fluoride contamination throughout India. *Environmental Science & Technology* **52**, 9889–9898 (2018).
15. Kom, K. P., Gurugnanam, B., Bairavi, S. & Chidambaram, S. Sources and geochemistry of high fluoride groundwater in hard rock aquifer of the semi-arid region. A special focus on human health risk assessment. *Total Environment Research Themes* **5**, 100026 (2023).
16. Yousefi, M. et al. Northwest of Iran as an endemic area in terms of fluoride contamina-tion: a case study on the correlation of fluoride concentration with physicochemical characteristics of groundwater sources in Showt. *DWT* **155**, 183–189 (2019).
17. Jali, B. R., Barick, A. K., Mohapatra, P. & Sahoo, S. K. A comprehensive review on quinones based fluoride selective colorimetric and fluorescence chemosensors. *Journal of Fluorine Chemistry* **244**, 109744 (2021).
18. World Health Organization. *Guidelines for Drinking-Water Quality: Incorporating the First and Second Addenda.* (World Health Organization, 2022).
19. Dar, F. A. & Kurella, S. Fluoride in drinking water: An in-depth analysis of its preva-lence, health effects, advances in detection and treatment. *Materials Today: Proceedings* (2023) doi:10.1016/j.matpr.2023.05.645.
20. Zhou, J., Sun, D. & Wei, W. Necessity to pay attention to the effects of low fluoride on human health: an overview of skeletal and non-skeletal damages in epidemiologic investigations and laboratory studies. *Biological Trace Element Research* **201**, 1627–1638 (2023).
21. Addison, M. J. et al. Fluoride occurrence in the lower East African Rift System, Southern Malawi. *Science of the Total Environment* **712**, 136260 (2020).
22. Strunecka, A. & Strunecky, O. Mechanisms of fluoride toxicity: From enzymes to underlying integrative networks. *Applied Sciences* **10**, 7100 (2020).
23. Waugh, D. T. Fluoride exposure induces inhibition of sodium-and potassium-activated adenosine triphosphatase (Na^+, K^+-ATPase) enzyme activity: Molecular mechanisms and implications for public health. *International Journal of Environmental Research and Public Health* **16**, 1427 (2019).
24. Chen, L. et al. Fluoride exposure disrupts the cytoskeletal arrangement and ATP syn-thesis of HT-22 cell by activating the RhoA/ROCK signaling pathway. *Ecotoxicology and Environmental Safety* **254**, 114718 (2023).
25. Johnston, N. R. & Strobel, S. A. Principles of fluoride toxicity and the cellular response: a review. *Archives of Toxicology* **94**, 1051–1069 (2020).
26. Cherukumilli, K., Maurer, T., Hohman, J. N., Mehta, Y. & Gadgil, A. J. Effective remediation of groundwater fluoride with inexpensively processed Indian bauxite. *Environmental Science & Technology* **52**, 4711–4718 (2018).

27. Mwiathi, N. F., Gao, X., Li, C. & Rashid, A. The occurrence of geogenic fluoride in shallow aquifers of Kenya Rift Valley and its implications in groundwater management. *Ecotoxicology and Environmental Safety* **229**, 113046 (2022).
28. Duggal, V. & Sharma, S. Fluoride contamination in drinking water and associated health risk assessment in the Malwa Belt of Punjab, India. *Environmental Advances* **8**, 100242 (2022).
29. Alkurdi, S. S. A., Al-Juboori, R. A., Bundschuh, J. & Hamawand, I. Bone char as a green sorbent for removing health threatening fluoride from drinking water. *Environment International* **127**, 704–719 (2019).
30. de Paula, N. et al. Application of ozone to enhance the defluoridation property of activated alumina: A novel approach. *Journal of Environmental Chemical Engineering* **11**, 110384 (2023).
31. Dhawane, S. H. et al. Insight into optimization, isotherm, kinetics, and thermodynamics of fluoride adsorption onto activated alumina. *Environmental Progress & Sustainable Energy* **37**, 766–776 (2018).
32. Araga, R. & Sharma, C. S. Amine functionalized electrospun cellulose nanofibers for fluoride adsorption from drinking water. *Journal of Polymers and the Environment* **27**, 816–826 (2019).
33. He, Y. et al. Enhanced fluoride removal from water by rare earth (La and Ce) modified alumina: Adsorption isotherms, kinetics, thermodynamics and mechanism. *Science of the Total Environment* **688**, 184–198 (2019).
34. Siauruševičiūtė, I. & Albrektienė, R. Removal of fluorides from aqueous solutions using exhausted coffee grounds and iron sludge. *Water* **13**, 1512 (2021).
35. Lee, J.-I. et al. Thermally treated *Mytilus coruscus* shells for fluoride removal and their adsorption mechanism. *Chemosphere* **263**, 128328 (2021).
36. Zhou, J., Liu, Y., Han, Y., Jing, F. & Chen, J. Bone-derived biochar and magnetic biochar for effective removal of fluoride in groundwater: Effects of synthesis method and coexisting chromium. *Water Environment Research* **91**, 588–597 (2019).
37. Wei, Y., Wang, L., Li, H., Yan, W. & Feng, J. Synergistic fluoride adsorption by composite adsorbents synthesized from different types of materials– a review. *Frontiers in Chemistry* **10**, 900660 (2022).
38. Habibi, N., Rouhi, P. & Ramavandi, B. Modification of tamarix hispida biochar by lanthanum chloride for enhanced fluoride adsorption from synthetic and real wastewater. *Environmental Progress & Sustainable Energy* **38**, S298–S305 (2019).
39. Jamwal, K. D. & Slathia, D. A review of defluoridation techniques of global and indian prominence. *Current World Environment* **17**, 41–57 (2022).
40. Mena, V. F. et al. Fluoride removal from natural volcanic underground water by an electrocoagulation process: Parametric and cost evaluations. *Journal of Environmental Management* **246**, 472–483 (2019).
41. Villaseñor-Basulto, D. L., Astudillo-Sánchez, P. D., del Real-Olvera, J. & Bandala, E. R. Wastewater treatment using *Moringa oleifera* Lam seeds: A review. *Journal of Water Process Engineering* **23**, 151–164 (2018).
42. Laney, B., Rodriguez-Narvaez, O. M., Apambire, B. & Bandala, E. R. Water defluoridation using sequentially coupled *Moringa oleifera* seed extract and electrocoagulation. *Groundwater Monitoring* **40**, 67–74 (2020).
43. Sandoval, M. A. et al. Simultaneous removal of fluoride and arsenic from groundwater by electrocoagulation using a filter-press flow reactor with a three-cell stack. *Separation and Purification Technology* **208**, 208–216 (2019).
44. Lacson, C. F. Z., Lu, M.-C. & Huang, Y.-H. Fluoride-containing water: A global perspective and a pursuit to sustainable water defluoridation management – An overview. *Journal of Cleaner Production* **280**, 124236 (2021).

45. Almuktar, S. A., Abed, S. N. & Scholz, M. Wetlands for wastewater treatment and subsequent recycling of treated effluent: a review. *Environmental Science and Pollution Research* **25**, 23595–23623 (2018).
46. Gao, J. et al. Effect of fluoride on photosynthetic pigment content and antioxidant system of *Hydrilla verticillata*. *International Journal of Phytoremediation* **20**, 1257–1263 (2018).
47. Chouhan, S., Tuteja, U. & Flora, S. J. S. Isolation, identification and characterization of fluoride resistant bacteria: possible role in bioremediation. *Applied Biochemistry and Microbiology* **48**, 43–50 (2012).
48. Shanker, A. S., Srinivasulu, D. & Pindi, P. K. A study on bioremediation of fluoride-contaminated water via a novel bacterium Acinetobacter sp. (GU566361) isolated from potable water. *Results in Chemistry* **2**, 100070 (2020).
49. Melak, F., Ambelu, A., Astatkie, H., Du Laing, G. & Alemayehu, E. Freeze desalination as point-of-use water defluoridation technique. *Applied Water Science* **9**, 1–10 (2019).
50. Xu, L. et al. Adsorptive removal of fluoride from drinking water using porous starch loaded with common metal ions. *Carbohydrate Polymers* **160**, 82–89 (2017).
51. Akafu, T., Chimdi, A. & Gomoro, K. Removal of fluoride from drinking water by sorption using diatomite modified with aluminum hydroxide. *Journal of Analytical Methods in Chemistry* 2019, 1–11 (2019).
52. Gitari, W. M., Izuagie, A. A. & Gumbo, J. R. Synthesis, characterization and batch assessment of groundwater fluoride removal capacity of trimetal Mg/Ce/Mn oxide-modified diatomaceous earth. *Arabian Journal of Chemistry* **13**, 1–16 (2020).
53. Xu, Y., Seward, P., Gaye, C., Lin, L. & Olago, D. O. Preface: Groundwater in Sub-Saharan Africa. *Hydrogeology Journal* **27**, 815–822 (2019).
54. Oladoja, N. A. & Helmreich, B. Reactive metal oxides in ceramic membrane formulation as a clue to effective point-of-use drinking water defluoridation. In (eds Inamuddin T.A., Mazumder M.A.J.) *Green Sustainable Process for Chemical and Environmental Engineering and Science* 173–196 (Elsevier, 2023).
55. Ahmad, S., Singh, R., Arfin, T. & Neeti, K. Fluoride contamination, consequences and removal techniques in water: a review. *Environmental Science: Advances Journal* **1**, 620–661 (2022).
56. Asefa, L., Ashenafi, A., Dhengesu, D., Roba, H. & Lemma, H. Household water treatment practice and associated factors among rural Kebeles (villages) in west Guji zone, southern Ethiopia: Community based cross-sectional study. *Clinical Epidemiology and Global Health* **22**, 101311 (2023).
57. Andreah, K., Tembo, M. & Manda, M. Community awareness of dental fluorosis as a health risk associated with fluoride in improved groundwater sources in Mangochi District, Malawi. *Journal of Water and Health* **21**, 192–204 (2023).
58. Puppala, H., Ahuja, J., Tamvada, J. P. & Peddinti, P. R. T. New technology adoption in rural areas of emerging economies: The case of rainwater harvesting systems in India. *Technological Forecasting and Social Change* **196**, 122832 (2023).
59. Gutierrez, L., Nocella, G., Ghiglieri, G. & Idini, A. Willingness to pay for fluoride-free water in Tanzania: disentangling the importance of behavioural factors. *International Journal of Water Resources Development* **39**, 294–313 (2023).
60. Idini, A. et al. Application of octacalcium phosphate with an innovative household-scale defluoridator prototype and behavioral determinants of its adoption in rural communities of the East African Rift Valley. *Integrated Environmental Assessment and Management* **16**, 856–870 (2020).
61. Nocella, G. et al. Insights to promote safe drinking water behavioural changes in zones affected by fluorosis in the East-African Rift Valley. *Groundwater for Sustainable Development* **19**, 100809 (2022).

62. Kumari, U., Swamy, K., Gupta, A., Karri, R. R. & Meikap, B. C. Global water challenge and future perspective. in *Green Technologies for the Defluoridation of Water* 197–212 (Elsevier, 2021).

63. Nyika, J. & Dinka, M. O. Management of water challenges in Sub-Saharan Africa. In (eds Nyika J., Dinka M.O.) *Water Challenges in Rural and Urban Sub-Saharan Africa and their Management* 57–75 (Springer, 2023).

64. Entele, B. R. & Lee, J. Estimation of household willingness to pay for fluoride-free water connection in the Rift Valley Region of Ethiopia: A model study. *Groundwater for Sustainable Development* **10**, 100329 (2020).

65. Burt, Z. et al. User preferences and willingness to pay for safe drinking water: experimental evidence from rural Tanzania. *Social Science & Medicine* **173**, 63–71 (2017).

13 Adoption of Ecohydrology Approaches for Urban Stormwater Management and Advancing the Circular Economy Concept

Yukun Ma, Erick R. Bandala,
Oscar M. Rodríguez-Narvaez, and
Ashantha Goonetilleke

13.1 INTRODUCTION

Urbanization converts natural surfaces to impervious surfaces leading to decreased stormwater infiltration and increased surface runoff in urban areas [1]. Consequent to accelerated urbanization and climate change impacts, recent decades have witnessed significant urban flooding, especially during extreme rainfall events in many cities across the globe [2]. It is commonly recognized that urban flooding is one of the most challenging issues and among the most significant climate change-linked impacts on cities [3]. In addition to the potential devastation caused by flooding, the urban water environment is also detrimentally affected by extensive anthropogenic activities due to urbanization [4]. The various anthropogenic activities common to urban areas such as industrial activities and traffic contribute a diversity of pollutants such as heavy metals, hydrocarbons, nutrients, and organic matter to the urban environment [5–8]. These pollutants are deposited on impervious surfaces during dry periods. During rainfall, the pollutants are washed off by stormwater runoff and finally discharged into the receiving waters [9]. This leads to urban water quality degradation and results in potential ecosystem and human health risks [10,11]. For example, Tian et al. [12] found that tire rubber residues carried by road runoff contained a high concentration of 6PPD-quinone which is an extremely toxic transformation by-product of 6PPD commonly used in tire rubber as an antioxidant and

DOI: 10.1201/9781003441007-16

was the main toxicant inducing acute mortality in coho salmon reproducing in urban creeks in the US Pacific Northwest. In response to the ever-increasing severity of urban flooding and water pollution, appropriate stormwater management strategies are essential for protecting the urban water environment.

From the perspective of sustainable development, stormwater is a potential water resource to overcome the serious water scarcity being experienced in urban areas in many parts of the world. Stormwater can be reused for various purposes such as for drinking, recreation, and building cooling after appropriate treatment. Capturing and reusing urban stormwater can form a key component of integrated urban water resources management [13,14]. It has been reported that the volume of stormwater runoff in major Australian cities typically exceeds the volume of water drawn from surface and groundwater sources for potable purposes [15]. Hence, the reuse of stormwater can provide an alternative water supply rather than directly discharging into receiving waters leading to water quality and quantity impacts. Consequently, the water demand from natural sources will be reduced while at the same time mitigating the adverse impacts associated with urban stormwater runoff due to reduced inflows to urban receiving waters. Therefore, sustainable stormwater management is essential to reduce the detrimental impacts associated with stormwater such as flooding and pollution of receiving waters.

Currently, sustainable urban stormwater management approaches primarily focus on mitigating detrimental environmental impacts rather than considering stormwater as an alternative water resource after appropriate treatment [16–18]. For example, nature-based solutions such as Low Impact Development (LID), Water Sensitive Urban Design (WSUD), or Best Management Practices (BMP) are commonly used measures aiming to mitigate stormwater pollution [19,20]. LID/WSUD/BMP measures focus on improving management performance such as runoff regulation and pollution removal to mitigate the detrimental influence of stormwater quality and quantity on receiving waters [21,22]. This linear management of stormwater leads to a growing demand for resources such as for the maintenance of treatment facilities on the one hand and waste production such as discharging pollutants to receiving waters on the other hand, generating high environmental and heavy economic burdens. Therefore, an innovative urban stormwater management approach is required to strengthen sustainable practices in the urban water environment and to recognize the value of stormwater as a resource.

13.2 ECOHYDROLOGY AS AN INNOVATIVE APPROACH FOR URBAN STORMWATER MANAGEMENT

Ecohydrology provides a multifunctional framework to address several environmental challenges associated with urban areas such as adaptation to climate change, water pollution mitigation, preventing biodiversity loss, and reversing the reduction of ecological services [23]. Due to rapid urbanization and the compounding impacts of climate change, aquatic ecosystems around the world are becoming significantly degraded. To alleviate these impacts, ecohydrology can provide an effective approach to the sustainable management of water resources [24]. Broadly, ecohydrology comprises three principles, namely, hydrological principle, ecological principle, and ecological engineering principle [25]. The hydrological principle refers to the quantification of the hydrological process for defining the cause–effect relationship.

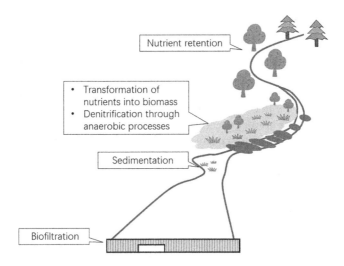

FIGURE 13.1   An example of an ecohydrological approach by applying a range of ecosystem practices and processes at the river basin scale for the restoration of a degraded waterway.

The ecological principle refers to the ecological identification of the potential ability of the biota to regulate the water cycle to enhance ecosystem sustainability. The ecological engineering principle is based on the first two principles referring to the dual regulation of hydrological and ecological processes for reducing water stress and enhancing ecosystem services [26]. Figure 13.1 illustrates the range of possible ecohydrology services that can be applied at the river basin scale to restore a degraded waterway [27]. In this example, the first stage is nutrient retention at the upper part of the basin through reforestation, the creation of ecotone buffer zones, and the optimization of agricultural practices. At the land and water interface, the buffer zones with shelter belts can increase nutrient uptake by cultivated land through the transformation of nutrients into biomass and reducing the supply of nutrients to the river. In the river valley, the wetlands can reduce the mineral sediments, organic matter, and nutrient load transported by the river during flood periods through sedimentation. Additionally, in some wetlands, the nitrogen load can be significantly reduced by regulating the water level to stimulate denitrification through anaerobic processes. Finally, the downstream, nutrients can be intercepted through biofiltration to reduce algae biomass.

The benefits of ecohydrology for urban stormwater management embody four aspects [28]. First, ecohydrology proposes the reduction of urban flooding to mitigate natural disasters and the reuse of urban stormwater to overcome water scarcity. Second, ecohydrology strives to remove pollutants in urban stormwater to minimize receiving water pollution. Third, ecohydrology can protect habitats to increase biodiversity and improve urban ecosystem services. Fourth, aesthetics is taken into consideration by ecohydrology to enhance the ecological environment in urban areas. These four aspects form a sustainable and integrated stormwater resources management framework that can combine water, biodiversity, economy, and human well-being in an urban area. These aspects are discussed in greater detail in the following sections.

13.2.1 FLOOD REDUCTION

Urban flooding due to the ever-increasing imperviousness of ground surfaces and inadequate water-carrying capacity of the aging gray infrastructure is one of the critical issues facing the urban water environment. Gray infrastructure refers to conventional concrete stormwater infrastructure such as pipes, gutters, kerbs, and channels. To minimize the flood risk in a city, ecohydrology approaches can be applied through enhanced stormwater retention and infiltration, thereby reducing flood peaks and flood volumes leading to a decrease in the severity of flooding. Further, urban flood mitigation can contribute to the reduction in economic costs associated with restoration and reconstruction after a flood disaster.

13.2.2 POLLUTION MINIMIZATION

Ecohydrology approaches can be employed to mitigate receiving water pollution by intercepting pollutants originating from source areas. Appropriate measures can be employed to protect urban water resources from point and non-point sources. Further, ecohydrology measures can be employed to enhance the resistance of aquatic ecosystems to anthropogenic stress by restoration of their homeostasis through hydrological regulation and ecological approaches. For example, ecohydrological techniques have been found to mitigate nutrient flow into rivers and largely minimize eutrophication [29]. Consequently, improved water quality can reduce ecosystem and public health risks associated with water pollution [30].

13.2.3 ENHANCED ECOSYSTEM SERVICES

Ecohydrology prefers low-cost, nature-based solutions such as swales, bioretention systems, and wetlands to mitigate urban stormwater pollution. These green spaces can play a pivotal role in ameliorating the negative ecological impacts of urbanization, thereby promoting human well-being and mitigating biodiversity loss common in urban areas [31,32]. Protecting biodiversity in green spaces will contribute to the preservation of related ecosystem services for urban residents including food security, human health, and access to clean water environments [33,34].

13.2.4 AESTHETICS OF WATERWAYS

Improving urban aquatic ecosystem services can enhance the aesthetic function of urban waterways. This will enhance human well-being in urban areas as well as promote urban water environmental protection through practical demonstration of the benefits of ecosystem protection.

13.3 THE LIMITATIONS IN CURRENT STORMWATER MANAGEMENT APPROACHES

Although ecohydrology measures have been employed for several decades for the management of urban stormwater, most of these technologies tend to drive urban water resources management in a linear direction. The focus has been on the safe disposal of stormwater after appropriate treatment to protect the receiving water environment with little thought given to its reuse potential and as a substitute resource for use for potable purposes in urban areas. On the other hand, freshwater resources are often imported from long distances and delivered to consumers for potable use and disposed of as residuals after use and appropriate treatment. For example, stormwater will flow into LID/WSUD/BMP structures and after treatment, the runoff in its entirety is commonly discharged into receiving waters. In this context, the treatment measures employed primarily focus on mitigating stormwater pollution and preventing the degradation of urban receiving waters rather than considering stormwater as a resource available to be reused. It is important to note that stormwater is the only uncommitted water resource available in urban areas whereas for the other surface and groundwater resources, there are always competing demands.

Unfortunately, very limited ecohydrological strategies are applied with consideration of enhancing the circularity of stormwater resources in urban areas such as stormwater reuse and nutrient recovery. This emphasizes the important need for ecohydrology to be applied in a circular approach to promote sustainable development. However, introducing ecohydrological measures could exert new problems in the functioning of existing infrastructure. This is because the quantity and quality characteristics of stormwater transported by grey infrastructure would be different from stormwater transported by infrastructure designed using ecohydrology concepts such as ecological engineering, green infrastructure, and urban green (and blue) spaces. For example, the coupling of stormwater pipes and LID/WSUD/BMP system needs to consider the consequent runoff regulation and pollution mitigation as well as cost-efficiency. Therefore, ecohydrology measures should be introduced using expert knowledge through further research, resulting not only in modifications but also in the use of hybrid systems to holistically contribute to sustainability [35].

13.4 APPLICATION OF THE CIRCULAR ECONOMY CONCEPT TO URBAN STORMWATER RESOURCES

The circular economy concept envisages ensuring economic growth without increasing the consumption of new resources and reducing the impact on the environment [36]. The major challenges to be addressed with the adoption of the circular economy for urban stormwater management are preserving natural freshwater resources by reducing their import and minimizing pollution discharges to receiving waters by reusing stormwater resources in cycles.

Freshwater resources are facing increasing pressure from climate change and population growth impacts. The current challenges faced by urban water resources due to extreme rainfall and rapid urbanization can be characterized as too much, too little, and too dirty [37]. "Too much" refers to the increased urban flooding risk. Due to accelerated urbanization, rapid expansion of impervious surfaces significantly decreases stormwater infiltration into the ground and increases surface runoff. Consequently, urban flooding has become one of the biggest challenges in urban areas. "Too little" represents the serious water scarcity faced by cities around the world. Currently, water supply for cities primarily depends on natural freshwater resources. Limited freshwater resources and a growing urban population compound the water supply stress. The contradiction between supply and demand has become a key factor limiting the sustainable development of cities. "Too dirty" means urban aquatic pollution from point and non-point sources poses a significant risk to water resources. Various anthropogenic activities generate large amounts of pollutants which are transported with stormwater runoff and finally discharged into urban receiving waters. Urban water problems can lead to serious aquatic environmental problems such as freshwater resource depletion, urban ecosystem degradation, and biodiversity loss. This contributes to a growing demand for freshwater resources on the one hand and pollution generation on the other hand. Therefore, water resources conservation is essential for the sustainability of urban areas.

The issues noted above in relation to water, also apply to other resources in demand in urban areas. It is in this context that the concept of the circular economy has been proposed. The concept of the circular economy is readily applicable to urban stormwater resources. The circular economy is based on three key principles, namely, "Regenerate natural capital," "Keep resources in use," and "Design out waste externalities" [38].

13.4.1 REGENERATE NATURAL CAPITAL

Regenerating natural capital ensures functional environmental flows and stocks, by reducing the use of resources, preserving, and enhancing ecosystems, and ensuring minimal disruptions from human interactions and use. In the circular economy context, water resources management requires the increase in retention and infiltration of stormwater through the application of ecohydrology measures. Consequently, urban water supply for people can be achieved through stormwater reuse and reducing the importation of natural freshwater resources.

13.4.2 KEEP RESOURCES IN USE

Keeping resources in use is aimed at closing material loops and minimizing materials and energy loss within the system. This is achieved by optimizing resource yields, minimizing energy and resource extraction, and maximizing their recycling and reuse. The circular economy shifts stormwater treatment and disposal from conventional management technologies to reclaim and reuse using ecohydrology approaches. Urban stormwater can contain significant amounts of nutrients such as nitrogen and phosphorous. Circular economy suggests retaining the nutrients in the locality through nutrient recovery. The retained nutrients can be used as fertilizer for plantings essential for nature-based solutions, thereby improving the cost-efficiency of stormwater treatment.

13.4.3 DESIGN OUT WASTE EXTERNALITIES

Designing out waste externalities focuses on the reduction and management of the remaining waste in the system, including economic efficiency. The cost of reducing waste by one unit is comparable to the economic and environmental benefits of having one less unit of waste. Circular economy proposes to keep pollutants in stormwater in the urban area where it is generated and to reduce pollution discharge into receiving waters for water quality improvement.

According to the three principles noted above, the circular economy focuses on shifting from a linear to the circular management of stormwater resources in urban areas. It is aimed at keeping and reusing resources within the locality and thus forms an essential approach for realizing urban sustainability. A circular economy approach for stormwater management is a model in which stormwater reuse can be adapted to maintain water security in an urban area [39]. Meanwhile, novel approaches such as ecohydrology and supplementing energy through stormwater reuse can provide nutrient-rich resources for urban farming to enhance ecosystem services [40]. In this context, resources can be utilized in a circular way and a reduced urban footprint can be achieved due to minimized water supply obtained from freshwater sources, reduced pollutants discharge to receiving waters, extended lifecycle of stormwater conveyance and treatment infrastructure, reduced import of nutrients for green space and increased cost-efficiency of stormwater management. In other words, urban areas will develop along a sustainable pathway because of the reduced demand for resources such as water and nutrients. Therefore, these benefits should provide the necessary impetus to adopt a circular economy approach for urban stormwater management.

13.5 HOW ECOHYDROLOGY CAN HELP TO ACHIEVE A CIRCULAR ECONOMY IN STORMWATER MANAGEMENT

To realize the goal of a circular economy, the appropriate application of ecohydrology concepts to urban stormwater management is needed to realign the current linear resources management approach to a circular management approach. Ecohydrology proposes stormwater management through several techniques as illustrated in Figure 13.2.

As illustrated in Figure 13.2, the key benefits of moving to the circular management of stormwater resources include:

- Restoring and maintaining the stormwater cycle;
- Stormwater treatment, recovery and reuse;
- Nutrient recovery and reuse.

13.5.1 RESTORING AND MAINTAINING THE STORMWATER CYCLE

Current urban stormwater drainage systems are not resilient to extreme rainfall events and consequently, urban areas have seen serious flooding and attendant environmental degradation. From a circular economy viewpoint, ecohydrology approaches can be adapted to provide greater retention and infiltration of stormwater into the subsurface by mimicking the natural processes and thus restoring and maintaining the natural water cycle processes.

13.5.2 STORMWATER TREATMENT, RECOVERY AND REUSE

As a potential non-potable water resource, urban stormwater can be recovered and reused to reduce the extraction of water from natural freshwater resources and thereby minimize the water importation footprint. It is estimated that almost 30% of the population in the world is currently affected by water scarcity [41]. In such a

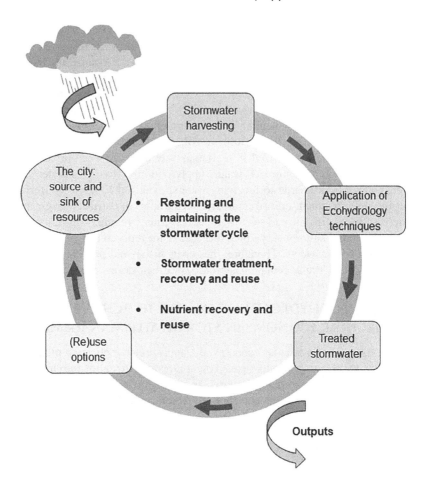

FIGURE 13.2   The approach for moving to the circular management of stormwater resources.

situation, urban stormwater reuse presents an obvious alternative resource. Currently, urban stormwater management primarily discharges the runoff after treatment, and most techniques such as LID/WSUD/BMP focus on low to medium rainfall events for treatment primarily due to the cost implications of undertaking treatment. On the other hand, ecohydrology solutions should enhance the ability to store and reuse stormwater resources including after high rainfall events. Therefore, further research is recommended to investigate the enhancement of runoff regulation and pollution treatment, particularly during extreme rainfall events for strengthening urban resilience and adaptation to climate change. In terms of stormwater reuse, current stormwater treatment approaches may not be effective in reducing pollution levels in stormwater to comply with the quality standards required for reuse. This demand can pose risks to human and ecosystem health. Therefore, further research should focus on the development of effective treatment strategies within the ecohydrology framework for the removal of toxic pollutants and emerging contaminants in stormwater. Additionally, more robust and accurate assessment frameworks are required to evaluate public and ecosystem health risks through stormwater reuse.

13.5.3 NUTRIENT RECOVERY AND REUSE

Excessive nutrients from stormwater runoff discharged into urban water bodies can cause serious degradation including eutrophication, leading to harmful algal blooms, water hypoxia, and even threatening human health [42]. Ecohydrology practices can improve stormwater quality through eco-solutions such as sorption and denitrification using biological processes to retain nutrients which can be recovered as fertilizer for urban greening [43,44]. Current knowledge about the adsorption and release of nutrients via ecohydrology approaches is limited. It is recommended that future studies should be undertaken to investigate further the dominant processes in relation to the adsorption or release of nitrogen and phosphorous by different ecohydrological measures, and how to improve the efficacy of these measures to reduce nutrients in the environment using passive methods.

The integration of urban stormwater management and ecohydrology can bring significant benefits to the sustainable development of urban areas. The multifunctionality of ecohydrology enables a whole range of implementation possibilities. Ecohydrological solutions for stormwater treatment have great potential for additional benefits such as the provision of healthy ecosystems, energy savings, biodiversity protection, and improved water and air quality. Ecohydrology brings nature-based solutions to urban areas to increase stormwater recovery and reduce freshwater demand, particularly for non-potable uses such as recreation, urban irrigation, and toileting. Besides, ecohydrological treatment of stormwater has the potential to recycle and recover resources including nutrients from stormwater, and transform organic material into bioenergy. Nutrient recovery and reuse play a major role in minimizing the use of manufactured nutrients and thus reducing related CO_2 emissions and energy demand, as well as contributing to improving water and air quality in urban areas. In short, the key message is that outputs from ecohydrology solutions can be transformed into valuable products and re-enter the value chain (e.g., reclaimed water, biomass, and organic fertilizers). Moreover, the inputs to ecohydrology solutions could be obtained from other local production chains (e.g., recycled materials).

The literature presents several case studies that have been conducted to evaluate the performance of ecohydrology integrated into urban stormwater management. Discussed below is an example of the application of ecohydrology to stormwater pollution mitigation and runoff regulation. An ecohydrological system was constructed in 2013 in Łódź city in Poland to retain and purify stormwater runoff from a street that flowed directly to a cascade of recreational reservoirs. The hybrid system consists of an underground separator system that is combined with a sequential sedimentation-biofiltration system (SSBS) (Figure 13.3). In the first 2 years of the system's operation, it effectively reduced pollution transported to the urban river

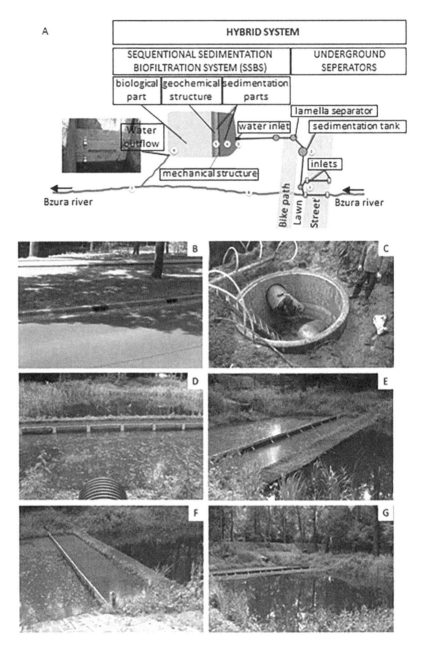

FIGURE 13.3 Location of the hybrid system (a) constructed for retention and purification of stormwater transported from the street to the river, with its purification stage sampling stations: st. (1) Street (an inlet of storm drainage system (B)), st. (2) Stormwater sedimentation tank (C), st. (3) Inflow to the SSBS (D), st. (4) Sedimentation zone (E, F) (part 1, above filtration grid), st. (5) Sedimentation zone (E, F) (part 2, above geochemical structure), st. (6) Biofiltration zone (G), st. (7) River below the hybrid system, st. 8–river above the hybrid system [45].

system by removing 86.0% of total suspended solids, 71.5% of total nitrogen (TN), 66.7% of total phosphorous (TP), and 58.1% of nitrate (NO_{3-}), respectively. In addition, the system was able to reduce the hydraulic stress induced by high discharges. The hybrid system is an example of a nature-based solution for mitigating the negative effects of nutrient transfer, eutrophication, and flooding in urbanized areas, as part of blue-green infrastructure.

13.6 CONCLUSIONS

Stormwater pollution and flooding due to extreme rainfall events and rapid urbanization create serious water environmental problems in urban areas. As an underutilized resource, urban stormwater can be reused to overcome the severe water shortages experienced in urban areas around the world. Sustainable management of stormwater is needed to reclaim stormwater for beneficial uses and to minimize the degradation of urban receiving waters. Ecohydrology provides a multifunctional framework to manage stormwater through the application of hydrological and ecological techniques. It proposes urban stormwater management primarily through retention, recovery, ecological treatment, and nutrient reuse. Ecohydrology can play an important role in the sustainable development of cities to achieve the goal of the circular economy. The circular economy shifts conventional resource management from a linear to a circular approach. From circular economy aspects, urban stormwater can be retained to minimize urban flood risk and reused for potable and non-potable purposes to reduce current urban water supply stress. Further, nutrients in stormwater can be recovered as organic fertilizers for urban farming to reduce the import of natural resources and to reduce receiving water pollution. In other words, stormwater resources can be used in a circular way for realizing urban sustainability.

REFERENCES

1. Hou, Y., Wang, S., Ma, Y., Shen, Z. & Goonetilleke, A., Influence of landscape patterns on nitrate and particulate organic nitrogen inputs to urban stormwater runoff. *J. Environ. Manage.* **348** 119190 (2023).
2. Bates, P., Uneven burden of urban flooding. *Nat. Sustain.* **69**, 119190 (2023).
3. Rowland, K., Flooding is the United Kingdom's biggest climate threat. *Nature.* https://www.nature.com/articles/nature.2012.9906#citeas (2012).
4. Ma, Y. et al., Influence of low impact development practices on urban diffuse pollutant transport process at catchment scale. *J. Clean. Prod.* **213** 357 (2019).
5. Ma, Y., Wang, S., Zhang, X. & Shen, Z., Transport process and source contribution of nitrogen in stormwater runoff from urban catchments. *Environ. Pollut.* **289** 117824 (2021).
6. Ma, Y., Gong, M., Zhao, H. & Li, X., Contribution of road dust from low impact development (LID) construction sites to atmospheric pollution from heavy metals. *Sci. Total Environ.* **698** 134243 (2020).
7. Ma, Y., Liu, A., Egodawatta, P., McGree, J. & Goonetilleke, A., Quantitative assessment of human health risk posed by polycyclic aromatic hydrocarbons in urban road dust. *Sci. Total Environ.* **575** 895 (2017).
8. Wang, S. et al., Nitrogen transport and sources in urban stormwater with different rainfall characteristics. *Sci. Total Environ.* **837** 155902 (2022).

9. Ma, Y. et al., Pollutant transport analysis and source apportionment of the entire non-point source pollution process in separate sewer systems. *Chemosphere* **211** 557 (2018).

10. Ma, Y. et al., Source quantification and risk assessment as a foundation for risk management of metals in urban road deposited solids. *J. Hazard. Mater.* **408** 124912 (2021).

11. Brodin, T., Fick, J., Jonsson, M. & Klaminder, J., Dilute concentrations of a psychiatric drug alter behavior of fish from natural populations. *Science* **339** 814 (2013).

12. Tian, Z. et al., A ubiquitous tire rubber-derived chemical induces acute mortality in coho salmon. *Science* **371** 185 (2021).

13. Goonetilleke, A. et al., Stormwater reuse, a viable option: Fact or fiction? *Econ. Anal. Policy* **56** 14 (2017).

14. Wijesiri, B., Liu, A. & Goonetilleke, A., Impact of global warming on urban stormwater quality: From the perspective of an alternative water resource. *J. Clean. Prod.* **262** 121330 (2020).

15. Commonwealth of Australia, Environment and Communications References Committee. https://www.aph.gov.au/parliamentary_business/committees/senate/environment_ and_communications/stormwater/~/media/Committees/ec_ctte/Stormwater/report.pdf 2015.

16. Ishaq, S., Rana, A., Hewage, K., Li, J. & Sadiq, R., Environmental impacts and cost analysis of low impact developments in arid climate: A case study of Okanagan Valley (Canada). *J. Clean. Prod.* **419** 138226 (2023).

17. Suresh, A., Pekkat, S. & Subbiah, S., Quantifying the efficacy of low impact developments (LIDs) for flood reduction in micro-urban watersheds incorporating climate change. *Sustain. Cities Soc.* **95** 104601 (2023).

18. Zhuang, Q., Li, M. & Lu, Z., Assessing runoff control of low impact development in Hong Kong's dense community with reliable SWMM setup and calibration. *J. Environ. Manage.* **345** 118599 (2023).

19. Xu, C., Jia, M., Xu, M., Long, Y. & Jia, H., Progress on environmental and economic evaluation of low-impact development type of best management practices through a life cycle perspective. *J. Clean. Prod.* **213** 1103 (2019).

20. Radcliffe, J. C., History of water sensitive urban desing/low impact development adoption in Australia and Internationally. In *Approaches to Water Sensitive Urban Design*, edited by Ashok K. Sharma, Ted Gardner and Don Begbie (Woodhead Publishing, 2019), p. 1.

21. Ma, Y. et al., Creating a hierarchy of hazard control for urban stormwater management. *Environ. Pollut.* **255** 113217 (2019).

22. Ma, Y. & Zhao, H., The role of spatial patterns of low impact development in urban runoff pollution control within parcel based catchments. *Front. Environ. Sci. Switz.* **10**, 926937, (2022).

23. Zalewski, M. et al., Low cost, nature-based solutions for managing aquatic resources: Integrating the principles of ecohydrology and the circular economy. *Ecohydrol. Hydrobiol.* **18** 309 (2018).

24. Fournier, F. Z. M., Janauer, G. A., Jolankai, G. & Pypaert, P., Ecohydrology a new paradigm for the sustainable use of aquatic resources. UNESCO: United Nations Educational, Scientific and Cultural Organisation. France. Retrieved from https:// policycommons.net/artifacts/10577581/ecohydrology/11482650/ on 25 Apr 2024. (1997).

25. Bridgewater, P. & Aricò, S., Turbo-charging the ecohydrology paradigm for the Anthropocene. *Ecohydrol. Hydrobiol.* **16** 74 (2016).

26. Schiemer, F., Building an eco-hydrological framework for the management of large river systems. *Ecohydrol. Hydrobiol.* **16** 19 (2016).

27. Zalewski, M. et al., Ecohydrology and adaptation to global change. *Ecohydrol. Hydrobiol.* **21** 393 (2021).

28. Li, C., Ecohydrology and good urban design for urban storm water-logging in Beijing, China. *Ecohydrol. Hydrobiol.* **12** 287 (2012).
29. Delibas, M. & Tezer, A., 'Stream Daylighting' as an approach for the renaturalization of riverine systems in urban areas: Istanbul-Ayamama Stream case. *Ecohydrol. Hydrobiol.* **17** 18 (2017).
30. Bertels, D., De Meester, J., Dirckx, G. & Willems, P., Estimation of the impact of combined sewer overflows on surface water quality in a sparsely monitored area. *Water Res* **244** 120498 (2023).
31. Bertram, C. & Rehdanz, K., The role of urban green space for human well-being. *Ecol. Econ.* **120** 139 (2015).
32. Lepczyk, C. A. et al., Biodiversity in the City: Fundamental questions for understanding the ecology of urban green spaces for biodiversity conservation. *Bioscience* **67** 799 (2017).
33. Xu, H. et al., Ensuring effective implementation of the post-2020 global biodiversity targets. *Nat. Ecol. Evol.* **5** 411 (2021).
34. Watson, K. B., Galford, G. L., Sonter, L. J. & Ricketts, T. H., Conserving ecosystem services and biodiversity: Measuring the tradeoffs involved in splitting conservation budgets. *Ecosyst. Serv.* **42** 101063 (2020).
35. Atanasova, N. et al., Nature-based solutions and circularity in cities. *Circ. Econ. Sustain.* *1*(1), 319–332, (2021).
36. Dale, G. et al., Education in ecological engineering – A need whose time has come. *Circ. Econ. Sustain.* **1** 333 (2021).
37. Wagner, I. & Breil, P., The role of ecohydrology in creating more resilient cities. *Ecohydrol. Hydrobiol.* **13** 113 (2013).
38. Atanasova, N., Castellar, J.A., Pineda-Martos, R. et al. Nature-Based Solutions and Circularity in Cities. Circ. Econ. Sust. 1, 319–332 (2021). https://doi.org/10.1007/s43615-021-00024-1.
39. Mbavarira, T. M. & Grimm, C., A systemic view on circular economy in the water industry: learnings from a Belgian and Dutch case. Sustainability (2021), Vol. 13(6), 3313.
40. Sušnik, J., Masia, S., Indriksone, D., Brēmere, I. & Vamvakeridou-Lydroudia, L., System dynamics modelling to explore the impacts of policies on the water–energy–food–land–climate nexus in Latvia. *Sci. Total Environ.* **775** 145827 (2021).
41. Díaz-Vázquez, D. et al., Characterization and multicriteria prioritization of water scarcity in sensitive urban areas for the implementation of a rain harvesting program: A case study for water-scarcity mitigation. *Urban Clim.* **51** 101670 (2023).
42. Wang, S. et al., Nitrogen transport and sources in urban stormwater with different rainfall characteristics. *Sci. Total Environ.* **837** 155902 (2022).
43. Costello, D. M., Hartung, E. W., Stoll, J. T. & Jefferson, A. J., Bioretention cell age and construction style influence stormwater pollutant dynamics. *Sci. Total Environ.* **712** 135597 (2020).
44. Li, J., Culver, T. B., Persaud, P. P. & Hathaway, J. M., Developing nitrogen removal models for stormwater bioretention systems. *Water Res.* **243** 120381 (2023).
45. Jurczak, T., Wagner, I., Kaczkowski, Z., Szklarek, S. & Zalewski, M., Hybrid system for the purification of street stormwater runoff supplying urban recreation reservoirs. *Ecol. Eng.* **110** 67 (2018).

14 Fundamental Understanding of Modified Biochar and the Current State-of-the-Art for Water Treatment

Alain S. Conejo-Dávila, Natyeli A. Ortiz-Tirado,
Brenda S. Morales-Verdín,
Gabriel Contreras-Zarazua,
Kattia A. Robles-Estrada,
and Oscar M. Rodríguez-Narvaez

NOMENCLATURE

ACRONYM

^{13}C-NMR	nuclear magnetic resonance of carbon
A	acidification potential of water and soil resources
AA	depletion potential of non-renewable material resources or depletion of abiotic resources
AA	abiotic resources depletion
AO	stratospheric ozone depletion potential
AOPs	advanced oxidation processes
BET	Brunauer–Emmett–Teller
BP	before pyrolysis
CAM	contact angle meter
CZP	carbamazepine
DCB	dichlorobenzene
DSC	differential scan calorimetric
EAD	freshwater ecotoxicity
EDS	energy-dispersive X-ray spectroscopy
EI99	eco-indicator 99
EPR	electron paramagnetic resonance

DOI: 10.1201/9781003441007-17

ESR	electron spin resonance spectroscopy
Eu	eutrophication potential
Fe-SEM	field emission scanning electron microscopy
FI	functionalization in situ
FPP	functionalization post-pyrolysis
FT-IR	Fourier transform infrared spectroscopy
GWP	global warming potential
HRTEM	high-resolution transmission electron microscopy
HT	human toxicity
IBI	international biochar initiative
IBP	ibuprofen
ICP-AES	inductively coupled plasma atomic emission spectroscopy
ICP-MS	inductively coupled plasma mass spectrometry
ICP-OES	inductively coupled plasma-optical emission spectroscopy
IR	infrared spectroscopy
ISO	International Organization for Standardization
LCA	life cycle assessment
LCA-E	environmental LCA
LCA-S	social LCA
LCSA	life cycle sustainability assessment
MPB	methyl paraben
OCG	oxygen-containing groups
OF	photochemical ozone-forming potential
PMS	peroxymonosulfate
SEM	scanning electron microscopy
SSA	specific surface areas
STEM	scanning transmission electron microscope
TCS	triclosan
TEM	transmission electron microscopes
TGA	thermo-gravimetric analysis
VOCs	volatile organic compound
XPS	X-ray photoelectron spectroscopy
XRD	X-ray powder diffraction
ZVI/nZVI	zero-valent iron

14.1   INTRODUCTION

The International Biochar Initiative (IBI) defines biochar as a solid material obtained from oxygen-limited thermochemical conversion of biomass which, depending on the feedstock (e.g., animal or vegetal waste) and operation conditions, possesses different chemical and physical properties [1,2]. Biochar has significant aromatization and anti-decomposition capabilities and, for that, it has been reported to have a high specific surface area and contain various negative surface functional groups (e.g., hydroxyl, carboxyl, amino) [3]. Therefore, it has been used in the last decade as a multifunctional material (e.g., absorbent and catalyst) with diverse environmental-, agricultural-, and energy-related applications [4–6].

Although biochar has essential properties, enhancing its capabilities is a growing research field using different methodologies to improve raw biomass' physical and chemical properties to generate cost-effective products that overcome the main drawbacks that currently limit its use. Some modifications involve biomass pretreatment (e.g., acid, alkaline, and oxidative) focusing on biopolymer degradation before pyrolysis [7], where the modification depends on the treatment. For example, hydrolysis reactions at acid conditions have been used to increase alcoholic and acidic functional groups present in biomass-producing oxygen-containing groups (OCGs) to increase surface area and promote microporous formation. In contrast, unsaturated OCGs are generated in alkaline conditions, increasing porosity but decreasing biochar polarity [7,8].

Biochar with and without modification has been intensely studied. Nevertheless, information is produced for specialized people, but for undergrad students or persons who are beginning biochar research development, the reported information is difficult to understand. Therefore, the present chapter is aimed at two topics: first, how to start the biochar research by explaining the production, which is the primary characterization and why each is used, and the life cycle analysis explanation for a possible scale-up process. Then, a mini-review of the state-of-the-art-modified biochar range in the last 4 years, highlighting the research gaps and further research lines to develop.

14.2   MODIFIED BIOCHAR

14.2.1   BIOCHAR MECHANISM

Biochar quality depends mainly on carbon content, pH, surface area, porous number, shape, size, and distribution [9], which is directly involved with the mechanics of biochar production, which generally happens in two steps; the first is the polymerization of biopolymers that constitute the biomass. Second, a reaction between functional groups of biomass oligomers generated by the hexagonal carbon–carbon structure (characteristics of carbon allotropes) [10,11]. Nonetheless, as different biopolymers are in the biomass (i.e., cellulose, hemicellulose, and lignin), several mechanism pathways are produced in a synergistic by secondary interaction that networks of carbon (Figure 14.1) [12,13].

The cellulose pyrolysis mechanism begins with the backbone's fragmentation to produce the first oligosaccharides. Then D-glucopyranose, through intramolecular rearrangement, generates the levoglucosan and then levoglucosenone by a dehydration reaction. Similarly, inter-, and intra-molecular condensations, decarboxylation, and aromatization reactions happen until the biochar network. Simultaneously, levoglucosan is transformed into hexose, producing D-glucopyranose and furan analogs by rearrangement and dehydration reactions. These aromatic by-products react through intermolecular condensation and rearrangement to generate mainly biochar networks, while bio-oil and syngas are at lower doses [14].

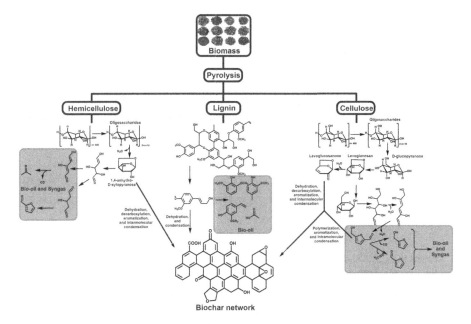

FIGURE 14.1   Biochar synthesis mechanisms.

Further, the hemicellulose transformation's first steps are like cellulose. Still, after the fragmentation and rearrangement steps, it is produced 1,4-anhydo-D-xylipyranose, which, as the aromatic cellulose by-products (i.e., D-glucopyranose and furan analogs) dehydration and decarboxylation reactions are generated to produce biochar. Finally, lignin is transformed into biochar by a straightforward reaction mechanism that produces phenolic-free radicals, and these react between them to produce biochar and continue the fragmentation to syngas generation [10,12].

14.2.2   BIOCHAR PRODUCTION TECHNOLOGIES

Biochar's chemical and physical properties are related to different parameters, where the most important are the feedstock (biomass), oxygen concentration, temperature, and heating rate [15]. Therefore, as Table 14.1 shows, several methods have been reported with several ranges of hating time, temperatures, and heating rates that generate different biochar production efficiency; for example, heating rate allows control of the thermal cracking of biomass and the chemical reaction that happens at pyrolysis [16,17]. Also, using the same techniques, depending on the parameter range (e.g., heat rate) used, various functionalization rates, biochar yield, porous size, and shape could be achieved.

Table 14.1 shows that the microwave method is the most efficient for biochar production (i.e., reaching 80%) because it replaces conventional heating methods (e.g., oven heat) due to producing hot spots by microwaves, which accelerate the reactions and the heating rates [18]. Nevertheless, it has several disadvantages for scale-up production, such as equipment and energy costs [1,15]. Therefore, although the laboratory scale has high biochar efficiency production, scaling up the process remains a crucial further research line to development, mainly on the cost/benefit analysis.

TABLE 14.1
Thermal Process Used for Biochar Production

Thermal Process	Conditions			Products/ By-Products, %			
	Temperature Range, °C	Residence Time	Heating Rate, °C/min	Biochar	Bio-oil	Syngas	References
Slow pyrolysis	300–700	Minutes-days	>30	~35	~30	~35	[13,29,30]
Intermediate pyrolysis	300–500	1–15 min	60–600	30–40	35–50	20–30	[31,32]
Fast pyrolysis	300–1,000	>2 s	60,000	5–38	50–75	13	[30,33]
Microwave	350–650 (400–2700 W)	1–60 min	25–50	15–80	8–70	12–60	[34,35]

From that, it is shown that even though new thermal treatments such as microwave assistance have been developed, slow pyrolysis remains used for biochar industrial production due to longer heating time and controlled heating rate increases the cracking reaction and produces less bio-oil than the other pyrolysis techniques (e.g., fast pyrolysis) [13,19]. Also, the slow pyrolysis generates that the biochar characteristics depend mainly on the feedstock chemical (e.g., cellulose, hemicellulose, and lignin content) and physical properties (e.g., particle size and molecular structure) [2,20–23].

14.3 MODIFIED BIOCHAR CHARACTERIZATION TECHNIQUES

Biochar modification by the different methodologies changes the physical and chemical properties such as structure, morphology, surface, and physicochemical properties. However, to identify the modification, different methodologies should perform the characterization, which could require more than one technique [24,25]. For example, structure characterizations focus on the composition and molecular arrangement of the biochar and biochar modified achieved. For that, elemental analysis (mainly, C, H, O, N, S, and P) should be performed to verify that the pyrolysis process has been carried out successfully and identify the OCG, aromatic, and unsaturated functional group rate. The most used techniques are the CHNS/O combustion analyzer [26], ICP-MS [27], and ICP-AES [28].

Once the element analysis methods are performed, thermal analysis should be used as a complementary method. It is used to identify the yield of organic material that does not change to carbon structure. The methodologies employed are thermos-gravimetric analysis (TGA) and differential scan calorimetric (DSC). TGA determines the weight loss of a sample as the temperature increases and, from that, biochar efficiency. Also, reaction intermediates are determined by identifying organic compounds of low molecular weight by their melting point lower than 200°C, and the biopolymers and carbon allotropes decomposition after 550°C and higher than 600°C, respectively [36]. Moreover, DSC indicates the endothermic and exothermic transition that the material has, showing the presence of biopolymer waste on the biochar and determining conversion efficiency [37].

Biochar structural analysis is performed through transmittance electron microscopy (TEM) and X-ray diffraction (XRD). TEM is employed in carbon allotropes to

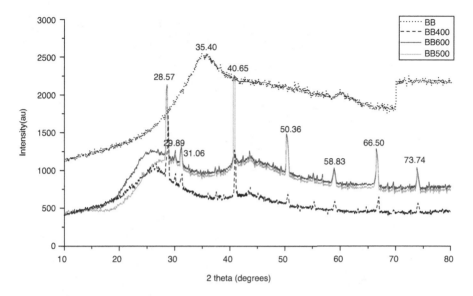

FIGURE 14.2 Diffractogram of pigeon pea stalk and its biochar [39].

identify crystal structures and orientation, for the biochar identifies the graphitization level with a nanoscale resolution. Also, it is used to study the porous shapes and the presence and localization of elements different from C, H, O, and N [10]. XRD indicates the different crystalline structures that graphitize carbon because despite biochar being an amorphous structure, it could show the lattice plane (1 0 0) characteristic of turbostratic carbon crystal between 40° and 48° [38] and plane (2 0 0) assigned to graphite structure between 20° and 25° [10]. Also, it can detect biomass remains through cellulose characteristic peaks at 15.7°, 22.4°, and 34.7° [38]. For example, Figure 14.2 shows the diffractogram of biochar produced from bamboo and its raw material reported by Swapna Sagarika Sahoo et al. [39]. The authors reported that a peak of 22.58° is assigned to the cellulose plane structure, and the peaks of 29.55°, 31.42°, 34.18°, and 39.43° should be to calcite structure, a product of minerals present in the plant a pyrolysis effect, highlighting that biochar remains with raw material structure after pyrolysis.

Molecular structure determination is complemented by nuclear magnetic resonance of carbon (^{13}C-NMR) and Raman spectroscopy. The ^{13}C-NMR identifies the different carbons in the structure, with different functional groups, hydration, and atoms linking. It is classified through chemical shifts, where the aliphatic carbons (-C-, -CH-, -CH$_2$-, -CH$_3$) are between 10 and 50 ppm, the C-O of OCG signals appear between 50 and 90 ppm, the aromatic and unsaturated carbons show around 120 ppm, and the signals more than 160 and less than 200 ppm correspond to carbonyl and its analogs functional groups [40].

Raman spectroscopy characterizes the carbon atoms with sp^2 and sp^3 hybridization materials, focusing on biochar; it could identify and interpret mainly the D and G bands. The G band is around 1,583 and 1,350 cm^{-1}, showing the graphitization level and imperfections and disorders in the graphene structure, respectively. For example, the disorder could be caused by the OCG, the sp^3 carbon, and the 5-carbon rings on

(a)   (b)

(c)   (d)

FIGURE 14.3   SEM images of (a) raw bamboo biomass and biochar produced at (b) 400°C, (c) 500°C, and (d) 600°C.

the structure. The D' band is looked at around 1,620 cm^{-1}. It shows the interaction of phonon (produced by graphene structure present in biochar) and non-carbon materials as doping agents, metal particles, or impurities [41,42].

Surface materials are primordial for complete characterization, and biochar is not an exception because different applications are reached depending on biochar surface characteristics. Scanning electron microscopy (SEM) and energy-dispersive X-ray (EDS) are commonly used for surface analysis. SEM identifies the morphology (i.e., the porous shape and size mainly) and topography (i.e., if it is a smooth or rough surface) [43]. Figure 14.3 is an SEM micrograph of pigeon pea biochar [40] showing the biochar porous shape and surface rugosity produced at different pyrolysis temperatures. Then, as complemented, EDS is performed to determine elemental composition. However, this is mainly for modified biochar by doping or metals/ceramic compound addition [44].

Finally, functional group characterization should be performed, and infrared spectroscopy (IR) and X-ray photoelectron spectroscopy (XPS) are the most reported techniques. IR studies the vibration of covalent bonds, mainly those that show a dipolar moment, and from that, the functional groups are identified.

XPS determines the different elements present on the surface and could identify functional groups as aromatic C (284.81–285.03 eV), carbonyl, C=O, (287.81–288.03 eV); carboxylic, COO, (287.81–288.03 eV); and aromatic C-O (288.60–289.55 eV) [45,46].

14.4   ENVIRONMENTAL IMPACT EVALUATION

Biochar production has a carbon footprint or environmental impact, which could increase or decrease depending on the production conditions [47] like, for example, biomass type used, reaction conditions, applications, and distance between the production plant, suppliers, and markets [48]. Therefore, this section describes strategies, tools, and methodologies for assessing, analyzing, and estimating the environmental impact using diverse reports [49].

First, biochar production and modification routes and boundaries to be analyzed should be defined to assess which and how many steps and equipment are required. For example, other studies [50] suggest that different processes are involved, including feedstock recollection and water influence on the system. Effectively setting system boundaries allows to reduce the amount of required and uploaded information. From the delimited system, it is possible to assess the environmental impact or determine the carbon footprint of biochar and hydrocarbon. A life cycle assessment (LCA) is the most common way to determine environmental impact and carbon footprint [49]. LCA is a tool used to evaluate and estimate the potential environmental, social, and economic impact through the life cycle of a product, system, or service, analyzing parameters such as raw material extraction, production, transportation, usage, and disposal [51].

There are different LCA types depending on the factor to be evaluated. The most used are environmental LCA (LCA-E), social LCA (LCA-S), and life cycle sustainability assessment (LCSA). However, this chapter will only focus on LCA-E for biochar applications. The methodology is based on the guidelines by the International Organization for Standardization (ISO, 14040 and 14044) as a systematic procedure for the evaluation of the environmental impacts of services and products, related to four fundamental stages [52–54]:

a. Goal definition and scope: The product or process is defined and described. Also, the boundaries and environmental effects of the assessment are identified and defined.
b. Inventory analysis: Identify and quantify the process requirements such as energy, water, and raw materials doses. Then, the environmental emissions include air emissions, solid waste disposal, and wastewater discharges.
c. Impact assessment: An environmental impact estimation is calculated using an impact model.
d. Interpretation: Results are analyzed during the inventory analysis and impact assessment steps according to the assumptions, boundaries, and uncertainties considered during previous phases. Then, the product, production route, or public policy is selected.

14.4.1   LCA METHODS

There is more than one method or metric for LCA evaluation. Therefore, selecting a suitable metric to analyze the environmental impact is important because some methods only evaluate certain aspects of the product or process. The method selection depends on the limits of the environmental impact analysis established in step two of the ISO standards. For this chapter, only the most common and suitable LCA methods for biochar will be discussed: the eco-indicator 99 and CML 2001. These methodologies quantify the environmental impact considering different biochar aspects, such as production technology, type and amount of raw material needed to produce them, energy required, and transportation.

Eco-indicator 99 (EI99). It is a methodology quantifies the environmental impact with a score scale using the Eco-point as a base unit [55]. Each Eco-point represents the 1,000th of the environmental load (i.e., damages) of an average European citizen annually [55–57]. The evaluation focuses on three major damage categories: (i) human health damage from environmental degradation such as reduced life expectancy and increased diseases and deaths. EI99 relates damage to climate change, depletion of the ozone layer, respiratory and carcinogenic effects, and ionizing radiation, (ii) ecosystem quality involving damage to species diversity, soil and water acidification, land use, and eutrophication, and (iii) resources depletion, including raw materials and energy sources depletion.

CML 2001. This method considers impact categories and evaluation methods, using a normalization step to sum all impacts without weighting. The basic unit is the equivalent kg of emissions [58]. Finally, a classification of impact indicators is used for three different categories: (i) obligatory impact categories, including common LCA impact categories such as ecotoxicity and resource depletion, (ii) additional impact categories including other categories for which indicators exist but are not always included, (iii) other indicators including impact categories for which no models are available like odor, noise, and drying.

LCA using the CML 2001 is performed considering mandatory categories shown in Table 14.2 [58]. Like EI99, CML 2001 has relevant disadvantages, highlighting the finding of adequate normalization factors and models for damage quantification.

TABLE 14.2
Damage Categories for CML Method Analysis

Category	Units
Global warming potential (GWP)	kg CO_2 equivalent
Acidification potential of water and soil resources (A)	kg SO_2 equivalent
Eutrophication potential (Eu)	kg PO_{4-3} equivalent
Stratospheric ozone depletion potential (AO)	kg of R11 equivalent
Depletion potential of non-renewable material resources or depletion of abiotic resources (AA)	kg of R11 equivalent
Abiotic resources depletion (AA)	kg of Sb
Human toxicity (HT)	kg of 1,4-dichlorobenzene (DCB) equivalent
Freshwater ecotoxicity (EAD)	kg of 1,4-dichlorobenzene (DCB) equivalent
Photochemical ozone-forming potential (OF)	kg of ethane equivalent
Ecotoxicity potential	kg of 1,4-dichlorobenzene (DCB) equivalent

14.5 STATE-OF-THE-ART OF MODIFIED BIOCHAR

14.5.1 WATER TREATMENT APPLICATION

Several studies have been published during the last decade on biochar applications [2,13,44]. However, in the last 4 years, biochar with impregnated metals and functionalized particles has been significantly reported (Table 14.3). Table 14.4 shows modified biochar wastewater treatment application, suggesting intensified use as catalysis for Fenton-like oxidation processes [1,13]. Some studies reported using peroxymonosulfate (PMS) as an oxidant which reacts with oxygen-containing groups (i.e., C=O and C-O) and generates sulfate-based radicals (Equations 14.1 and 14.2) [59–61].

$$C = C = 0 + HSO_5^- \rightarrow SO_4^- + C = C - O^+ + OH^+ \tag{14.1}$$

$$C = C - 0^+ + HSO_5^- \rightarrow SO_4^- + C = C = 0 + SO_5^- + H^+ \tag{14.2}$$

Photocatalysts supported on carbon-based materials produced heterojunction-based photocatalysts (e.g., Bi, TiO_2) and improved photocatalytic efficiency [62,63]. However, few photocatalysts have been used, generating a knowledge gap worth exploring. For example, the plasmonic nanostructures (e.g., silver and gold) stand out because surface plasmons of these materials are used to improve their optical properties [64] and enhance photocatalytic activity through plasmonic energy transfer from the metal nanostructure to the semiconductor, which decreases the band gap and extends the light absorption range [65,66]. Also, as with the Fenton-like reaction, the biochar photocatalytic activities and interaction with the organic compound photodegradation have not been sufficiently studied [63].

14.5.2 LCA ANALYSIS

Biochar has been reported for different high-efficiency applications (e.g., adsorbent) [78,79], and scale-up analysis has been developed, such as LCA [47,50,80]. However, few analyses are available for modified biochar and, of these, most are focused on metal [75,81] and microwave assistance [76] environmental impact. Therefore, LCA analysis on the effect of different biochar modifications on its environmental impact is a further research avenue pending development, mainly to determine scale-up potential. In that sense, some points have been identified for the life cycle analysis of modified biochar. First, boundary delimitation should use broader limits (e.g., extraction to disposal) because LCA accuracy heavily depends on boundary delimitation. Second, raw materials availability because feedstock could generate an efficient adsorbent, but transporting the raw material to the industrial plant increases GHE generation. Third, evaluating other environmental effects such as material extraction, inventory, or other forms of damage. Finally, a supply chain assessment should be considered before LCA because up to 40% of the environmental damage may be related to transportation [48].

TABLE 14.3

Reported Biochar Modifications

Modification/Treatment	Feedstock	Pyrolysis	Notes	Characterization	Reference
Fe$_3$O$_4$ and FeO/FI	Anaerobic digestion sludge	600°C–1,000°C, 90 min, N$_2$ = 100 mL/min	PDS-ZVI pretreatment	SEM, EDS, HRTEM, FT-IR, XRD, XPS, ZP, SSA	[67]
FeCu/FPP	Dewatering sludge	500°C, 4 h	[FeSO$_4$·7H$_2$O]:0.2 M [CuSO$_4$·5H$_2$O]:0.1 M 500°C, 4 h	SEM, EDS, FT-IR, XRD, XPS. SSA (BET), Raman, ESR	[68]
nZVI/FPP	Killed pine trees	550°C–600°C	[FeCl$_3$·6H$_2$O]:0.027–0.162 M [NaBH$_4$]:0.135–0.81 M pH: 5–11, Ar	SEM, STEM, FT-IR, XPS, SSA (BET), CAM, ICP-OES	[69]
Co/FI	Spent coffee ground	700°C, 120 min Ar:100 mL/min	[NaOH]:4 M [CoCl$_2$·6H$_2$O]:4%	SEM, EDS, TEM, HRTEM, XRD, SSA (BET)	[70]
N/FI	Corncob	700°C, 2 h	Urea	SEM, EDS, XRD, XPS, Raman,	[71]
Fe$_3$O$_4$ and Fe$_2$O$_3$/BP	Sugarcane bagasse and steel pickling waste liquor	300°C–500°C, 2 h	[Fe]: 12 g/L	SEM, TEM, FT-IR, XRD, XPS, PPMS-9	[72]
Bi and Bi$_2$O$_3$/FI	Rice straws	500°C, 60 min	Bi(NO$_3$)$_3$·5H$_2$O [HNO$_3$]: 50, 100, 500 mL	SEM, EPR, EDS, TEM, FT-IR, XRD, XPS, SSA (BET)	[63]
TiO$_2$/FI	*Salvinia molesta*	350°C, pH: 7	TTiP or TiOSO$_4$ NH$_4$OH	SEM, TEM, FT-IR, XRD, XPS, SSA (BET)	[62]
TiO$_2$/FPP	*Salvinia molesta*	350°C	[TTiP or TiOSO$_4$]		
Bentonite/FI	Rice husk	450°C, 2 h	Raw Bentonite [MgCl$_2$]:1.25 M [NaOH]: 2.25 M	FE-SEM, EDS, FT-IR, XRD, SSA	[73]
nZVI/FPP	Corn stalk	250°C–700°C, 2 h	[FeCl$_3$·6H$_2$O] [NaBH$_4$]:0.36 M	SEM, EDS, FT-IR, XRD, XPS	[74]
ZnCl$_2$/FPP	Waste sawdust	500°C, 1 h	ZnCl$_2$, 350–650°C, 30–120 min	SEM, EDS, FT-IR, XRD, SSA (BET), CHNS/O	[75]
Fe/FPP	Raw canola straw	200–650–1,000 W	[FeCl$_3$]:0.1,0.25, 0.5 M, 9 min	SEM, EDS, XRD, SSA (BET)	[76]

TABLE 14.4

Water Treatment Process Efficacy Using Modified Biochar

Modification	Process	Pollutant	Efficiency %	Notes	References
Bentonite	Adsorption	Humic acid	52	[Humic acid]: 20 mg/L, [Ammonium]: 60 mg/L, [Phosphates]: 60 mg/L, [Adsorbent]: 0.3 g/L, 4 h, 25°C	[73]
		Ammonium ions	29		
		Phosphate ions	63		
nZVI	Adsorption	Cd(II)	28	[Cd(II)]: 30 mg/L, [As(III)]: 80 mg/L, [Adsorbent]: 0.25 g/L, pH=2–8, 25°C	[74]
		As(III)	46		
ZnCl$_2$	Adsorption	Methyl orange	34–92	[Pollutant]: 10–200 mg/L, [Absorbent]: 0.2–1 g/L, 30–120 min, 25°C–45°C	[75]
Fe	Adsorption	As(V)	94	[Pollutant]: 1 mg/L, [Absorbent]: 1 g/L, pH: 3–7, 15°C–35°C	[76]
		As(III)	89		
TiO$_2$	Adsorption	Acid orange 7 dye	46	[Pollutant]: 20 mg/L, [Absorbent]: 0.1 g/L	[62]
Fe$_3$O$_4$ and FeO	Fenton-Like	Sulfamethazine	33–100	[Pollutant]: 20 mg/L [Catalyst]: 0.2 g/L, 90 min [PDS]: 6 mM	[67]
		Orange G	100		
		Nitrobenzene	100		
		Malachite green	100		
		Methoxazole	85		
		Benzoic acid	79		
FeCu	Fenton-Like	Tetracycline	92	[Pollutant]: 5, 10, 20 mg/L [Catalyst]: 0.05 g/L, pH: 5, 6.67, 9.03	[68]
nZVI	Fenton-Like	Trichloroethylene	88	[Pollutant]: 40 µg/L, [Catalyst]: 250 mg/L, pH: 3, 20 min	[69]
Co	Fenton-Like	Tetracycline	100	[Pollutant]: 10–300 mg/L [Catalyst]: 100 mg/L, [PMS]: 0.6 mM, pH: 7, 25°C	[70]
N	Fenton-Like	Sulfadiazine	96	[Pollutant]: 1–50 µM, [Catalyst]: 0.5–3 g/L, [PDS]: 0.25–3 mM, pH=3–11, 25°C	[71]
Fe$_3$O$_4$ and Fe$_2$O$_3$	Fenton-Like	Metronidazole	8–100	[Pollutant]: 40 mg/L, [Catalyst]: 1 g/L, [H$_2$O$_2$]: 5 mM, 120 min, 30°C	[77]
Fe$_3$O$_4$ and Fe$_2$O$_3$	Fenton-Like	Metronidazole	93–100	[Pollutant]: 40 mg/L [Catalyst]: 0.3 g/L, [H$_2$O$_2$]: 5 mM, pH: 5.6, 2 h, 30°C	[72]
Bi and Bi$_2$O$_3$	Photocatalysis	Estrone	92–96	[Pollutant]: 10.4 µmol/L [Photocatalyst]: 1 g/L, pH: 6–8, 500 W, λ=350–700 nm	[63]
TiO$_2$	Photocatalysis	Acid Orange 7 dye	58–90	[Pollutant]: 20 mg/L, [Photocatalyst]: 0.1 g/L, 10 mW/m, 380–480 nm	[62]

14.6 CONCLUSIONS

From the analysis of modified biochar, the following are our main findings:

- Despite biochar being widely studied, other carbon-based materials are commonly confused with biochar (e.g., hydrochar). Therefore, parameters and characteristics related to biochar should be standardized.
- The last decade's analysis shows different biochar modification affects its chemical and physical properties. For example, an acid pretreatment for increasing surface area. Therefore, production scaling-up analysis is an important research line focused on application.
- Characterization techniques need to be selected carefully to avoid duplicity. For example, ^{13}C NMR is not necessary if a Raman analysis is performed. Also, XPS analysis is not necessary when FT-IR is available. However, elemental analysis (e.g., EDS), XRD, and microscopy (e.g., SEM or TEM) are characterizations of unavoidable characterization techniques.
- Applications of modified biochar for water treatment are shifting from adsorption to catalysts because of knowledge gaps included in review studies. In the same way, scaling up analysis to justify its carbon footprint and contribution to the circular economy is attracting interest. The lack of information on modified biochar suggests that more research is needed to highlight if these modifications lead to a lower carbon footprint.

ACKNOWLEDGMENTS

All the authors would also like to thank CONACyT for the fellowship to undertake this research study. Also, IDEAG for the grant for this project (grants: CONVO/045/2021 and CONVO/090/2021).

REFERENCES

1. Rodriguez-Narvaez, O. M., Peralta-Hernandez, J. M. J. M., Goonetilleke, A. & Bandala, E. R. Biochar-supported nanomaterials for environmental applications. *J. Ind. Eng. Chem.* **78**, 21–33 (2019).
2. Rodríguez-Narvaez, O. M., Medina-Orendain, D. A. & Mendez-Alvarado, L. N. Functionalized green carbon-based nanomaterial for environmental application. In (eds Koduru J.R., Karri R.R., Mubarak N.M., Bandala E.R.) *Sustainable Nanotechnology for Environmental Remediation* 347–382 (Elsevier, 2022). doi:10.1016/B978-0-12-824547-7.00005-9.
3. Wang, B., Gao, B. & Fang, J. Recent advances in engineered biochar productions and applications. *Crit. Rev. Environ. Sci. Technol.* **47**, 2158–2207 (2017).
4. Tan, X. et al. Application of biochar for the removal of pollutants from aqueous solutions. *Chemosphere* **125** 70–85 (2015).
5. Inyang, M. I. et al. A review of biochar as a low-cost adsorbent for aqueous heavy metal removal. *Crit. Rev. Environ. Sci. Technol.* **46** 406–433 (2016).
6. Rodriguez-Narvaez, O. M., Peralta-Hernandez, J. M., Goonetilleke, A. & Bandala, E. R. Treatment technologies for emerging contaminants in water: A review. *Chem. Eng. J.* **323**, 361–380 (2017).

7. Zhou, Y. et al. Modulating hierarchically microporous biochar via molten alkali treatment for efficient adsorption removal of perfluorinated carboxylic acids from wastewater. *Sci. Total Environ.* **757**, 143719 (2021).

8. Chu, G. et al. Phosphoric acid pretreatment enhances the specific surface areas of biochars by generation of micropores. *Environ. Pollut.* **240**, 1–9 (2018).

9. Yao, Z., You, S., Ge, T. & Wang, C. H. Biomass gasification for syngas and biochar co-production: Energy application and economic evaluation. *Appl. Energy* **209**, 43–55 (2018).

10. Ghodake, G. S. et al. Review on biomass feedstocks, pyrolysis mechanism and physicochemical properties of biochar: State-of-the-art framework to speed up vision of circular bioeconomy. *J. Clean. Prod.* **297**, 126645 (2021).

11. Clurman, A. M. et al. Influence of surface hydrophobicity/hydrophilicity of biochar on the removal of emerging contaminants. *Chem. Eng. J.* **402**, 126277 (2020).

12. Zhang, Z., Zhu, Z., Shen, B. & Liu, L. Insights into biochar and hydrochar production and applications: A review. *Energy* **171**, 581–598 (2019).

13. Li, S., Chan, C. Y., Sharbatmaleki, M., Trejo, H. & Delagah, S. Engineered biochar production and its potential benefits in a closed-loop water-reuse agriculture system. *Water*, **12**, 2847 (2020).

14. Liu, W. J., Jiang, H. & Yu, H. Q. Development of biochar-based functional materials: Toward a sustainable platform carbon material. *Chem. Rev.* **115**, 12251–12285 (2015).

15. Ethaib, S., Omar, R., Kamal, S. M. M., Biak, D. R. A. & Zubaidi, S. L. Microwave-assisted pyrolysis of biomass waste: A mini review. *Processes*, **8**, 1190 (2020).

16. Safarian, S., Rydén, M. & Janssen, M. Development and comparison of thermodynamic equilibrium and kinetic approaches for biomass pyrolysis modeling. *Energies*, **15**, 3999 (2022).

17. Bridgwater, A. Fast pyrolysis of biomass for the production of liquids. *Biomass Combust. Sci. Technol. Eng.* 130–171 (2013) doi:10.1533/9780857097439.2.130.

18. Mubarak, N. M., Sahu, J. N., Abdullah, E. C. & Jayakumar, N. S. Plam oil empty fruit bunch based magnetic biochar composite comparison for synthesis by microwave-assisted and conventional heating. *J. Anal. Appl. Pyrolysis* **120**, 521–528 (2016).

19. Roy, P. & Dias, G. Prospects for pyrolysis technologies in the bioenergy sector: A review. *Renew. Sustain. Energy Rev.* **77**, 59–69 (2017).

20. Hernandez-Mena, L. E., Pécora, A. A. B. & Beraldo, A. L. Slow pyrolysis of bamboo biomass: Analysis of biochar properties. *Chem. Eng. Trans.*, **37**, 115–120 (2014).

21. Yu, J., Wang, D. & Sun, L. The pyrolysis of lignin: Pathway and interaction studies. *Fuel* **290**, 120078 (2021).

22. Mohan, D., Sarswat, A., Ok, Y. S. & Pittman, C. U. Organic and inorganic contaminants removal from water with biochar, a renewable, low cost and sustainable adsorbent – A critical review. Bioresour. Technol. **160**, 191–202 (2014).

23. Rajapaksha, A. U. et al. Engineered/designer biochar for contaminant removal/immobilization from soil and water: Potential and implication of biochar modification. *Chemosphere* **148**, 276–291 (2016).

24. Chee H. Chia, Adriana Downie, P. Munroe Characteristics of biochar: Physical and structural properties. In (eds Lehmann J., Joseph S.) *Biochar for Environmental Management* 1–2 (Taylor & Francis, 2019). doi:10.4324/9780203762264-12.

25. Li, Y. et al. Characterization of modified biochars derived from bamboo pyrolysis and their utilization for target component (furfural) adsorption. *Energy Fuels* **28**, 5119–5127 (2014).

26. Bakshi, S., Banik, C. & Laird, D. A. Estimating the organic oxygen content of biochar. *Sci. Rep.* **10**, 1–12 (2020).

27. Bachmann, H. J. et al. Toward the standardization of biochar analysis: The COST action TD1107 interlaboratory comparison. *J. Agric. Food Chem.* **64**, 513–527 (2016).

28. Mašek, O., Bogush, A., Jayakumar, A., Wurzer, C. & Peters, C. Biochar characterization methods. *Biochar* 5-1–5-19 (2020) doi:10.1088/978-0-7503-2660-5CH5.
29. Suliman, W. et al. Influence of feedstock source and pyrolysis temperature on biochar bulk and surface properties. *Biomass Bioenergy* **84**, 37–48 (2016).
30. Safarian, S. Performance analysis of sustainable technologies for biochar production: A comprehensive review. *Energy Rep.* **9**, 4574–4593 (2023).
31. Kisiki Nsamba, H., Hale, S. E., Cornelissen, G., Bachmann, R. T. & Nsamba, H. K. Sustainable technologies for small-scale biochar production – a review. *J. Sustain. Bioenergy Syst.* **05**, 10–31 (2015).
32. Jung, S. H. & Kim, J. S. Production of biochars by intermediate pyrolysis and activated carbons from oak by three activation methods using CO_2. *J. Anal. Appl. Pyrolysis* **107**, 116–122 (2014).
33. Enaime, G., Baçaoui, A., Yaacoubi, A. & Lübken, M. Biochar for wastewater treatment-conversion technologies and applications. *Appl. Sci.* **10**, 3492 (2020).
34. Kumar, A., Saini, K. & Bhaskar, T. Hydochar and biochar: Production, physicochemical properties and techno-economic analysis. *Bioresour. Technol.* **310**, 123442 (2020).
35. Mutsengerere, S., Chihobo, C. H., Musademba, D. & Nhapi, I. A review of operating parameters affecting bio-oil yield in microwave pyrolysis of lignocellulosic biomass. *Renew. Sustain. Energy Rev.* **104**, 328–336 (2019).
36. Torquato, L. D. M., Crnkovic, P. M., Ribeiro, C. A. & Crespi, M. S. New approach for proximate analysis by thermogravimetry using CO_2 atmosphere: Validation and application to different biomasses. *J. Therm. Anal. Calorim.* **128**, 1–14 (2017).
37. Meri, N. H. et al. Effect of chemical washing pre-treatment of empty fruit bunch (EFB) biochar on characterization of hydrogel biochar composite as bioadsorbent. *IOP Conf. Ser. Mater. Sci. Eng.* **358**, 012018 (2018).
38. Zhang, C. et al. Evolution of the functionalities and structures of biochar in pyrolysis of poplar in a wide temperature range. *Bioresour. Technol.* **304**, 123002 (2020).
39. Sahoo, S. S., Vijay, V. K., Chandra, R. & Kumar, H. Production and characterization of biochar produced from slow pyrolysis of pigeon pea stalk and bamboo. *Clean. Eng. Technol.* **3**, 100101 (2021).
40. Mao, J., Cao, X. & Chen, N. Characterization of biochars using advanced solid-state 13C nuclear magnetic resonance spectroscopy. In: Lee, J. (ed.) *Advanced Biofuels Bioproducts* 47–55. https://doi.org/10.1007/978-1-4614-3348-45 (Springer, New York, NY, 2013).
41. Li, Z., Deng, L., Kinloch, I. A. & Young, R. J. Raman spectroscopy of carbon materials and their composites: Graphene, nanotubes and fibres. *Prog. Mater. Sci.* **135**, 101089 (2023).
42. Xu, J. et al. Raman spectroscopy of biochar from the pyrolysis of three typical Chinese biomasses: A novel method for rapidly evaluating the biochar property. *Energy* **202**, 117644 (2020).
43. Zhang, K., Mao, J. & Chen, B. Reconsideration of heterostructures of biochars: Morphology, particle size, elemental composition, reactivity and toxicity. *Environ. Pollut.* **254**, 113017 (2019).
44. Manyà, J. J. Pyrolysis for biochar purposes: A review to establish current knowledge gaps and research needs. *Environ. Sci. Technol.* 46 7939–7954 (2012).
45. Singh, B., Fang, Y., Cowie, B. C. C. & Thomsen, L. NEXAFS and XPS characterisation of carbon functional groups of fresh and aged biochars. *Org. Geochem.* **77**, 1–10 (2014).
46. Santhosh, C. et al. Synthesis and characterization of magnetic biochar adsorbents for the removal of Cr(VI) and Acid orange 7 dye from aqueous solution. *Environ. Sci. Pollut. Res.* **27**, 32874–32887 (2020).
47. Matuštík, J., Hnátková, T. & Kočí, V. Life cycle assessment of biochar-to-soil systems: A review. *J. Clean. Prod.* **259**, 120998 (2020).

48. Contreras-Zarazúa, G., Martin, M., Ponce-Ortega, J. M. & Segovia-Hernández, J. G. Sustainable design of an optimal supply chain for furfural production from agricultural wastes. *Ind. Eng. Chem. Res.* **60**, 14495–14510 (2021).
49. Peña, C. et al. Using life cycle assessment to achieve a circular economy. *Int. J. Life Cycle Assess.* **26**, 215–220 (2021).
50. Choudhary, V. & Philip, L. Sustainability assessment of acid-modified biochar as adsorbent for the removal of pharmaceuticals and personal care products from secondary treated wastewater. *J. Environ. Chem. Eng.* **10**, 107592 (2022).
51. Curran, M.A. Life Cycle Assessment: Principles and Practice. *National Risk Management Research Laboratory Office of Research and Development U.S. Environmental Protection Agency* (2006).
52. Solé, A., Miró, L. & Cabeza, L. F. Environmental approach. In *High-Temperature Thermal Storage Systems Using Phase Change Materials* 277–295 (Elsevier Ltd., 2018). doi:10.1016/B978-0-12-805323-2.00010-2.
53. ISO14040:2006. *Environmental Management Life Cycle Assessment Principles and Framework.* Technical Committee ISO/TC, Geneva. (2022).
54. ISO 14044:2006. *Environmental Management-Life Cycle Assessment-Requirements and Guidelines.* International Organization for Standardization (ISO). (2022).
55. Goedkoop, M. J.V. The Eco-Indicator 99: A Damage Oriented Method for Life Cycle Impact Assessment Methodology Report. (1999).
56. Guillen-Gosalbez, G., Caballero, J. A., Esteller, L. J. & Gadalla, M. Application of life cycle assessment to the structural optimization of process flowsheets. *Comput. Aided Chem. Eng.* **24**, 1163–1168 (2007).
57. Dincer, I. & Bicer, Y. Life cycle assessment of energy. In Dincer, I. ed., *Comprehensive Energy Systems* 1–5 1042–1084 (Elsevier, 2018).
58. Frischknecht, R. et al. Implementation of life cycle impact assessment methods. *Am. Midl. Nat.* **150**, 1–151 (2007).
59. Xiao, R. et al. Activation of peroxymonosulfate/persulfate by nanomaterials for sulfate radical-based advanced oxidation technologies. *Curr. Opin. Chem. Eng.* **19**, 51–58 (2018).
60. Duan, X. et al. Unveiling the active sites of graphene-catalyzed peroxymonosulfate activation. *Carbon N. Y.* **107**, 371–378 (2016).
61. Wang, S. & Wang, J. Activation of peroxymonosulfate by sludge-derived biochar for the degradation of triclosan in water and wastewater. *Chem. Eng. J.* **356**, 350–358 (2019).
62. Silvestri, S. et al. TiO_2 supported on Salvinia molesta biochar for heterogeneous photocatalytic degradation of Acid Orange 7 dye. *J. Environ. Chem. Eng.* **7**, 102879 (2019).
63. Zhu, N. et al. Bismuth impregnated biochar for efficient estrone degradation: The synergistic effect between biochar and Bi/Bi_2O_3 for a high photocatalytic performance. *J. Hazard. Mater.* **384** (2020).
64. Sánchez-Cid, P., Jaramillo-Páez, C., Navío, J. A., Martín-Gómez, A. N. & Hidalgo, M. C. Coupling of Ag_2CO_3 to an optimized ZnO photocatalyst: Advantages vs. disadvantages. *J. Photochem. Photobiol. A Chem.* **369**, 119–132 (2019).
65. Kale, M. J., Avanesian, T. & Christopher, P. Direct photocatalysis by plasmonic nanostructures. *ACS Catal.* **4** 116–128 (2014).
66. Wu, N. Plasmonic metal–semiconductor photocatalysts and photoelectrochemical cells: A review. *Nanoscale* **10** 2679–2696 (2018).
67. Chen, Y. di et al. Magnetic biochar catalysts from anaerobic digested sludge: Production, application and environment impact. *Environ. Int.* **126**, 302–308 (2019).
68. Liu, J. et al. The biochar-supported iron-copper bimetallic composite activating oxygen system for simultaneous adsorption and degradation of tetracycline. *Chem. Eng. J.* **402**, 126039 (2020).

69. Mortazavian, S. et al. Heat-treated biochar impregnated with zero-valent iron nanoparticles for organic contaminants removal from aqueous phase: Material characterizations and kinetic studies. *J. Ind. Eng. Chem.* **76**, 197–214 (2019).

70. Nguyen, V.-T. et al. Cobalt-impregnated biochar (Co-SCG) for heterogeneous activation of peroxymonosulfate for removal of tetracycline in water. *Bioresour. Technol.* **292**, 121954 (2019).

71. Wang, H. et al. Edge-nitrogenated biochar for efficient peroxydisulfate activation: An electron transfer mechanism. *Water Res.* **160**, 405–414 (2019).

72. Yi, Y., Tu, G., Eric Tsang, P. & Fang, Z. Insight into the influence of pyrolysis temperature on Fenton-like catalytic performance of magnetic biochar. *Chem. Eng. J.* **380**, 122518 (2020).

73. Jing, H. P., Li, Y., Wang, X., Zhao, J. & Xia, S. Simultaneous recovery of phosphate, ammonium and humic acid from wastewater using a biochar supported Mg(OH)₂/bentonite composite. *Environ. Sci. Water Res. Technol.* **5**, 931–943 (2019).

74. Yang, D. et al. Simultaneous adsorption of Cd(II) and As(III) by a novel biochar-supported nanoscale zero-valent iron in aqueous systems. *Sci. Total Environ.* **708**, 134823 (2020).

75. Maiti, P., Siddiqi, H., Kumari, U., Chatterjee, A. & Meikap, B. C. Adsorptive remediation of azo dye contaminated wastewater by ZnCl₂ modified bio-adsorbent: Batch study and life cycle assessment. *Powder Technol.* **415**, 118153 (2023).

76. Norberto, J., Zoroufchi Benis, K., McPhedran, K. N. & Soltan, J. Microwave activated and iron engineered biochar for arsenic adsorption: Life cycle assessment and cost analysis. *J. Environ. Chem. Eng.* **11**, 109904 (2023).

77. Yi, Y., Tu, G., Zhao, D., Tsang, P. E. & Fang, Z. Pyrolysis of different biomass pre-impregnated with steel pickling waste liquor to prepare magnetic biochars and their use for the degradation of metronidazole. *Bioresour. Technol.* **289**, 121613 (2019).

78. Thompson, K. A. et al. Environmental comparison of biochar and activated carbon for tertiary wastewater treatment. *Environ. Sci. Technol.* **50**, 11253–11262 (2016).

79. Uddin, M. M. & Wright, M. M. *Life Cycle Analysis of Biochar Use in Water Treatment plants. Sustainable Biochar for Water and Wastewater Treatment* (Elsevier Inc., 2022). doi:10.1016/b978-0-12-822225-6.00012-9.

80. James, A., Sánchez, A., Prens, J. & Yuan, W. Biochar from agricultural residues for soil conditioning: Technological status and life cycle assessment. *Curr. Opin. Environ. Sci. Heal.* **25**, 100314 (2022).

81. Gallego-Ramírez, C., Chica, E., Rubio-Clemente, A. Life cycle assessment of raw and Fe-modified biochars: Contributing to circular economy. *Materials* **16**, 6059 (2023).

Index

Printed in the United States
by Baker & Taylor Publisher Services